SYSTEMS, CONTROL, MODELING AND OPTIMIZATION

IFIP – The International Federation for Information Processing

IFIP was founded in 1960 under the auspices of UNESCO, following the First World Computer Congress held in Paris the previous year. An umbrella organization for societies working in information processing, IFIP's aim is two-fold: to support information processing within its member countries and to encourage technology transfer to developing nations. As its mission statement clearly states,

> *IFIP's mission is to be the leading, truly international, apolitical organization which encourages and assists in the development, exploitation and application of information technology for the benefit of all people.*

IFIP is a non-profitmaking organization, run almost solely by 2500 volunteers. It operates through a number of technical committees, which organize events and publications. IFIP's events range from an international congress to local seminars, but the most important are:

• The IFIP World Computer Congress, held every second year;
• Open conferences;
• Working conferences.

The flagship event is the IFIP World Computer Congress, at which both invited and contributed papers are presented. Contributed papers are rigorously refereed and the rejection rate is high.

As with the Congress, participation in the open conferences is open to all and papers may be invited or submitted. Again, submitted papers are stringently refereed.

The working conferences are structured differently. They are usually run by a working group and attendance is small and by invitation only. Their purpose is to create an atmosphere conducive to innovation and development. Refereeing is less rigorous and papers are subjected to extensive group discussion.

Publications arising from IFIP events vary. The papers presented at the IFIP World Computer Congress and at open conferences are published as conference proceedings, while the results of the working conferences are often published as collections of selected and edited papers.

Any national society whose primary activity is in information may apply to become a full member of IFIP, although full membership is restricted to one society per country. Full members are entitled to vote at the annual General Assembly, National societies preferring a less committed involvement may apply for associate or corresponding membership. Associate members enjoy the same benefits as full members, but without voting rights. Corresponding members are not represented in IFIP bodies. Affiliated membership is open to non-national societies, and individual and honorary membership schemes are also offered.

SYSTEMS, CONTROL, MODELING AND OPTIMIZATION

Proceedings of the 22nd IFIP TC7 Conference held from July 18-22, 2005, in Turin, Italy

Edited by

F. Ceragioli
Politecnico di Torino, Torino, Italy

A. Dontchev
AMS and University of Michigan, Ann Arbor, USA

H. Furuta
Kansai University, Osaka, Japan

K. Marti
Federal Armed Forces University, Munich, Germany

L. Pandolfi
Politechnico di Torino, Torino, Italy

Systems, Control, Modeling and Optimization
Edited by F. Ceragioli, A. Dontchev, H. Furuta, K. Marti, and L. Pandolfi

p. cm. (IFIP International Federation for Information Processing, a Springer Series in Computer Science)

ISSN: 1571-5736 / 1861-2288 (Internet)

ISBN-13: 978-1-4419-4155-8 e-SBN-13: 978-0-387-33882-8
Printed on acid-free paper

9 8 7 6 5 4 3 2 1
springer.com

Contents

Preface

We publish in this volume a selection of the papers presented at the 22nd Conference on System Modeling and Optimization, held at the Politecnico di Torino in July 2005. The conference has been organized by the Mathematical Department of the Politecnico di Torino.

The papers presented in this volume mostly concern stochastic and distributed systems, their control/optimization and inverse problems.

IFIP is a multinational federation of professional and technical organizations concerned with information processes. It was established in 1959 under the auspices of UNESCO. IFIP still mantains friendly connections with specialized agencies of the UN systems. It consists of Technical Committees. The Seventh Technical Committee, established in 1972, was created in 1968 by A.V. Balakrishnan, J.L. Lions and G.I. Marchuk with a joint conference held in Sanremo and Novosibirsk.

The present edition of the conference is dedicated to Camillo Possio, killed by a bomb during the last air raid over Torino, in the sixtieth anniversary of his death. The special session "On the Possio equation and its special role in aeroelasticity" was devoted to his achievements. The special session "Shape Analysis and optimization" commemorates the 100th anniversary of Pompeiu thesis.

All the fields of interest for the seventh Technical Committee of the IFTP, chaired by Prof. I. Lasiecka, had been represented at the conference: Optimization; Optimization with PDE constraints; structural systems optimization; algorithms for linear and nonlinear programming; stochastic optimization; control and game theory; combinatorial and discrete optimization. identification and inverse problems; fault detection; shape identification. complex systems; stability and sensitivity analysis; neural networks; fractal and chaos; reliability. computational techniques in distributed systems and in information processing environments; transmission of information in complex systems; data base design. Applications of optimization techniques and of computational methods to scientific and technological areas (such as medicine, biology, economics, finances, aerospace and aeronautics etc.).

Over 300 researchers took part to the conference, whose organization was possible thanks to the help and support of the Department of Mathematics of the Politecnico di Torino. We would like to thank Istituto Boella, Fondazione Cassa di Risparmio Torino, Unicredit Banca and the Regione Piemonte for financial support. We also acknowledge support from Politecnico di Torino and INRIA.

We would like to thank the following colleagues, who organized special and invited sessions, greatly contributing to the success of the conference:

V. Agoshkov, O. Alvarez, G. Avalos, A. Bagchi , U. Boscain, A.V. Balakrishnan, M. Bardi, F. Bucci, J. Cagnol, P. Cardaliaguet, P. Cannarsa , M.C. Delfour, D. Dentcheva, A. Dontchev, H. Furuta, K. Juszczyszyn, P. Kall, A. Kalliauer, R. Katarzyniak, D. Klatte, B. Kummer, M. A. Lòpez, V. Maksimov, K. Marti, J. Mayer, S. Migorski, Z. Naniewicz , N.T. Nguyen, J. Outrata, B. Piccoli, S. Pickenhain, M. Polis, E. Priola, A. Ruszczynski, I.F. Sivergina, H. Scolnik, J. Sobecki, J. Sokolowski, G. Tessitore, D. Tiba, F. Troeltzsch , S. Vessella, J. Zabczyk, T. Zolezzi, J.P. Zolesio.

L. PANDOLFI

Contributing Authors

N. Ahmed University of Ottawa, Ottawa, Canada

M. Badra University Paul Sabatier, Toulouse, France

J.W. Barret Imperial College, London, U.K.

M Battipede Politecnico di Torino, Torino, Italy

L. Blanchard INRIA, Sophia Antipolis, France

S. Bonaccorsi Università di Trento, Trento, Italy

B. Bonnard UMR CNRS, Dijon, France

K. Bredies University of Bremen, Bremen, Germany

J. Cagnol Pôle Universitaire Lèonard de Vinci, Paris, France

J.-B. Caillau ENSEEIHT-IRIT, UMR CNRS, Toulouse, France

P. Cannarsa ,Università di Roma "Tor Vergata", Roma ,Italy

E. Casas Universidad de Cantabria, Santander, Spain

M. Chyba University of Hawaii, Honolulu, U.S.A.

D. Lorenz University of Bremen, Bremen, Germany

P. Maass University of Bremen, Bremen, Germany

V. Maksimov Russian Academy of Sciences, Ekaterinburg, Russia

E. Mamontov Gothenburg University, Gothenburg, Sweden

P. Maponi Università di Camerino, Camerino, Italy

R. Marchand Slippery Rock University, Slippery Rock PA, U.S.A.

M. Mateos Alberdi Universidad de Oviedo, Gijòn, Spain

H. Maurer Wilhelms-Universität, Münster, Germany

G.A. Meles Università di Milano, Milano, Italy

A. Morassi Università di Udine, Udine, Italy

A. Myslinski Polish Academy of Sciences, Warszaw, Poland

G. Naldi Università di Milano, Milano, Italy

N.P. Osmolovskii Polish Academy of Sciences, Warszaw, Poland

G. Parravicini Università di Milano, Milano, Italy

G. Ponzini Università di Milano, Milano, Italy

L. Prigozhin Ben Gurion University, Sede-Boqer, Israel

E. Rosset Università di Trieste, Trieste, Italy

Y. Shi Shandong University, Jinan, China

OPTIMAL STOCHASTIC CONTROL OF MEASURE SOLUTIONS ON HILBERT SPACE

N.U. Ahmed [1]

[1] *University of Ottawa, SITE & Department of Mathematics, ahmed@site.uottawa.ca*

Abstract This paper is concerned with optimal control of semilinear stochastic evolution equations on Hilbert space driven by stochastic vector measure. Both continuous and discontinuous (measurable) vector fields are admitted. Due to nonexistence of regular solutions, existence and uniqueness of generalized (or measure valued) solutions are proved. Using these results, existence of optimal feedback controls from the class of bounded Borel measurable maps are proved for several interesting optimization problems.

keywords: Stochastic Differential Equations, Hilbert Space, Measurable Vector Fields, Finitely Additive Measure Solutions, Optimal Feedback Controls.

1. Introduction

For motivation let us consider the deterministic evolution equation

$$\dot{x} = Ax + F(x), \ t \geq 0, \ x(0) = x_0, \tag{1}$$

in a Hilbert space H where A is the infinitesimal generator of a C_0-semigroup, $S(t), t \geq 0$, on H and $F : H \longrightarrow H$ is a continuous map. It is well known that if H is finite dimensional, mere continuity of F is good enough to prove existence of local solutions with possibly finite blow up time. If H is an infinite dimensional Hilbert space continuity no longer guarantees existence of even local solutions unless the semigroup $S(t), t > 0$, is compact. Because of this, the very notion of solutions required a major generalization to cover continuous as well as discontinuous vector fields [11], [1],[2],[3],[4],[5]. Using the general concept of measure solutions one can completely avoid standard assumptions such as local Lipschitz property and linear growth for both the drift and the diffusion operators as often used in [8] and [11].

Many physical models are described by differential equations with discontinuous and possibly unbounded vector fields without satisfying any of the standard regularity properties including monotonicity or accretivity etc. In this

Please use the following format when citing this chapter:

Ahmed, N.U., 2006, in IFIP International Federation for Information Processing, Volume 202, Systems, Control, Modeling and Optimization, eds. Ceragioli, F., Dontchev, A., Furuta, H., Marti, K., Pandolfi, L., (Boston: Springer), pp. 1-12.

situation there is no solution in the sense of any one of the standard notions (classical,strong,mild,weak). Even the weak solution, if one exists, is highly irregular. However under fairly general assumptions one is able to prove existence of measure valued solutions. Functionals (physical observables) of such measure solutions, such as

$$\Upsilon(\mu) \equiv E \int_I f(\mu_t(\varphi_1), \mu_t(\varphi_2). \cdots, \mu_t(\varphi_m))\rho(dt),$$

where $f \in BC(R^m)$, $\varphi_i \in BC(H)$ and ρ is a countably additive bounded positive measure on I having bounded total variation, are of practical interest. For example, in the study of hydrodynamic turbulence, the solution of Euler equation, as the limit of the Navier-Stokes equation with the Reynolds number increasing towards infinity, can be considered as a measure solution. Its impact on mixing is an important physical variable.
Let $\{H, \Xi, E\}$ be any three Hilbert spaces relating the stochastic system governed by an evolution equation of the form

$$\begin{aligned} dx(t) &= Ax(t)dt + F(x(t))dt + \Gamma(x(t))u(t, x(t))dt \\ &\qquad + G(x(t-))M(dt), t \geq 0, \\ x(0) &= x_0. \end{aligned} \tag{2}$$

Here A and F are as described above, and $G : H \longrightarrow \mathcal{L}(E, H)$ is a continuous map and $\Gamma : H \longrightarrow \mathcal{L}(\Xi, H)$ is Borel measurable map and M is an E-valued stochastic vector measure defined on the sigma algebra $\mathcal{B}_0$ of Borel subsets of $R_0 \equiv [0, \infty)$.

The last term on the right hand side of equation (2) gives rise to a well defined stochastic integral since the integrand is adapted and therefore previsible or equivalently measurable with respect to the sigma algebra of predictable sets. An intuitive physical meaning is as follows. Given that $\{t\}$ is an atom of the measure M, the state at time t is uniquely determined by the state $x(t-)$ (just before the jump) and the jump size $M(\{t\})$ of the measure M.

For simplicity of presentation, we have considered the operators F, G, Γ to be independent of time. However the results presented here can be extended to the time varying case without any difficulty.
The paper is organized as follows. In section 2 we recall some important facts from analysis sufficient to serve our needs. In section 3 and 4, we consider system (2) and present without proof two results on existence of measure valued solutions and their regularity properties. Using these results, in section 5, we consider control problems and present several results on the question of existence of optimal feedback controls.

2. Basic Facts from Analysis

Recently the author dealt with the question of existence of measure valued solutions for semilinear stochastic differential equations with continuous but unbounded nonlinearities driven by cylindrical Brownian motion [3]. Here we admit Borel measurable, possibly unbounded, vector fields and replace the Brownian motion by a more general stochastic vector measure. Properties of the stochastic vector measure are stated in the sequel.

Radon Nikodyme Property & Lifting:

For any normal topological space Z, let $BC(Z)$ and $B(Z)$ denote the vector spaces of bounded continuous and bounded Borel measurable functions on Z respectively. Furnished with the supnorm topology these are Banach spaces. It follows from a well known result [10] that the corresponding duals are given by $\Sigma_{rba}(Z)(\Sigma_{ba}(Z))$ which are regular bounded (bounded) finitely additive measures on the algebra of sets determined closed subsets of Z. Note that the dual pairs $\{BC(Z), \Sigma_{rba}(Z)\}$ and $\{B(Z), \Sigma_{ba}(Z)\}$ do not satisfy Radon-Nikodym (RNP) property [8]. Hence, for any finite measure space $(S, \mathcal{S}, \gamma)$, it follows from the theory of lifting that the dual of $L_1(S, BC(Z))$ is given by $L^w_\infty(S, \Sigma_{rba}(Z))$. These are weak star measurable measure valued functions. To study the question of existence, we need these spaces.

Special Vector Spaces Used:

Now we are prepared to introduce the vector spaces used in the paper. Let H, E be two separable Hilbert spaces and $(\Omega, \mathcal{F}, \mathcal{F}_t \uparrow, t \geq 0, P)$ a complete filtered probability space, $M(J), J \in \mathcal{B}_0$, an E valued $\mathcal{F}_t$ adapted vector measure in the sense that for any $J \in \mathcal{B}_0$ with $J \subset [0, t]$, $M(J)$ or more precisely $e^*(M(J))$ is $\mathcal{F}_t$ measurable for every $e^* \in E^* = E$. For the purpose of this paper we consider $\mathcal{F}_t \equiv \mathcal{F}_t^M \vee \sigma(x_0)$, where $\mathcal{F}_t^M, \sigma(x_0)$ are the smallest sigma algebras with respect to which the measures M and the initial state x_0 respectively are measurable. Let $I \times \Omega$ be furnished with the predictable σ-field with reference to the filtration $\mathcal{F}_t, t \in I$ and $M^w_{\infty,2}(I \times \Omega, \Sigma_{rba}(H)) \subset L^w_{\infty,2}(I \times \Omega, \Sigma_{rba}(H))$ denote the vector space of $\Sigma_{rba}(H)$ valued random processes $\{\lambda_t, t \in I\}$, which are $\mathcal{F}_t$-adapted and w^*-measurable in the sense that, for each $\phi \in BC(H)$, $t \longrightarrow \lambda_t(\phi)$ is a bounded $\mathcal{F}_t$ measurable random variable possessing finite second moments. We furnish this space with the w^* topology. Clearly this is the dual of the Banach space

$$M_{1,2}(I \times \Omega, BC(H)) \subset L_{1,2}(I \times \Omega, BC(H))),$$

where the later space is furnished with the natural topology induced by the norm given by

$$\| \varphi \| \equiv \int_I \left(\mathcal{E}(\sup\{|\varphi(t, \omega, \xi)|, \xi \in H\})^2 \right)^{1/2} dt.$$

Similarly, one can verify that $M^{w}_{\infty,2}(I \times \Omega, \Sigma_{ba}(H))$ is the dual of the Banach space $M_{1,2}(I \times \Omega, B(H))$. We will have occasion to use both these spaces.

Basic properties of M :

(M1): $\{M(J), M(K), J \cap K = \emptyset, J, K \in \mathcal{B}_0\}$ are pair wise independent E-valued random variables (vector measures) satisfying $\mathcal{E}\{(M(J), \xi)\} = 0, J \in \mathcal{B}_0, \xi \in E$, where $\mathcal{E}(z) \equiv \int_\Omega zP(d\omega)$.

(M2): There exists a countably additive bounded positive measure $\pi \in M_c(\mathcal{R}_0)$, having bounded total variation on bounded sets, such that for every $\xi, \zeta \in E$,

$$\mathcal{E}\{(M(J), \xi)(M(K), \zeta)\} = (\xi, \zeta)_E \, \pi(J \cap K).$$

Clearly, it follows from this last property that, for any $\xi \in E, \mathcal{E}\{(M(J), \xi)^2\} = |\xi|^2_E \pi(J)$, and that the process N, defined by

$$N(t) \equiv \int_0^t M(ds), t \geq 0,$$

is a square integrable E-valued $\mathcal{F}_t$-martingale. Gyongi and Krylov [12] have also considered general noise for a special class of finite dimensional systems.

3. EXISTENCE OF MEASURE SOLUTIONS

In recent years a notion of generalized solution, which consists of regular finitely additive measure valued functions, has been extensively used in the study of semi linear and quasi linear systems with vector fields which are merely continuous and bounded on bounded sets; see [1],[2],[3], [11] and the references therein. Existence of solutions for deterministic systems, such as (1), was proved in [1,2,11] with varying degrees of generality. Recently [6] existence of measure solutions for stochastic system (2), generalizing a previous result of the author [3], has been proved. These latest results cover Borel measurable drift and diffusion assumed to be merely bounded on bounded sets. Our main objective here is to prove existence of optimal feedback controls for these class of systems.

Since the measure solutions may not be fully supported on the original state space H, it is useful to extend the state space to a compact Hausdorff space containing H as a dense subspace. Since every metric space is a Tychonoff space, H is a Tychonoff space. Hence $H^+ \equiv \beta H$, the Stone-Čech compactification of H, is a compact Hausdorff space and consequently bounded continuous functions on H can be extended to continuous functions on H^+. In view of this we shall often use H^+ in place of H and the spaces $M_{1,2}(I \times \Omega, BC(H^+))$ with dual $M^{w}_{\infty,2}(I \times \Omega, \Sigma_{rba}(H^+)) \supset M^{w}_{\infty,2}(I \times \Omega, \Pi_{rba}(H^+))$. Here $M^{w}_{\infty,2}(I \times \Omega, \Pi_{rba}(H^+))$ is the set of all finitely additive probability measure valued processes, a subset of the vector space $M^{w}_{\infty,2}(I \times \Omega, \Sigma_{rba}(H^+))$. Note that, since

H^+ is a compact Hausdorff space, $\Sigma_{rba}(H^+) = \Sigma_{rca}(H^+)$. In view of the fact that the measure solutions of stochastic evolution equations restricted to H are only finitely additive, we prefer to use the notation $\Sigma_{rba}(H^+)$ to emphasize this fact though they are countably additive on H^+.

Without further notice, throughout this paper we use $D\phi$ and $D^2\phi$ to denote the first and second Frechet derivatives of the function ϕ whenever they exist. We denote by Ψ the class of test functions as defined below:

$$\Psi \equiv \{\phi \in BC(H) : D\phi, D^2\phi \text{ exist, continuous and bounded on } H\}.$$

Define the operators $\mathcal{A}$ $\mathcal{B}$ and $\mathcal{C}$ with domains given by

$$\begin{aligned}
&\mathcal{D}(\mathcal{A}) \equiv \{\phi \in \Psi : \mathcal{A}\phi \in BC(H^+)\} \\
&\mathcal{D}(\mathcal{B}) \equiv \{\phi \in \Psi : D\phi \in D(A^*) \;\&\; \mathcal{B}\phi \in BC(H^+)\},
\end{aligned}$$

where

$$\begin{aligned}
&(\mathcal{A}\phi)(\xi) = (1/2)Tr(D^2\phi\, GG^*)(\xi), \phi \in \mathcal{D}(\mathcal{A}) \\
&\mathcal{B}\phi = (A^* D\phi(\xi), \xi) + (F(\xi), D\phi(\xi)) \;\text{ for }\; \phi \in \mathcal{D}(\mathcal{B}) \\
&\mathcal{C}\phi(\xi) \equiv G^*(\xi) D\phi(\xi).
\end{aligned} \tag{3}$$

First we consider the uncontrolled system

$$\begin{aligned}
&dx(t) = Ax(t)dt + F(x(t))dt + G(x(t-))M(dt), \\
&x(0) = x_0,
\end{aligned} \tag{4}$$

and use the notion of measure (generalized) solutions introduced in [3] and finally add modifications necessary for the control system.

DEFINITION 1 *A measure valued random process $\mu \in M^w_{\infty,2}(I \times \Omega, \Pi_{rba}(H^+))$ is said to be a measure (or generalized) solution of equation (4) if for every $\phi \in \mathcal{D}(\mathcal{A}) \cap \mathcal{D}(\mathcal{B})$ and $t \in I$, the following equality holds*

$$\begin{aligned}
\mu_t(\phi) \;\; = \;\; & \phi(x_0) + \int_0^t \mu_s(\mathcal{A}\phi)\, \pi(ds) + \int_0^t \mu_s(\mathcal{B}\phi)\, ds \\
& + \int_0^t < \mu_{s-}(\mathcal{C}\phi), M(ds) >_E \;\; P - a.s.
\end{aligned} \tag{5}$$

where $\mu_t(\psi) \equiv \int_{H^+} \psi(\xi)\mu_t(d\xi), t \in I$.

Again the stochastic integral is well defined since the integrand is predictable and the process $\mu_t, t \in I$, is cadlag in the weak star sense. This follows from the assumption that the compensating measure π has bounded total variation and hence M can have at most a countable set of atoms.

REMARK 2 *Note that equation (5) can be written in the differential form as follows:*

$$d\mu_t(\phi) = \mu_t(\mathcal{A}\phi)\pi(dt) + \mu_t(\mathcal{B}\phi)dt + < \mu_{t-}(\mathcal{C}\phi), M(dt) >$$

with $\mu_0(\phi) = \phi(x_0)$*. This is in fact the weak form of the stochastic evolution equation*

$$d\mu_t = \mathcal{A}^*\mu_t\pi(dt) + \mathcal{B}^*\mu_t dt + < \mathcal{C}^*\mu_{t-}, M(dt) >_E, \ \mu_0 = \delta_{x_0}, \quad (6)$$

on the state space $\Sigma_{rba}(H)$ *where* $\{\mathcal{A}^*, \mathcal{B}^*, \mathcal{C}^*\}$ *are the duals of the operators* $\{\mathcal{A}, \mathcal{B}, \mathcal{C}\}$.

In case the system (2) has a weak solution in the classical sense, the measure solution degenerates into a Dirac measure concentrated along the path and the expression (5) defines the weak solution.

To proceed further we shall need the following

ASSUMPTION 3 *(A1): there exists a sequence* $\{F_n, G_n\}$ *with* $F_n(x) \in D(A)$ *and* $G_n(x) \in \mathcal{L}(E, D(A))$*, for each* $x \in H$*, and*

$$F_n(x) \xrightarrow{\tau_{wuc}} F(x) \text{ in } H \text{ and } G_n(x) \xrightarrow{\tau_{souc}} G(x) \text{ in } \mathcal{L}(E, H),$$

where $\tau_{wuc}(\tau_{souc})$ *denotes the topology of weak convergence (convergence in strong operator topology) uniformly on compacts.*

(A2): there exists a pair of sequence of real numbers $\{\alpha_n, \beta_n > 0\}$*, possibly* $\alpha_n, \beta_n \to \infty$ *as* $n \to \infty$*, so that both* F_n, G_n *are Lipschitz having linear growth with coefficients* α_n, β_n *respectively.*

We note that under the very relaxed assumptions used here, nonlinearities having polynomial growth are also admissible.

Following result generalizes our previous result ([3], Theorem 3.2).

THEOREM 4 *Suppose A is the infinitesimal generator of a* C_0*-semigroup in H and the maps* $F : H \longrightarrow H$, $G : H \longrightarrow \mathcal{L}(E, H)$ *are continuous, and bounded on bounded subsets of H, satisfying the approximation properties (A1) and (A2); and M is the vector measure satisfying (M1) and (M2). Then, for every* x_0 *for which* $P\{\omega \in \Omega : |x_0|_H < \infty\} = 1$*, the evolution equation (4) has at least one measure valued solution*

$$\lambda^0 \in M^w_{\infty,2}(I \times \Omega, \Sigma_{rba}(H^+))$$

in the sense of Definition 1. Further, $\lambda^0 \in M^w_{\infty,2}(I \times \Omega, \Pi_{rba}(H^+))$.

Proof. Proof will appear in [6].

Supports of measure solutions may leak out to infinity in the sense that $\lambda_t^0(H) < 1$ while $\lambda_t^0(H^+) = 1$ with probability one. For any $\beta > 0$ and $t \in I$, the operator Q_t^β as given below is well defined

$$(Q_t^\beta \xi, \xi) \equiv E \int_{H^+} (x, \xi)^2 \exp - \{\beta |x|^2\} \, \lambda_t^0(dx).$$

It follows from Milnos-Sazanov theorem that if $\limsup_{\beta \downarrow 0} Tr(Q_t^\beta) < \infty$ then we are assured of H being the support of the measure solution. If λ_t^o is not supported on H, we can still define an H-valued process given by

$$X_t^\beta \equiv \int_{H^+} x \exp - \{\beta |x|_H^2\} \, \lambda_t^0(dx)$$

which can serve as the approximate path process becoming highly irregular with decreasing β and possibly escaping to H^+ as $\beta \downarrow 0$.

REMARK 5 *In view of Theorem 4, F, G are required to be merely continuous and bounded on bounded sets and hence they may have polynomial growth [3]. In contrast, for standard mild solutions it is usually assumed that F, G are locally Lipschitz having at most linear growth [8].*

The following corollary is an immediate consequence of Theorem 4.

COROLLARY 6 *Consider the forward Kolmogorov equation,*

$$d\vartheta_t = \mathcal{A}^* \vartheta_t \pi(dt) + \mathcal{B}^* \vartheta_t dt, \; \vartheta(0) = \nu_0, \tag{7}$$

where $\mathcal{A}^*$, $\mathcal{B}^*$ are the duals of the operators $\mathcal{A}$, $\mathcal{B}$ respectively (see equation 3) with F, G satisfying the assumptions of Theorem 4. Then, for every $\nu_0 \in \Pi_{rba}(H)$, equation (7) has at least one weak solution $\nu \in L_\infty^w(I, \Pi_{rba}(H^+)) \subset L_\infty^w(I, \Sigma_{rba}(H^+))$ in the sense that for each $\phi \in D(\mathcal{A}) \cap \mathcal{D}(\mathcal{B})$ and $t \in I$, the following equality holds

$$\nu_t(\phi) = \nu_0(\phi) + \int_0^t \nu_s(\mathcal{A}\phi) \, \pi(ds) + \int_0^t \nu_s(\mathcal{B}\phi) ds. \tag{8}$$

Proof. Proof will appear in [6].

REMARK 7 *Note that Corollary 6 proves existence of (measure) solutions of Kolmogorov equation (7) with unbounded coefficients generalizing similar results of [7] for parabolic and elliptic equations on finite dimensional spaces.*

The following corollary asserts uniqueness.

COROLLARY 8 (UNIQUENESS) *Suppose the assumptions of Corollary 6 hold. Then the solution (weak solution) of the evolution equation (7) is unique.*

Proof. Proof will appear in [6].

REMARK 9 *Using this result we can prove the uniqueness of mild and hence weak solution of the stochastic measure equation (6).*

4. MEASURABLE VECTOR FIELDS

In many applications, F, G and Γ may not be even continuous. However, assuming that they are bounded Borel measurable, it is possible to prove existence results similar to those of deterministic evolutions [5].

Consider the control system

$$dx(t) = Ax(t)dt + F(x(t))dt + \Gamma(x(t))\, u(t,x)\, dt + G(x(t-))M(dt)$$
$$x(0) = x_0, \tag{9}$$

where $\Gamma : H \longrightarrow \mathcal{L}(\Xi, H)$ is a bounded Borel measurable map with Ξ being another separable Hilbert space and $u : I \times H \longrightarrow \Xi$ is any bounded Borel measurable function representing the control. Let $BM(I \times H, \Xi)$ denote the class of bounded Borel measurable functions from $I \times H$ to Ξ. Furnished with the uniform norm topology,

$$\| u \| \equiv \sup\{|u(t,x)|_{\Xi}, (t,x) \in I \times H\},$$

it is a Banach space. We present here a result analogous to that of Theorem 4with the major exception that in the present case the measure solutions are no longer regular. They are bounded finitely additive measure valued processes.

THEOREM 10 *Consider the system (9). Suppose $\{A, M\}$ satisfy the assumptions of theorem 4, $F : H \longrightarrow H$, $G : H \longrightarrow \mathcal{L}(E, H)$ and $\Gamma : H \longrightarrow \mathcal{L}(\Xi, H)$ are Borel measurable maps bounded on bounded sets. Then, for every x_0 for which $P\{\omega \in \Omega : |x_0|_H < \infty\} = 1$, statistically independent of M, and $u \in BM(I \times H, \Xi)$, the evolution equation (9) has a unique measure solution*

$$\lambda^0 \in M^w_{\infty,2}(I \times \Omega, \Pi_{ba}(H^+)).$$

Proof. Proof will appear in [6].

5. OPTIMAL FEEDBACK CONTROLS

Consider the control system (9). For admissible controls, we use a weaker topology and introduce the following class of functions. Let U be a closed bounded (possibly convex) subset of Ξ and

$$\mathcal{U} \equiv \{u \in BM(I \times H, \Xi) : u(t,x) \in U\, \forall\, (t,x)\}.$$

On $BM(I \times H, \Xi)$, we introduce the topology of weak convergence in Ξ uniformly on compact subsets of $I \times H$ and denote this topology by τ_{wu}. In other words, a sequence $\{u_n\} \subset BM(I \times H, \Xi)$ is said to converge to $u_0 \in BM(I \times H, \Xi)$ in the topology τ_{wu} if, for every $v \in \Xi$,

$$(u_n(t,x), v)_{\Xi} \longrightarrow (u_0(t,x), v)_{\Xi}$$

uniformly in (t, x) on compact subsets of $I \times H$. We assume that $\mathcal{U}$ has been furnished with the relative τ_{wu} topology and $\mathcal{U}_{ad}$ any τ_{wu} sequentially compact (possibly) convex subset of $\mathcal{U}$ and choose this set for admissible controls.

We consider the Lagrange problem $P1$: Find $u^o \in \mathcal{U}_{ad}$ that minimizes the cost functional

$$J(u) \equiv \mathcal{E} \int_0^T \ell(t, x(t))dt, \tag{10}$$

where ℓ is any real valued Borel measurable function on $I \times H$ which is bounded on bounded sets. Since, under the general assumptions of Theorem 4 and Theorem 10, the control system (9) has only measure solutions, the control problem as stated above must be reformulated in terms of measure solutions. For this purpose we introduce the operator $\mathcal{B}_u$ associated with the control u as follows. Define, for $(t, \xi) \in I \times H$,

$$(\mathcal{B}_u \phi)(t, \xi) \equiv (u(t, \xi), \Gamma^*(\xi) D\phi(\xi))_{\Xi},$$

where $\Gamma^*(\xi) \in \mathcal{L}(H, \Xi)$ is the adjoint of the operator $\Gamma(\xi)$. Clearly the operator $\mathcal{B}_u$ is well defined on $D(\mathcal{A}) \cap D(\mathcal{B})$. Then the correct formulation of the original control problem is given by $(P1)$: find $u^o \in \mathcal{U}_{ad}$ that minimizes the functional

$$J(u) \equiv \mathcal{E} \int_0^T \int_H \ell(t, \xi) \lambda_t^u(d\xi)dt \tag{11}$$

where λ^u is the (weak) solution of equation

$$\begin{aligned} d\lambda_t &= \mathcal{A}^* \lambda_t \pi(dt) + \mathcal{B}^* \lambda_t dt + \mathcal{B}_u^* \lambda_t dt + < \mathcal{C}^* \lambda_{t-}, M(dt) >_E, \\ \lambda_0 &= \delta_{x_0} \,. \end{aligned} \tag{12}$$

Note that the initial measure need not be a Dirac measure, it suffices if $\lambda_0 = \pi_0 \in \Pi_{ba}(H)$.

For convenience of reference we identify this problem as P_1. We need the following result on continuous dependence of solutions on control.

LEMMA 11 *Consider the system (12) with admissible controls $\mathcal{U}_{ad}$ as defined above, and suppose the assumptions of Theorem 10 hold and that $\Gamma : H \longrightarrow \mathcal{L}(\Xi, H)$ is a bounded Borel measurable map. Then, for every $u \in \mathcal{U}_{ad}$, the system (12) has a unique weak solution $\lambda^u \in M^w_{\infty,2}(I \times \Omega, \Pi_{ba}(H^+))$ and further, the control to solution map $u \longrightarrow \lambda^u$ from $\mathcal{U}_{ad}$ to $M^w_{\infty,2}(I \times \Omega, \Sigma_{ba}(H^+))$ is (sequentially) continuous with respect to the topologies τ_{wu} on $\mathcal{U}_{ad}$ and weak star topology on $M^w_{\infty,2}(I \times \Omega, \Sigma_{ba}(H^+))$.*

Proof. Detailed proof will appear in [6].

Now we consider the control problem $P1$.

THEOREM 12 *Consider the system (12) and the Lagrange problem (11) with admissible controls $\mathcal{U}_{ad}$. Suppose the assumptions of Lemma 11 hold and that ℓ is a Borel measurable real valued function defined on $I \times H$ and bounded on bounded sets and that there exists a function $\ell_0 \in L_1(I)$ such that $\ell(t,\xi) \geq \ell_0(t)\ \forall\ \xi \in H$. Then there exists an optimal control for the problem $P1$.*

Proof. Since ℓ is bounded from below by an integrable function ℓ_0, we have $J(u) > -\infty$, $\forall\ u \in \mathcal{U}_{ad}$. Clearly if $J(u) = +\infty$ for all $u \in \mathcal{U}_{ad}$, there is nothing to prove. So suppose the contrary. Define $\inf\{J(u), u \in \mathcal{U}_{ad}\} = m$, and let $\{u^n\} \subset \mathcal{U}_{ad}$ be a minimizing sequence. Since $\mathcal{U}_{ad}$ is τ_{wu} sequentially compact, there exists a subsequence, relabeled as the original sequence, and a control $u^o \in \mathcal{U}_{ad}$ such that $u^n \xrightarrow{\tau_{wu}} u^o$. Then by virtue of Lemma 11, along a further subsequence if necessary, we have $\lambda^{u^n} \xrightarrow{w^*} \lambda^{u^o}$. Note that the functional (11) is linear in λ^u and bounded (since $\{u^n\}$ is a minimizing sequence) and hence continuous along the minimizing sequence $\{\lambda^{u^n}\}$. Thus $\lim_{n\to\infty} J(u^n) = J(u^o) = m$ and u^o is the optimal control. •

Next we consider the control problem $P2$:

$$J(u) \equiv \mathcal{E}\int_{I\times H}\{\ell(t,\xi) + \rho(\xi)|u(t,\xi)|_{\Xi}\}\lambda_t^u(d\xi)dt \longrightarrow \inf, \quad (13)$$

where ρ is a nonnegative bounded Borel measurable function on H with compact support and λ^u is the weak solution of the stochastic evolution equation (12) corresponding to control u.

THEOREM 13 *Consider the Lagrange problem $P2(13)$ subject to the dynamics of system (12) with admissible controls $\mathcal{U}_{ad}$. Suppose ℓ satisfies the conditions as in Theorem 12, and ρ is any nonnegative bounded Borel measurable function on H having compact support. Then there exists an optimal control for the problem $P2$.*

Proof. Again by virtue of the assumption on ℓ, we have $J(u) > -\infty$. If $J(u) \equiv +\infty$ for all $u \in \mathcal{U}_{ad}$ there is nothing to prove. So we may assume the contrary. Let $\{u^n\}$ be a minimizing sequence so that

$$\lim_{n\to\infty} J(u^n) = \inf\{J(u), u \in \mathcal{U}_{ad}\} \equiv \tilde{m}.$$

We show that the second term of the objective functional (13), denoted by J_2, is τ_{wu} lower semi continuous on $\mathcal{U}_{ad}$. Since $\mathcal{U}_{ad}$ is τ_{wu} sequentially compact, the sequence $\{u^n\}$ contains a subsequence, relabeled as the original sequence, which converges in τ_{wu} topology to an element $u^o \in \mathcal{U}_{ad}$. Consider the value of J_2 at u^o,

$$J_2(u^o) \equiv \mathcal{E}\int_{I\times H}\rho(\xi)|u^o(t,\xi)|_{\Xi}\lambda_t^{u^o}(d\xi)dt. \quad (14)$$

Since $u^o(t,\xi)$ is a Ξ valued bounded Borel measurable function, by Hahn-Banach theorem there exists a $B_1(\Xi)$ valued bounded measurable function η^o on $I \times H$ such that

$$|u^o(t,\xi)|_\Xi = (u^o(t,\xi), \eta^o(t,\xi))_\Xi, \ \forall (t,\xi) \in I \times H.$$

In fact one can take $\eta^o(t,\xi) = u^o(t,\xi)/|u^o(t,\xi)|_\Xi$. Using Lemma 11 and some functional analytic arguments one can verify that

$$J_2(u^o) \le \liminf_{n\to\infty} J_2(u^n). \tag{15}$$

Thus J_2 is τ_{wu} lower semicontinuous and it follows from continuity of the first term that J is τ_{wu} lower semicontinuous. The existence now follows from τ_{wu} sequential compactness of $\mathcal{U}_{ad}$. •

Another interesting control problem, identified as $P3$, consists of maximizing the functional:

$$J(u) = f(\mathcal{E}\lambda^u_{t_1}(\varphi_1), \cdots, \mathcal{E}\lambda^u_{td}(\varphi_d))\} \longrightarrow \sup$$

where $f : R^d \longrightarrow R$ is a function, and $\{\varphi_i\} \in B(H)$ is a finite set of bounded real valued Borel measurable functions on H.

THEOREM 14 *Consider the system (12) with admissible controls $\mathcal{U}_{ad}$ and suppose the assumptions of Lemma 11 hold. Further, suppose the stochastic vector measure M is nonatomic and the associated quadratic variation measure π is absolutely continuous with respect to the Lebesgue measure and the function f is upper semicontinuous from R^d to R and $\{\varphi_i\} \in B(H)$ are real valued bounded Borel measurable functions. Then the Problem P3 has a solution.*

Proof. For proof see [6].•

A fourth interesting problem, identified as $P4$, can be stated as follows. Let $\Psi \in B(H)$ and $g \in C_b(R)$ be given. The problem is to find a control that minimizes (maximizes) the functional

$$J(u) \equiv \mathcal{E}g(\lambda^u_T(\Psi)). \tag{16}$$

THEOREM 15 *Consider the system (12) with admissible controls $\mathcal{U}_{ad}$ and suppose the assumptions of Lemma 11 hold. Further, suppose $\{M, \pi\}$ satisfy the assumptions of Theorem 14 and $g \in C_b(R)$ and $\Psi \in B(H)$. Then the Problem P4 has a solution.*

Proof. Proof will appear in [6].

REMARK 16 *For construction of optimal controls, it is essential to develop necessary and sufficient conditions for optimality. We leave this as an open problem for the readers.*

REMARK 17 *In section 5, the set $\mathcal{U}_{ad}$ is assumed to be τ_{wu} sequentially compact. The author feels that simple τ_{wu} compactness is sufficient. In general topological spaces compactness and continuity can all be described in terms of nets. Lemma 11 can be proved using nets and since all the results following Lemma 11 depend only on τ_{wu} continuity and compactness, they remain valid under this relaxed condition.*

References

[1] N.U. Ahmed. Measure Solutions for Semi-linear Evolution Equations with Polynomial Growth and Their Optimal Controls. *Discussiones Mathematicae, Differential Inclusions.*, 17, 5-27, 1997.

[2] N.U. Ahmed. Measure Solutions for Semilinear Systems with Unbounded Nonlinearities. *Nonlinear Analysis: Theory, Methods and Applications*, 35, 487-503, 1999.

[3] N.U. Ahmed. Relaxed Solutions for Stochastic Evolution Equations on Hilbert Space with Polynomial Nonlinearities. *Publicationes Mathematicae, Debrecen*, 54 (1-2), 75-101, 1999.

[4] N.U. Ahmed. Measure Solutions for Evolution Equations with Discontinuous Vector Fields. *Nonlinear Functional analysis and Applications.* 9(3), 467-484, 2004.

[5] N.U. Ahmed. Measure Solutions for Impulsive Evolution Equations with Measurable Vector Fields. *Journal of Math. Annal. and Appl*, 2004 (submitted).

[6] N.U. Ahmed. Measure Valued Solutions for Stochastic Evolution Equations on Hilbert Space and Their Feedback Control. *Discussiones Mathematicae, DICO, 2005*, 25 (to appear).

[7] S. Cerrai. Elliptic and parabolic equations in R^n with coefficients having polynomial growth. *Preprints di Matematica, n.9*, Scuola Normale Superiore, Pisa, 1995.

[8] G. Da Prato and J. Zabczyk. *Stochastic Equations in Infinite Dimensions*. Encyclopedia of Mathematics and its Applications, 44, Cambridge University Press, 1992.

[9] J. Diestel and J.J.Uhl,Jr.. *Vector Measures*. Math. Surveys Monogr. 15, AMS, Providence, RI, 1977.

[10] N. Dunford and J.T. Schwartz. *Linear Operators, Part 1.* Interscience Publishers, Inc., New York, 1958.

[11] H. Fattorini. A Remark on Existence of solutions of Infinite Dimensional Non-compact Optimal Control Problems, *SIAM J. Control and Optim.* 35(4), 1422-1433, 1997.

[12] I. Gyongy and N.V. Krylov. On Stochastic Equations with Respect to Semimartingales, *Stochastics* no. 1, 1-21, 1980/81.

FEEDBACK STABILIZATION OF THE 3-D NAVIER-STOKES EQUATIONS BASED ON AN EXTENDED SYSTEM

M. Badra[1]

[1]*Laboratoire MIP, UMR CNRS 5640, Université Paul Sabatier, 31062 Toulouse Cedex 4, France, badra@mip.ups-tlse.fr*

Abstract We study the local exponential stabilization of the 3D Navier-Stokes equations in a bounded domain, around a given steady-state flow, by means of a boundary control. We look for a control so that the solution to the Navier-Stokes equation be a strong solution. In the 3D case, such solutions may exist if the Dirichlet control satisfies a compatibility condition with the initial condition. In order to determine a feedback law satisfying such a compatibility condition, we consider an extended system coupling the Navier-Stokes equations with an equation satisfied by the control on the boundary of the domain. We determine a linear feedback law by solving a linear quadratic control problem for the linearized extended system. We show that this feedback law also stabilizes the nonlinear extended system.

Keywords: Navier-Stokes equation, Feedback stabilization, Riccati equation.

1. Introduction

Let $\mathcal{O}$ and $\mathcal{B}$ two regular bounded domains of class C^∞ in $\mathbf{R}^3$ such that $\overline{\mathcal{B}} \subset \mathcal{O}$, $\Omega = \mathcal{O}\backslash\overline{\mathcal{B}}$, $\Gamma_e = \partial\mathcal{O}$ and $\Gamma_i = \partial\mathcal{B}$. We have $\Gamma_i \cap \Gamma_e = \emptyset$ and $\partial\Omega = \Gamma_i \cup \Gamma_e$. We consider the motion of an incompressible fluid around the bounded body $\mathcal{B}$ in Ω which is described by the couple (z_e, p_e), the velocity and the pressure, solution to the stationary Navier-Stokes equations

$$\begin{aligned} &-\Delta z_e + (z_e \cdot \nabla) z_e + \nabla p_e = 0 \text{ and } \nabla \cdot z_e = 0 \text{ in } \Omega, \\ &z_e = 0 \text{ on } \Gamma_i, \quad z_e = v_\infty \text{ on } \Gamma_e. \end{aligned}$$

According to [4], if $v_\infty \in H^{\frac{3}{2}}(\Gamma_i; \mathbf{R}^3)$ obeys $\int_{\Gamma_e} v_\infty \cdot n = 0$, such a stationary solution exists in $H^2(\Omega, \mathbf{R}^3) \times H^1(\Omega)/\mathbf{R}$. For an initial condition of the form $z_e + z_0$ and a Dirichlet boundary control u on Γ_i such that $\int_{\Gamma_i} u(t) \cdot n = 0$, the pair $(z + z_e, p + p_e)$ satisfies the instationary Navier-Stokes equations, and

Please use the following format when citing this chapter:

Badra, M., 2006, in IFIP International Federation for Information Processing, Volume 202, Systems, Control, Modeling and Optimization, eds. Ceragioli, F., Dontchev, A., Furuta, H., Marti, K., Pandolfi, L., (Boston: Springer), pp. 13-24.

(z,p) obeys:

$$\partial_t z - \Delta z + (z\cdot\nabla)z_e + (z_e\cdot\nabla)z + (z\cdot\nabla)z + \nabla p = 0 \text{ in } Q, \quad (1)$$

$$\nabla\cdot z = 0 \text{ in } Q,\ \ z = u \text{ on } \Sigma_i,\ \ z = 0 \text{ on } \Sigma_e,\ \ z(0) = z_0. \quad (2)$$

In this setting $Q = \Omega\times(0,\infty)$, $\Sigma_i = \Gamma_i\times(0,\infty)$, $\Sigma_e = \Gamma_e\times(0,\infty)$ and n denotes the unit normal vector to Γ_i, exterior to Ω. We assume that z_e is an unstable solution of (1)-(2) corresponding to $z_0 = z_e$. Our goal is to find a Dirichlet boundary control u on Γ_i which stabilizes the instationary Navier-Stokes system (1)-(2) for initial data z_0 small enough in an appropriate functional space. To achieve this goal, the three dimensional case is hightly demanding in terms of velocity regularity: we need that $z\in L^2(0,\infty;\mathbf{H}^{\frac{3}{2}}(\Omega))$ to obtain a stabilization result. Therefore, we look for a control u regular enough to fit the expected smoothness of z and in particular, the initial compatibility condition $u(0) = z_0|_{\Gamma_i}$ should be satisfied. A way to obtain such compatibility condition is to characterize the trace u as the first component of $(u,\sigma)\in L^2_{loc}([0,\infty);L^2(\Gamma_i;\mathbf{R}^3))\times L^2_{loc}([0,\infty))$, where (u,σ) is the solution to the time dependent equation:

$$\begin{aligned} \partial_t u &= \Delta_b u + \sigma\, n + g \quad \text{on } \Sigma_i,\\ u(0) &= z_{0|\Gamma_i}, \quad \text{and } \textstyle\int_{\Gamma_i} u(t)\cdot n = 0. \end{aligned}$$

Here Δ_b is a Laplace Beltrami operator and $g\in L^2(\Sigma_i;\mathbf{R}^3)$ is such that $g(t)$ obeys $\int_{\Gamma_i} g(t)\cdot n = 0$. Thus the state (z,u) now satisfy an extended system of two coupled equations with a distributed control g on Γ_i:

$$\partial_t z - \Delta z + (z\cdot\nabla)z_e + (z_e\cdot\nabla)z + (z\cdot\nabla)z + \nabla p = 0 \text{ in } Q, \quad (3)$$

$$\nabla\cdot z = 0 \text{ in } Q, \quad z = u \text{ on } \Sigma_i, \quad z = 0 \text{ on } \Sigma_e, \quad z(0) = z_0, \quad (4)$$

$$\partial_t u - \Delta_b u - \sigma\, n = g \text{ in } \Sigma_i, \quad \int_{\Gamma_i} u(t)\cdot n = 0, \quad u(0) = z_0|_{\Gamma_i}. \quad (5)$$

In a first step, we consider the linear problem derived from this last coupled system by dropping the nonlinear term $(z\cdot\nabla)z$. We introduce the velocity space $V^0_n(\Omega) = \{y\in L^2(\Omega;\mathbf{R}^3) \mid \nabla\cdot y = 0 \text{ in } \Omega, y\cdot n = 0 \text{ on } \partial\Omega\}$, and the orthogonal projector P from $L^2(\Omega;\mathbf{R}^3)$ into $V^0_n(\Omega)$. Next we rewrite the extended system as an evolution equation (see section **2.2**) involving a linear unbounded operator $\mathcal{A}$ which is studied in section **4**. Then we state a linear quadratic optimal control problem (see section **2.3**) which provides a distributed feedback controller for the extended system (see section **5**). Finally, we apply the feedback controller to the initial nonlinear system (see section **6**) and we show a local stabilization result.

2. Extended system and optimal control problem

2.1 Functional framework

Let us define the spaces of free divergence functions

$$\begin{aligned} V^s(\Omega) &= \{y \in H^s(\Omega; \mathbf{R}^3) \mid \nabla \cdot y = 0 \text{ in } \Omega,\ \int_{\partial\Omega} y \cdot n = 0\}, \quad s \geq 0, \\ V^s_n(\Omega) &= \{y \in H^s(\Omega; \mathbf{R}^3) \mid \nabla \cdot y = 0 \text{ in } \Omega,\ y \cdot n = 0 \text{ on } \partial\Omega\}, \quad s \geq 0, \end{aligned}$$

and the corresponding trace spaces with a free mean normal component

$$V^s(\Gamma_i) = \{y \in H^s(\Gamma_i; \mathbf{R}^3) \mid \int_{\Gamma_i} y \cdot n = 0\}, \quad V^{-s}(\Gamma_i) = V^s(\Gamma_i)', \quad s \geq 0.$$

We denote by $V^s_0(\Omega)$ the interpolation space $[V^2(\Omega) \cap H^1_0(\Omega; \mathbf{R}^3)\,,\ V^0_n(\Omega)]_{1-s/2}$ for $0 \leq s \leq 2$ and $V^{-s}(\Omega) = V^s_0(\Omega)'$ its dual counterpart with respect to the pivot space $V^0_n(\Omega)$. It is well known that

$$\begin{aligned} V^s_0(\Omega) &= V^s_n(\Omega), \quad 0 \leq s < \tfrac{1}{2}, \\ V^{\frac{1}{2}}_0(\Omega) &= \{\, y \in V^{\frac{1}{2}}_n(\Omega) \mid \int_\Omega \rho(x)^{-1} |y|^2 < +\infty \,\}, \\ V^s_0(\Omega) &= \{\, y \in V^s_n(\Omega) \mid y = 0 \text{ on } \partial\Omega \,\}, \quad \tfrac{1}{2} < s \leq 2, \end{aligned}$$

where $\rho(x)$ is the distance from x to $\partial\Omega$. Notice that, according to the above definition, we have $V^s_0(\Omega) = V^s(\Omega) \cap H^1_0(\Omega; \mathbf{R}^3)$ for $1 \leq s \leq 2$. Finally, for $0 < T \leq \infty$, and X_1 and X_2 two Banach spaces, we introduce the function space

$$W(0, T; X_1, X_2) = L^2(0, T; X_1) \cap H^1(0, T; X_2).$$

2.2 Abstract formulation of the extended system

In this section, we state an abstract weak formulation for the system

$$\partial_t z - \Delta z + (z \cdot \nabla) z_e + (z_e \cdot \nabla) z + \kappa (z \cdot \nabla) z + \nabla p = 0 \text{ in } Q, \quad (6)$$

$$\nabla \cdot z = 0 \text{ in } Q,\ z = u \text{ on } \Sigma_i,\ z = 0 \text{ on } \Sigma_e,\ z(0) = z_0 \in V^0(\Omega), \quad (7)$$

$$\partial_t u - \Delta_b u - \sigma n = g \text{ in } \Sigma_i, \quad u(0) = u_0 \in V^{-\frac{1}{2}}(\Gamma_i). \quad (8)$$

Equation (6) corresponds to the Oseen equation if $\kappa = 0$, and to the Navier-Stokes equation if $\kappa = 1$. Equation (6), the left hand side of (8) and (7) are satisfied in the sense of distributions. We observe that σ plays the role of the Lagrange multiplier associated with the constraint $\int_{\Gamma_i} u \cdot n$.
By using transformations developed in [6] we are going to give an equivalent formulation of (6)-(7)-(8). First we define the unbounded operator $(\mathcal{D}(A), A) = (V^2_0(\Omega), A(x, \partial))$ where $A(x, \partial) = P\Delta - P(\nabla z_e) - P(z_e \cdot \nabla)$. We choose $\lambda_0 > 0$ such that $\langle (\lambda_0 - A) y | y \rangle \geq \frac{1}{2} \|y\|_{V^1_0(\Omega)}$ for all $y \in V^1_0(\Omega)$, and we

introduce the Dirichlet operator $D \in \mathcal{L}(V^0(\Gamma_i), V^0(\Omega))$ associated with $\lambda_0 - A$: $Du = w$ where $u \in V^{\frac{1}{2}}(\Omega)$ satisfies

$$\lambda_0 w - A(x,\partial)w = 0,\ \nabla \cdot w = 0 \text{ in } \Omega,\ w = u \text{ on } \Gamma_i \text{ and } w = 0 \text{ on } \Gamma_e.$$

Thus, for $z \in L^{12/5}(\Omega; \mathbf{R}^3)$ we define $b(z,z) \in L^2(0,T;\mathcal{D}(A^*)')$ by

$$\langle b(z,z)|v\rangle = \int_\Omega (\nabla v)\, z \cdot z, \quad \forall v \in \mathcal{D}(A^*).$$

Finally, we define the unbounded operator $(\mathcal{D}(A_b), A_b) = (V^2(\Gamma_i), P_b\Delta_b)$ in $V^0(\Gamma_i)$, where $P_b \in \mathcal{L}(L^2(\Gamma_i, \mathbf{R}^3), V^0(\Gamma_i))$ denotes the orthogonal projector from $L^2(\Gamma_i; \mathbf{R}^3)$ into $V^0(\Gamma_i)$.

DEFINITION 1 *We shall say that* $(z,u) \in L^2(0,T;V^0(\Omega)) \times L^2(0,T;V^{-\frac{1}{2}}(\Gamma_i))$ *if* $\kappa = 0$, *or* $(z,u) \in L^2(0,T;V^0(\Omega) \cap L^{12/5}(\Omega, \mathbf{R}^3)) \times L^2(0,T;V^{-\frac{1}{2}}(\Gamma_i))$ *if* $\kappa = 1$, *is a weak solution to (6)-(7)-(8) if and only if it obeys the system:*

$$(Pz)' = APz + (\lambda_0 - A)PDu + \kappa b(z,z) \in L^2(0,T;\mathcal{D}(A^*)'), \quad (9)$$

$$u' = A_b u + g \in L^2(0,T;\mathcal{D}(A_b)'), \quad (10)$$

$$Pz(0) = Pz_0 \in V_n^0(\Omega), \quad u(0) = u_0 \in V^{-\frac{1}{2}}(\Gamma_i), \quad (11)$$

$$(I-P)z = (I-P)Du \in L^2(0,T;V^0(\Omega)), \quad u_0 \cdot n = z_0 \cdot n. \quad (12)$$

THEOREM 2 *Let* (z,p,u,σ) *be an element of* $W(0,T;V^1(\Omega),V^{-1}(\Omega)) \times L^2(0,T;L^2(\Omega)/\mathbf{R}) \times W(0,T;V^{\frac{1}{2}}(\Gamma_i),V^{-\frac{3}{2}}(\Gamma_i)) \times L^2(0,T)$. *Then* (z,p,u,σ) *satisfies (6)-(7)-(8) if and only if* (z,u) *satisfy (9)-(10)-(11)-(12).*

According to [6], the right hand side of (12) is equivalent to $(I-P)z_0 = (I-P)Du_0$. Then (12) ensures that the couple (z,z_0) is entirely determined by its projected part (Pz,Pz_0) and the boundary values (u,u_0). In the following we only consider the new 'extended' state $Y = (Pz,u)$ and the initial condition $Y_0 = (Pz_0,u_0)$. We define $\mathcal{H}^0 = V_n^0(\Omega) \times V^{-\frac{1}{2}}(\Gamma_i)$ and an adequate unbounded operator $(\mathcal{D}(\mathcal{A}),\mathcal{A})$ in $\mathcal{H}^0$ - $\mathcal{A}$ is defined by (19), (20) and studied in section **4**. We introduce the bilinear operator B

$$B(Y_1,Y_2) = \begin{pmatrix} b(y_1 + (I-P)Du_1, y_2 + (I-P)Du_2 \\ 0 \end{pmatrix}. \quad (13)$$

THEOREM 3 *Let* $(z,p,u,\sigma) \in W(0,T;V^1(\Omega),V^{-1}(\Omega)) \times L^2(0,T;L^2(\Omega)/\mathbf{R}) \times W(0,T;V^{\frac{1}{2}}(\Gamma_i),V^{-\frac{3}{2}}(\Gamma_i)) \times L^2(0,T)$. *Then* (z,p,u,σ) *satisfies (6)-(7)-(8) if and only if (12) holds true and the state* $Y = (Pz,u)$ *satisfies*

$$Y' = \mathcal{A}Y + \kappa B(Y,Y) + \begin{pmatrix} 0 \\ g \end{pmatrix} \in L^2(0,T;\mathcal{D}(\mathcal{A}^*)'), \quad Y(0) = Y_0 \in \mathcal{H}^0. \quad (14)$$

2.3 The extended system and the linear quadratic control problem

The feedback control law is obtained by studying the control problem

$$(\mathcal{Q})^{\infty}_{z_0,u_0} \quad \inf\{\, \mathcal{I}(g) \mid g \in L^2(0,\infty; V^0(\Gamma_i))\,\},$$

where

$$\mathcal{I}(g) = \frac{1}{2}\int_0^\infty \int_\Omega |\nabla z_g|^2 + \frac{1}{2}\int_0^\infty \int_{\Gamma_i} |g|^2,$$

and $(z_g, u_g) \in W(0,\infty; V^1(\Omega) \times V^{\frac{1}{2}}(\Gamma_i), V^{-1}(\Omega) \times V^{-\frac{3}{2}}(\Gamma_i))$ satisfies (6)-(7)-(8). If we introduce $\mathcal{C}Y = \nabla(y + (I-P)Du) \in \mathcal{Z}$ with $\mathcal{Z} = L^2(\Omega; \mathbf{R}^9)$ - $\mathcal{C}$ is studied in section **4** lemma 9 - we can rewrite $(\mathcal{Q})^{\infty}_{z_0,u_0}$ in the form:

$$(\mathcal{P}^{\infty}_{Y_0}) \quad \inf\{\mathcal{J}(g) \mid g \in L^2(0,\infty; V^0(\Gamma_i))\},$$

where

$$\mathcal{J}(g) = \frac{1}{2}\int_0^\infty \|\mathcal{C}Y_g\|^2_{\mathcal{Z}} + \frac{1}{2}\int_0^\infty \int_{\Gamma_i} |g|^2,$$

and Y_g satisfies (14) for $\kappa = 0$.

3. Main result

THEOREM 4 *Let Π_2 and Π_3 be the operator defined in (35). Consider the following coupled system,*

$$\partial_t z - \Delta z + (z\cdot\nabla)z_e + (z_e\cdot\nabla)z + (z\cdot\nabla)z + \nabla p = 0 \; in\; Q, \quad (15)$$
$$\nabla\cdot z = 0 \; in\; Q, \quad z = u \quad in\; \Sigma, \quad z = 0 \quad in\; \Sigma_e, \quad (16)$$
$$\partial_t u - \Delta_b u + \Pi_3 u - \sigma\, n = -\Pi_2 P z \; in\; \Sigma, \quad (17)$$
$$z(0) = z_0 \in V^{\frac{1}{2}}(\Omega), \quad u(0) = u_0 \in V^0(\Gamma_i). \quad (18)$$

There exists $c_0 > 0$ and $\mu_0 > 0$ such that, if $\delta \in (0,\mu_0)$ and

$$\begin{aligned}(z_0,u_0) \in \mathcal{W}_\delta \;=\; & \{(z_0,u_0) \in V^{\frac{1}{2}}(\Omega) \times V^0(\Gamma_i) \mid z_0 - Du_0 \in V_0^{\frac{1}{2}}(\Omega), \\ & \|u_0\|_{V^0(\Gamma_i)} + \|Pz_0\|_{V^{\frac{1}{2}}(\Omega)} \le c_0\delta\},\end{aligned}$$

then, (15)-(16)-(17)-(18) admit a unique solution in the set

$$\begin{aligned}\mathcal{D}_\delta \;=\; & \Big\{ (z,p,u,\sigma) \in W(0,+\infty; V^{\frac{3}{2}}(\Omega), V^{-\frac{1}{2}}(\Omega)) \times L^2(0,\infty; H^{\frac{1}{2}}(\Omega)/\mathbf{R}) \\ & \times W(0,+\infty; V^1(\Gamma_i), V^{-1}(\Gamma_i)) \times L^2(0,\infty) \mid \\ & \|z\|_{L^2(0,+\infty;V^{\frac{3}{2}}(\Omega))} + \|u\|_{L^2(0,+\infty;V^1(\Gamma_i))} + \|\sigma\|_{L^2(0,\infty)} \le \delta, \\ & \|p\|_{L^2(0,+\infty;H^{\frac{1}{2}}(\Omega))} \le \delta(1+\delta) \Big\}.\end{aligned}$$

Moreover, (z,u) obeys

$$\|z(t)\|_{V^{\frac{1}{2}}(\Omega)} + \|u(t)\|_{V^0(\Gamma_i)} \leq C(\|u_0\|_{V^0(\Gamma_i)} + \|Pz_0\|_{V^{\frac{1}{2}}(\Omega)})\, e^{-\eta t}, \quad t \geq 0.$$

4. The operator $\mathcal{A}$

The goals of this section are:

- to give a definition of the unbounded operator $(\mathcal{D}(\mathcal{A}), \mathcal{A})$ in $\mathcal{H}^0$.
- to characterize the function spaces for which the mapping $Y \mapsto (Y' - \mathcal{A}Y, Y(0))$ is an isomorphism, in order to have optimal regularity results for the extended system (14) when $\kappa = 0$.
- to characterize the functional spaces for which the mapping $Q \mapsto (-Q' - \mathcal{A}^*Q, Q(T))$ is an isomorphism, in order to study the backward adjoint equation which appears in the characterization of the solution to $(\mathcal{P}^{\infty}_{Y_0})$ - see part **4**.

THEOREM 5
We define the unbounded operator $(\mathcal{D}(\mathcal{A}), \mathcal{A})$ in $\mathcal{H}^0 = V^0_n(\Omega) \times V^{-\frac{1}{2}}(\Gamma_i)$ by

$$\mathcal{D}(\mathcal{A}) = \Big\{(y,u) \in V^2_n(\Omega) \times V^{\frac{3}{2}}(\Gamma_i) \mid (y - PDu) \in V^2_0(\Omega)\Big\}, \quad (19)$$

$$\mathcal{A} = \begin{pmatrix} A(x,\partial) & (\lambda_0 - A(x,\partial))PD \\ 0 & P_b\Delta_b \end{pmatrix}. \quad (20)$$

The domain $\mathcal{D}(\mathcal{A})$ is dense in $\mathcal{H}^0$, and $\mathcal{A}$ generates an analytic semigroup in $\mathcal{H}^0$. Moreover, for $0 \leq \theta \leq 1$, the identifications below hold

$$\begin{aligned}\mathcal{D}((\lambda_0 - \mathcal{A})^{\theta}) &= [\mathcal{D}(\mathcal{A}), \mathcal{H}^0]_{1-\theta}, \quad (21)\\ &= \{(y,u) \in V^{2\theta}_n(\Omega) \times V^{2\theta-\frac{1}{2}}(\Gamma_i) \mid (y - PDu) \in V^{2\theta}_0(\Omega)\}.\end{aligned}$$

The unbounded operator $(\mathcal{D}(\mathcal{A}^*), \mathcal{A}^*)$ in $\mathcal{H}^0_* = V^0_n(\Omega) \times V^{\frac{1}{2}}(\Gamma_i)$ is defined by

$$\mathcal{D}(\mathcal{A}^*) = V^2_0(\Omega) \times V^{\frac{5}{2}}(\Gamma_i),$$

$$\mathcal{A}^* = \begin{pmatrix} A^*(x,\partial) & 0 \\ D^*(\lambda - A^*(x,\partial)) & P_b\Delta_b \end{pmatrix}.$$

It is the adjoint of $(\mathcal{D}(\mathcal{A}), \mathcal{A})$ with respect to the pivot space $V^0_n(\Omega) \times V^0(\Gamma_i)$. The domain $\mathcal{D}(\mathcal{A}^*)$ is dense in $\mathcal{H}^0_*$ and $\mathcal{A}^*$ generates an analytic semigroup in $\mathcal{H}^0_*$. Finally, for $0 \leq \theta \leq 1$, the identifications below hold

$$\mathcal{D}((\lambda_0 - \mathcal{A}^*)^{\theta}) = [\mathcal{D}(\mathcal{A}^*), \mathcal{H}^0_*]_{1-\theta} = V^{2\theta}_0(\Omega) \times V^{\frac{1}{2}+2\theta}(\Gamma_i). \quad (22)$$

Moreover, for $0 \leq \theta \leq 1$, we define the function spaces

$$\mathcal{H}^{2\theta} = [\mathcal{D}(\mathcal{A}), \mathcal{H}^0]_{1-\theta}, \; \mathcal{H}_*^{2\theta} = [\mathcal{D}(\mathcal{A}^*), \mathcal{H}_*^0]_{1-\theta},$$
$$\mathcal{H}^{-2\theta} = (\mathcal{H}_*^{2\theta})', \; \mathcal{H}_*^{-2\theta} = (\mathcal{H}^{2\theta})'.$$

Then, as a consequence of the analyticity of the semigroups $(e^{\mathcal{A}t})_{t\geq 0}$ and $(e^{\mathcal{A}^*t})_{t\geq 0}$ respectively in $\mathcal{H}^0$ and in $\mathcal{H}_*^0$, we can state a general isomorphism Theorem (see [1], Chap.3, Thm 2.2, p.166):

THEOREM 6 *For every $0 \leq \theta \leq 1$, the mappings below are isomorphisms:*

$$\begin{array}{rcl} W(0,T;\mathcal{H}^{2\theta},\mathcal{H}^{2(\theta-1)}) & \to & L^2(0,T;\mathcal{H}^{2(\theta-1)}) \times [\mathcal{H}^{2\theta},\mathcal{H}^{2(\theta-1)}]_{\frac{1}{2}}, \\ Y & \mapsto & (Y' - \mathcal{A}Y, Y(0)), \end{array}$$

$$\begin{array}{rcl} W(0,T;\mathcal{H}_*^{2\theta},\mathcal{H}_*^{2(\theta-1)}) & \to & L^2(0,T;\mathcal{H}_*^{2(\theta-1)}) \times [\mathcal{H}_*^{2\theta},\mathcal{H}_*^{2(\theta-1)}]_{\frac{1}{2}}, \\ Q & \mapsto & (-Q' - \mathcal{A}^*Q, Q(T)). \end{array}$$

Next we determine the spaces $[\mathcal{H}^{2\theta},\mathcal{H}^{2(\theta-1)}]_{\frac{1}{2}}$ of initial conditions.

LEMMA 7 *For all $0 \leq \theta \leq 1$ the following characterization holds:*

$$[\mathcal{H}^{2\theta},\mathcal{H}^{2(\theta-1)}]_{\frac{1}{2}} = \mathcal{H}^{2\theta-1}. \tag{23}$$

Finally, a direct application of Theorem 6 with (23) ensures the existence of a unique solution Y to the extended linear system (14) when $\kappa = 0$.

THEOREM 8 *Let $g \in L^2(0,T;V^0(\Gamma_i))$ and $Y_0 \in \mathcal{H}^0$. There exists a unique solution $Y \in L^2(0,T;\mathcal{H}^0)$ to the extended system*

$$Y' = \mathcal{A}Y + \begin{pmatrix} 0 \\ g \end{pmatrix} \in L^2(0,T;\mathcal{D}(\mathcal{A}^*)'), \quad Y(0) = Y_0 \in \mathcal{H}^0. \tag{24}$$

Moreover, Y belongs to $W(0,T;\mathcal{H}^1,\mathcal{H}^{-1})$. More generally, if we assume that $Y_0 \in \mathcal{H}^{2\theta}$, $0 \leq \theta \leq \frac{1}{2}$, then Y belongs to $W(0,T;\mathcal{H}^{2\theta+1},\mathcal{H}^{2\theta-1})$.

We now treat the backward adjoint equation which appears in the characterization of the solution to $(\mathcal{P}_{Y_0}^{\infty})$.

LEMMA 9 *Let us define $\mathcal{C} \in \mathcal{L}(\mathcal{H}^1,\mathcal{Z})$ with $\mathcal{Z} = L^2(\Omega,\mathbf{R}^9)$ by $\mathcal{C} : Y \to \nabla(y + (I-P)Du)$. Then*

$$\|\mathcal{C}.\|_{\mathcal{Z}} \sim \|.\|_{\mathcal{H}^1} \quad \textit{and} \quad \mathcal{C}^*\mathcal{C} \in \mathcal{L}(\mathcal{H}^{\theta+1},\mathcal{H}_*^{\theta-1}), \quad 0 \leq \theta \leq 1. \tag{25}$$

Then Theorem 6 with (25) leads to the following theorem.

THEOREM 10 *Let* $Y \in L^2(0,T;\mathcal{H}^1)$. *There exists a unique solution* $Q \in L^2(0,T;\mathcal{H}^0_*)$ *to the backward equation*

$$-Q' = \mathcal{A}^*Q + C^*CY, \quad Q(T) = 0. \tag{26}$$

Moreover, Q belongs to $W(0,T;\mathcal{H}^1_*,\mathcal{H}^{-1}_*)$. *More generally, if we assume that* $Y \in L^2(0,T;\mathcal{H}^{2\theta})$, $\frac{1}{2} \le \theta \le 1$, *then Q belongs to* $W(0,T;\mathcal{H}^{2\theta}_*,\mathcal{H}^{2(\theta-1)}_*)$.

5. Resolution of the optimal control problem

5.1 The finite time horizon case

Let $0 < T < \infty$ be a finite time horizon. To deal with the optimal control problem $(\mathcal{P}^T_{Y_0})$ we first study the following problem:

$$(\mathcal{P}^T_\xi) \quad \inf\{\mathcal{J}_T(g) \mid g \in L^2(0,T;V^0(\Gamma_i))\}, \tag{27}$$

where

$$\mathcal{J}_T(g) = \frac{1}{2}\int_0^T \|CY_g\|^2_{\mathcal{Z}} + \frac{1}{2}\int_0^T \int_{\Gamma_i} |g|^2,$$

and $Y_g \in W(0,T;\mathcal{H},\mathcal{H}^{-1})$ is the solution to

$$Y' = \mathcal{A}Y + \begin{pmatrix} 0 \\ g \end{pmatrix}, \quad Y(0) = \xi \in \mathcal{H}^0. \tag{28}$$

We introduce the projection operator $\Lambda : (f,g) \in \mathcal{H}^0_* \mapsto (0,g)$. The problem $(\mathcal{P}^T_\xi)$ admits a unique solution $(0,g_{\xi,T})$ where $(0,g_{\xi,T}) = -\Lambda Q_{\xi,T}$ and $(Y_{\xi,T},Q_{\xi,T})$ is the unique solution to the system

$$(S^T) \quad \begin{cases} Y' &= \mathcal{A}Y - \Lambda Q, \qquad Y(0) = \xi \in \mathcal{H}^0, \\ -Q' &= \mathcal{A}^*Q + C^*CY, \quad Q(T) = 0. \end{cases}$$

Finally, we denote by $\Pi(T) \in \mathcal{L}(\mathcal{H}^0,\mathcal{H}^0_*)$, the mapping

$$\Pi(T) : \xi \mapsto Q_{\xi,T}(0).$$

5.2 The infinite time horizon case

Since, for every ξ and $0 < T < \infty$, the solution to $(\mathcal{P}^T_\xi)$ has been characterized, we are in position to study the optimal control problem $(\mathcal{P}^\infty_\xi)$ and the regularity of its solution in function of the regularity of ξ. The problem $(\mathcal{P}^\infty_\xi)$ is defined by

$$(\mathcal{P}^\infty_\xi) \quad \inf\{\mathcal{J}(g) \mid g \in L^2(0,\infty;V^0(\Gamma_i))\}, \tag{29}$$

where the following functional satisfies (28):

$$\mathcal{J}(g) = \frac{1}{2}\int_0^\infty \|CY_g\|_{\mathcal{Z}}^2 + \frac{1}{2}\int_0^\infty \int_{\Gamma_i} |g|^2, \quad Y_g \in W(0,\infty;\mathcal{H},\mathcal{H}^{-1}).$$

Using a null controllability result stated in [3] we can show that there exists a control $g \in L^2(0,\infty;V^0(\Gamma_i))$ such that $\mathcal{J}(g) < +\infty$. This gives us the existence of a unique solution g_ξ to $(\mathcal{P}_\xi^\infty)$.

THEOREM 11 *The problem* $(\mathcal{P}_\xi^\infty)$ *admits a unique solution* g_ξ *where* $g_\xi = -\psi_\xi$ *and* $(Y_\xi, Q_\xi) = ((y_\xi, u_\xi), (\Phi_\xi, \psi_\xi)) \in W(0,\infty;\mathcal{H}^1,\mathcal{H}^{-1}) \times W(0,\infty;\mathcal{H}_*^1,\mathcal{H}_*^{-1})$ *is the unique solution to the system:*

$$(S)\begin{cases} Y' &= \mathcal{A}Y - \Lambda Q, \qquad Y(0) = \xi \in \mathcal{H}^0, \\ -Q' &= \mathcal{A}^*Q + C^*CY, \qquad Q(\infty) = 0. \end{cases}$$

Moreover, there exists $\Pi \in \mathcal{L}(\mathcal{H}^*, \mathcal{H}_*^0)$, $\Pi^* = \Pi$ *such that*

$$Q_\xi = \Pi Y_\xi, \quad \mathcal{J}(g_\xi) = \frac{1}{2}\langle \Pi\xi|\xi\rangle_{\mathcal{H}_*^0,\mathcal{H}^0}. \tag{30}$$

The control g_ξ has been calculated as the limit of $g_{\xi,T}$ when $T \to \infty$, where $g_{\xi,T}$ is the unique solution to $(\mathcal{P}_\xi^T)$.
We now focus on the properties of Π. First, from the limit $\Pi(T) \to \Pi$ as $T \to \infty$ we can show that Π satisfies an algebraic Riccati equation. Next, Theorems 8 and 10 for (S) lead to sharp regularity result for Π.

THEOREM 12 Π *satisfies a Riccati equation:* $\forall(\xi,\zeta) \in \mathcal{H}^1 \times \mathcal{H}^1$,

$$\langle \Pi\xi|\mathcal{A}\zeta\rangle_{\mathcal{H}_*^1,\mathcal{H}^{-1}} + \langle \mathcal{A}\xi|\Pi\zeta\rangle_{\mathcal{H}^{-1},\mathcal{H}_*^1} + (C\xi|C\zeta)_{\mathcal{Z}} - (\Lambda\Pi\xi|\Lambda\Pi\zeta)_{V^0(\Gamma_i)} = 0, \tag{31}$$

and the regularizing property $\quad \Pi \in \mathcal{L}(\mathcal{H}^{2\theta}, \mathcal{H}_*^{2\theta}), \quad 0 \le \theta \le \frac{1}{2}$.

Next we set $\mathcal{A}_\Pi = (\Lambda\Pi - \mathcal{A})$, so that the optimal trajectory Y_ξ satisfy

$$Y' + \mathcal{A}_\Pi Y = 0, \quad Y(0) = \xi \in \mathcal{H}^0. \tag{32}$$

From a classical result due to Datko in [5, Chap 4, Thm 4.1], we show the exponential stability of $e^{-\mathcal{A}_\Pi t}Y_0$. Let $\alpha > 0$, the positivity of $\mathcal{A}_\Pi$ allows us to define its fractional power $\mathcal{A}_\Pi^\alpha$ and we respectively identify $\mathcal{D}(\mathcal{A}_\Pi^\alpha)$ and $\mathcal{D}(\mathcal{A}_\Pi^{*\alpha})$ with $\mathcal{D}((\lambda_0 - \mathcal{A})^\alpha)$ and $\mathcal{D}((\lambda_0 - \mathcal{A}^*)^\alpha)$. Thus (21) and (22) ensures a characterization $\mathcal{D}(\mathcal{A}_\Pi^\alpha)$ and $\mathcal{D}(\mathcal{A}_\Pi^{*\alpha})$.

THEOREM 13 *The unbounded operator* $(\mathcal{D}(-\mathcal{A}_\Pi), -\mathcal{A}_\Pi)$ *generates an analytic exponentially stable semigroup in* $\mathcal{H}^0$ *and the characterizations below hold*

$$\mathcal{D}(\mathcal{A}_\Pi^\theta) = \mathcal{H}^{2\theta}, \quad \mathcal{D}(\mathcal{A}_\Pi^{*\theta}) = \mathcal{H}_*^{2\theta}, \quad 0 \le \theta \le 1.$$

Next we define new norms in the spaces $\mathcal{H}^\alpha$ and $\mathcal{H}^{1+\alpha}$ which are essential to study the stabilization of the Navier-Stokes equation (see section **6**).

DEFINITION 14 *We define* $\Pi^{(\alpha)} \in \mathcal{L}(\mathcal{H}^\alpha, \mathcal{H}_*^{-\alpha})$ *by* $\Pi^{(\alpha)} = {\mathcal{A}_\Pi}^{*\frac{\alpha}{2}}\, \Pi\, {\mathcal{A}_\Pi}^{\frac{\alpha}{2}}$.

THEOREM 15 *The operator* $\Pi^{(\alpha)}$ *has the regularizing property:*

$$\Pi^{(\alpha)} \in \mathcal{L}(\mathcal{H}^{2\theta+\alpha}, \mathcal{H}_*^{2\theta-\alpha}), \quad 0 \le \theta \le \frac{1}{2}. \tag{33}$$

DEFINITION 16 *We define the two mappings* $\mathcal{N}_\alpha$ *and* $\mathcal{R}_{1+\alpha}$ *by*

$$\begin{aligned}
\mathcal{N}_\alpha(\xi) &= (\langle \Pi^{(\alpha)}\xi | \xi \rangle_{\mathcal{H}_*^{-\alpha}, \mathcal{H}^\alpha})^{\frac{1}{2}}, \quad \xi \in \mathcal{H}^\alpha, \\
\mathcal{R}_{1+\alpha}(\xi) &= (\langle \mathcal{A}_\Pi \xi | \Pi^{(\alpha)}\xi \rangle_{\mathcal{H}^{-1+\alpha}, \mathcal{H}_*^{1-\alpha}})^{\frac{1}{2}}, \quad \xi \in \mathcal{H}^{1+\alpha}.
\end{aligned}$$

THEOREM 17 $\mathcal{N}_\alpha$ *and* $\mathcal{R}_{1+\alpha}$ *define norms respectively on* $\mathcal{H}^\alpha$ *and* $\mathcal{H}^{1+\alpha}$,

$$\mathcal{N}_\alpha(\cdot) \sim \|\cdot\|_{\mathcal{H}^\alpha}, \quad \mathcal{R}_{1+\alpha}(.) \sim \|\cdot\|_{\mathcal{H}^{1+\alpha}}. \tag{34}$$

We shall point out that the expression of $\mathcal{R}_{1+\alpha}(\xi)$ is explicitly given by

$$\langle \mathcal{A}_\Pi \xi | \Pi^{(\alpha)}\xi \rangle_{\mathcal{H}^{-1+\alpha}, \mathcal{H}_*^{1-\alpha}} = \frac{1}{2}\|C{\mathcal{A}_\Pi}^{\frac{\alpha}{2}}\xi\|_{\mathcal{Z}}^2 + \frac{1}{2}\|\Lambda\Pi{\mathcal{A}_\Pi}^{\frac{\alpha}{2}}\xi\|_{V^0(\Gamma_i)}^2, \quad \forall \xi \in \mathcal{H}^{1+\alpha},$$

which follows from (31) in which we have replaced ξ and ζ by $\mathcal{A}_\Pi^{\frac{\alpha}{2}}\xi$.
We finish this section by giving the PDE formulation of the closed loop system (32).

THEOREM 18 *The operator* Π *can be rewritten as follows,*

$$\Pi = \begin{pmatrix} \Pi_1 & \Pi_2^* \\ \Pi_2 & \Pi_3 \end{pmatrix}, \quad \begin{array}{l} \Pi_1 \in \mathcal{L}(V_n^0(\Omega)), \\ \Pi_2 \in \mathcal{L}(V_n^0(\Omega), V^{\frac{1}{2}}(\Gamma_i)), \\ \Pi_3 \in \mathcal{L}(V^{-\frac{1}{2}}(\Gamma_i), V^{\frac{1}{2}}(\Gamma_i)), \end{array} \tag{35}$$

where Π_1 *and* Π_3 *are positive, definite and self-adjoint operators. Then* $Y = (Pz, u) \in W(0,\infty; \mathcal{H}^1, \mathcal{H}^{-1})$ *satisfies (32) if and only if the element* (z, p, u, σ) *of* $W(0,\infty; V^1(\Omega), V^{-1}(\Omega)) \times L^2(0,\infty; L^2(\Omega)/\mathbf{R}) \times W(0,\infty; V^{\frac{1}{2}}(\Gamma_i), V^{-\frac{3}{2}}(\Gamma_i)) \times L^2(0,\infty)$ *satisfies*

$$\begin{aligned}
&\partial_t z - \Delta z + (z\cdot\nabla)z_e + (z_e\cdot\nabla)z + \nabla p = 0, \quad \nabla\cdot z = 0 \text{ in } Q, \\
&\partial_t u - \Delta_b u + \Pi_3 u - \sigma n = -\Pi_2 Pz, \quad z = u \text{ in } \Sigma_i, \quad z = 0 \text{ in } \Sigma_e, \\
&z(0) = z_0 \in V^0(\Omega), \quad u(0) = u_0 \in V^{-\frac{1}{2}}(\Gamma_i), \quad z_0\cdot n = u_0\cdot n.
\end{aligned}$$

6. Stabilization of the Navier-Stokes equations

We now come back to the stabilization of the Navier-Stokes system. We now consider the nonlinear system (15)-(16)-(17)-(18) which we can rewrite in the abstract formulation

$$Y' + \mathcal{A}_{\Pi} Y = B(Y,Y), \quad Y(0) = Y_0 \in \mathcal{H}^{\frac{1}{2}}. \tag{36}$$

Finally, Theorem 4 is a direct consequence of the following result.

THEOREM 19 *There exists $c_0 > 0$ and $\mu_0 > 0$ such that, if $\delta \in (0, \mu_0)$ and $Y_0 \in \mathcal{V}_\delta = \{Y \in \mathcal{H}^{\frac{1}{2}} | \ \|Y\|_{\mathcal{H}^{\frac{1}{2}}} < c_0\delta\}$ then, (36) admit a unique solution in the set $\mathcal{S}_\delta = \{ Y \in W(0,\infty; \mathcal{H}^{\frac{3}{2}}, \mathcal{H}^{-\frac{1}{2}}) | \ \|Y\|_{W(0,\infty;\mathcal{H}^{\frac{3}{2}},\mathcal{H}^{-\frac{1}{2}})} \leq \delta \}$. Moreover, there exists $\eta > 0$ such that*

$$\|Y(t)\|_{\mathcal{H}^{\frac{1}{2}}} \leq C \|Y_0\|_{\mathcal{H}^{\frac{1}{2}}} e^{-\eta t}. \tag{37}$$

Proof. Here, we give a brief sketch of the proof of the stability result. We multiply the left hand side of (36) by $\Pi^{(\frac{1}{2})} Y(t)$. According to (34) with $\alpha = \frac{1}{2}$ we obtain

$$\frac{1}{2}\frac{d}{dt}\mathcal{N}^2_{\frac{1}{2}}(Y(t)) + \mathcal{R}^2_{\frac{3}{2}}(Y(t)) = \langle B(Y(t),Y(t)) | \Pi^{(\frac{1}{2})} Y(t)\rangle.$$

Then we invoke a classical estimation - see [2, Chap.6, 6.9 and 6.10] - to obtain $|\langle B(Y(t),Y(t)) | \Pi^{(\frac{1}{2})} Y(t)\rangle| \leq C \|Y(t)\|_{\mathcal{H}^{\frac{1}{2}}} \|Y(t)\|_{\mathcal{H}^{\frac{1}{2}}} \|\Pi^{(\frac{1}{2})} Y(t)\|_{\mathcal{H}^{\frac{1}{2}}}$. According to (33) with $\alpha = \theta = \frac{1}{2}$ it yields

$$|\langle B(Y(t),Y(t)) | \Pi^{(\frac{1}{2})} Y(t)\rangle| \leq C \|Y(t)\|_{\mathcal{H}^{\frac{1}{2}}} \|Y(t)\|^2_{\mathcal{H}^{\frac{3}{2}}}.$$

Thus (34) gives us $C_0 > 0$ such that

$$\frac{d}{dt}\mathcal{N}^2_{\frac{1}{2}}(Y(t)) + 2(1 - C_0 \mathcal{N}_{\frac{1}{2}}(Y(t)))\mathcal{R}^2_{\frac{3}{2}}(Y(t) \leq 0. \tag{38}$$

It is obvious to see that if $\mathcal{N}_{\frac{1}{2}}(Y_0)$ is small enough, we can choose $\mathcal{N}_{\frac{1}{2}}(Y_0) < \frac{1}{4C_0}$, so that the mapping $t \to \mathcal{N}_{\frac{1}{2}}(Y(t))$ be a nonincreasing function with values less than $\frac{1}{4C_0}$. This gives the existence of δ, and (37) follows from (38).

References

[1] A. Bensoussan, G. Da Prato, M.C. Delfour, S.K. Mitter *Representation and Control of Infinite Dimentional Systems, Volume I.* Birkhauser, Boston, 1992.

[2] P. Constantin, C. Foias. *Navier-Stokes Equations.* University of Chicago Press, Chicago Lectures in Mathematics, 1988.

[3] E. Fernández-Cara, S. Guerrero, O. Yu. Imanuvilov, J.-P. Puel. Local exact controllability of the Navier-Stokes system. to appear in *J. Math. Pures et Appliquées.*.

[4] G.P. Galdi. *An Introduction to the Mathematical Theory of the Navier-Stokes Equations, Vol. II., Nonlinear Steady Problems* . Springer, New York, 1994.

[5] A. Pazy. *Semigroups of Linear Operators and Applications to Partial Differential Equations.* Springer-Verlag, New York,1983.

[6] J.-P. Raymond. Stokes and Navier-Stokes Equations with Nonhomogeneous Boundary Conditions, preprint, 2005.

SANDPILES AND SUPERCONDUCTORS: DUAL VARIATIONAL FORMULATIONS FOR CRITICAL-STATE PROBLEMS

J.W. Barrett [1] and L. Prigozhin [2]

[1] *Dept. of Mathematics, Imperial College, London SW7 2AZ, UK, jwb@ic.ac.uk*

[2] *Blaustein Institute for Desert Research, Ben-Gurion University, Sede-Boqer Campus 84990, Israel, leonid@cs.bgu.ac.il*

Abstract Similar evolutionary variational inequalities appear as variational formulations of continuous models for sandpile growth, magnetization of type-II superconductors, and evolution of some other dissipative systems characterized by the multiplicity of metastable states, long-range interactions, avalanches, and hysteresis. Such formulations for sandpile and superconductor models are, however, convenient for modeling only some of the variables (evolving pile shape and magnetic field for sandpile and superconductor models, respectively). The conjugate variables (the surface sand flux and the electric field) are also of interest in various applications. Here we derive dual variational formulations, similar to mixed variational inequalities in plasticity, for the sandpile and superconductor models. These formulations are used in numerical simulations and allow us to approximate simultaneously both the primary and dual variables.

keywords: variational inequalities, critical-state problems, duality, numerical solution.

1. Introduction

Sandpiles and type-II superconductors are examples of spatially extended open dissipative systems which have infinitely many metastable states but, driven by the external forces, tend to organize themselves into a marginally stable "critical state" and are then able to demonstrate almost instantaneous long-range interactions. The evolution of such systems is often accompanied by sudden collapses, like sandpile avalanches, and hysteresis. Although these are dissipative systems of a different nature, their continuous models are equivalent to similar variational (or quasivariational) inequalities (see [1] and the references therein). The origin of this similarity is that these models are quasistationary models of equilibrium and the multiplicity of metastable states is a consequence of a unilateral equilibrium condition. Typically, dynamics of

Please use the following format when citing this chapter:

Barrett, J.W., and Prigozhin, L., 2006, in IFIP International Federation for Information Processing, Volume 202, Systems, Control, Modeling and Optimization, eds. Ceragioli, F., Dontchev, A., Furuta, H., Marti, K., Pandolfi, L., (Boston: Springer), pp. 25-29.

such a system occur at the border of equilibrium and the system in a marginally stable state, often called critical. The rate with which these systems adjust themselves to the changing external conditions is determined implicitly and appears in the model as a Lagrange multiplier. Typically, the multiplier is eliminated in transition to a variational formulation written in terms of a "primary" variable (surface of a sandpile, magnetic field in a superconductor, stress tensor in elastoplasticity, etc.) In many situations, however, the Lagrange multiplier or, equivalently, a "dual" variable (sand flux upon the pile surface, electric field, and strain tensor, respectively) also has to be found. We present, for both the sandpile and the superconductivity problem, the variational formulations written for the dual variables. On discretization these dual formulations yield an efficient algorithm to compute the dual and primal variables simultaneously. Only the simplest version of each problem is considered.

2. Sandpiles

Let sand be poured out onto a rigid support surface, $y = h_0(x)$, given in a domain $\Omega \subset R^2$ with boundary $\partial\Omega$. If the support boundary is open, a model for pile surface evolution can be written as

$$\partial_t h + \nabla \cdot \underline{q} = w, \quad h|_{t=0} = h_0, \quad h|_{\partial\Omega} = h_0|_{\partial\Omega},$$

where h is the pile surface, $w \geq 0$ is the given source density, $\underline{q}$ is the horizontal projection of the flux of sand pouring down the pile surface. If the support has no slopes steeper than the sand angle of repose, $|\nabla h_0| \leq \gamma = \tan\alpha$, the simplest constitutive relations for this model read: (i) the flux $\underline{q}$ is directed towards the steepest decent of the surface, (ii) the surface slope cannot exceed the critical angle α, and (iii) the flux is zero upon subcritical slopes. Equivalently, one can write $\underline{q} = -m\nabla h$ and show that $m(x,t) \geq 0$ is the Lagrange multiplier related to the constraint $|\nabla h| \leq \gamma$. The model can be rewritten as a variational inequality for h,

$$h(.,t) \in K : (\partial_t h - w, \varphi - h) \geq 0 \ \forall \varphi \in K, \ h|_{t=0} = h_0, \tag{1}$$

where $K = \{\varphi \in H^1(\Omega) : |\nabla\varphi| \leq \gamma \text{ a.e.}, \varphi|_{\partial\Omega} = h_0|_{\partial\Omega}\}$ and $(\cdot,\cdot)$ is the standard $L^2(\Omega)$ inner product. Simple analytical solutions of this inequality describe piles generated on the support $h_0 \equiv 0$. For the point source $w = \delta(x - x_0)$, a conical pile with critical slopes grows until its base touches the domain boundary. Then there appears a runway connecting the cone apex with the boundary and the pile growth stops: all additional sand just follows the runway and leaves the system. On the other hand, if $w > 0$ everywhere in Ω, the final stationary shape of the pile is different: $h(x) = \gamma \text{dist}(x, \partial\Omega)$. For $w \geq 0$, the general stationary solution and an integral representation formula for the corresponding Lagrange multiplier m, determining the surface sand flux

$\underline{q}$, have also been obtained recently (see [2] and the references therein). Note, however, that it is not easy to compute the Lagrange multiplier using this formula. In the non-stationary case, determining the surface flux $\underline{q}$ remains difficult even if the unique solution h to (1) is found. To compute both these variables simultaneously, we derive a dual variational formulation of the evolutionary problem.

Let $\{h, \underline{q}\}$ satisfy the model relations (i)-(iii). Then, for any test field $\underline{\psi}$, $\nabla h \cdot (\underline{\psi} - \underline{q}) \geq -|\nabla h||\underline{\psi}| - \nabla h \cdot \underline{q} = -|\nabla h||\underline{\psi}| + \gamma|\underline{q}| \geq -\gamma|\underline{\psi}| + \gamma|\underline{q}|$. Hence, $(\nabla h, \underline{\psi} - \underline{q}) \geq \phi(\underline{q}) - \phi(\underline{\psi})$, where $\phi(\underline{q}) = \gamma \int_\Omega |\underline{q}|$. Since $(\nabla h, \underline{\psi} - \underline{q}) = \oint_{\partial\Omega} h_0\{\underline{\psi}_n - \underline{q}_n\} - (h, \nabla \cdot \{\underline{\psi} - \underline{q}\})$, where $\underline{\psi}_n$ is the normal component of $\underline{\psi}$ on $\partial\Omega$, we have $\phi(\underline{\psi}) - \phi(\underline{q}) - (h, \nabla \cdot \{\underline{\psi} - \underline{q}\}) + \oint_{\partial\Omega} h_0\{\underline{\psi}_n - \underline{q}_n\} \geq 0$. Noting that $h = h_0 + \int_0^t w\, dt - \nabla \cdot \{\int_0^t \underline{q}\, dt\}$ we finally obtain

$$\underline{q}(.,t) \in V: \left(\nabla \cdot \left\{\int_0^t \underline{q}\, dt\right\}, \nabla \cdot \{\underline{\psi} - \underline{q}\}\right) + \mathcal{F}(\underline{\psi} - \underline{q}) + \phi(\underline{\psi}) - \phi(\underline{q}) \geq 0 \tag{2}$$

for any $\underline{\psi} \in V$. Here $\mathcal{F}(\underline{u}) = \oint_{\partial\Omega} h_0 \underline{u}_n - (h_0 + \int_0^t w\, dt, \nabla \cdot \underline{u})$ and we define $V = \{\underline{\psi} \in [\mathcal{M}(\Omega)]^2 \ : \ \nabla \cdot \underline{\psi} \in L^2(\Omega)\}$, where $\mathcal{M}(\Omega)$ is the Banach space of bounded Radon measures. Existence of a solution to problem (2) is proved in [3].

To approximate (2) numerically, we smoothed the non-differentiable functional ϕ by introducing $|\underline{q}|_\varepsilon = (|\underline{q}|^2 + \varepsilon^2)^{1/2}$, discretized the regularized equality problem in time, employed Raviart-Thomas finite elements of lowest order with vertex sampling on the nonlinear term, and solved the resulting nonlinear algebraic system at each time level iteratively using a form of successive over-relaxation (see Fig. 1 for an example of a numerical simulation).

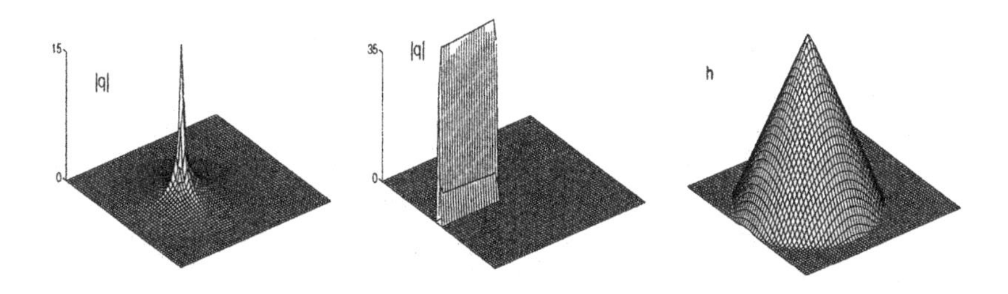

Figure 1. Point source above square support with $h_0 = 0$. The cone grows until its base touches the support boundary and the runway appears. Shown: 1) sand flux $|\underline{q}|$ before and after this moment; 2) the final pile shape computed as $h = h_0 + \int_0^t w\, dt - \nabla \cdot \{\int_0^t \underline{q}\, dt\}$.

3. Superconductors

Phenomenologically, the magnetic field penetration into type-II superconductors can be understood as a nonlinear eddy current problem. Let a long cylindrical superconductor with a simply connected cross-section Ω be placed into a non-stationary uniform external magnetic field $\underline{h}_e(t)$ parallel to the cylindrical generators. According to Faraday's law, time variations of this field induce in a conductor an electric field $\underline{e}$ leading to a current $\underline{j}$ parallel to the cross-section plane; this current induces a magnetic field $\underline{h}(x,t)$ parallel to $\underline{h}_e$. Omitting the displacement current in Maxwell's equations and scaling the magnetic permeability to be unity, we obtain the following model,

$$\partial_t(h+h_e)+\operatorname{curl}\underline{e}=0,\ \ \mathbf{curl}\, h=\underline{j},\ \ h|_{t=0}=h_0(x),\ \ h|_{\partial\Omega}=0,$$

where $\operatorname{curl}\underline{u}=\partial_{x_1}u_2-\partial_{x_2}u_1$ and $\mathbf{curl}\, u=(\partial_{x_2}u,-\partial_{x_1}u)$. Instead of the usual Ohm law, a multivalued current-voltage relation (the Bean model) is often employed for type-II superconductors. It is postulated that (i) the electric field $\underline{e}$ and the current density $\underline{j}$ have the same direction, (ii) the current density cannot exceed some critical value, j_c, and (iii) if the current is subcritical, the electric field is zero. One can write $\underline{e}=\rho\underline{j}$ and show that the effective resistivity, $\rho(x,t)\geq 0$, is the Lagrange multiplier related to current density constraint $|\underline{j}|\leq j_c$. Using conditions (i)-(iii) we can eliminate the electric field from the model. This yields the variational inequality,

$$h(.,t)\in K:\ (\partial_t\{h+h_e\},\varphi-h)\geq 0\ \ \forall\varphi\in K,\ \ h|_{t=0}=h_0,$$

where $K=\{\varphi\in H_0^1(\Omega)\ :\ |\nabla\varphi|\leq\gamma\ a.e.\}$. This inequality for h can be approximated numerically and often even solved analytically. However, computing the electric field $\underline{e}$ may be difficult [4]. As for the sandpile model, a dual variational formulation can be derived to find both variables simultaneously:

$$\underline{e}(.,t)\in W:\ \left(\operatorname{curl}\left\{\int_0^t \underline{e}\,dt\right\},\operatorname{curl}\{\underline{\psi}-\underline{e}\}\right)+\mathcal{F}(\underline{\psi}-\underline{e})+\phi(\underline{\psi})-\phi(\underline{e})\geq 0 \tag{3}$$

for any $\underline{\psi}\in W$, where $W=\{\underline{\psi}\in[\mathcal{M}(\Omega)]^2\ :\ \operatorname{curl}\underline{\psi}\in L^2(\Omega)\}$, $\phi(\underline{u})=\int_\Omega j_c|\underline{u}|$, and $\mathcal{F}(\underline{u})=(h_e(t)-h_e(0)-h_0,\operatorname{curl}\underline{u})$. The primary variable, h, is found as

$$h=h_0+h_e(0)-h_e(t)-\operatorname{curl}\{\textstyle\int_0^t\underline{e}\,dt\}.$$

The simple transformation $R:\underline{e}=(e_1,e_2)\to(e_2,-e_1)$ maps W to V and enables us to use the same Raviart-Thomas finite element as in the previous case. To model the magnetization of a superconductor with a multiply connected cross-section (Fig. 2) we "filled" the hole and set $j_c=0$ there.

References

[1] L. Prigozhin. Variational inequalities in critical-state problems. *Physica D* 167:197-210, 2004.

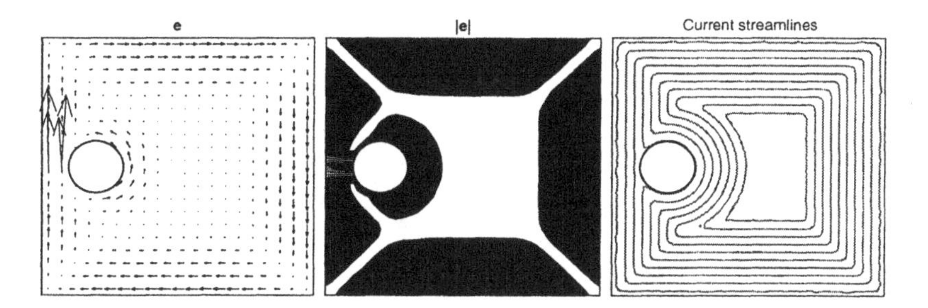

Figure 2. Cylindrical superconductor in a growing field (square cross-section with a hole, zero initial state). Shown at the same moment in time: the electric field $\underline{e}$, level contours of $|\underline{e}|$, and the current streamlines plotted as levels of h. Note the "runway" (red region in the $|\underline{e}|$ contour plot) through which the magnetic field penetrates the hole and where the electric field is the strongest.

[2] P. Cannarsa, P. Cardaliaguet. Representation of equilibrium solutions to the table problem for growing sandpiles. *J. Eur. Math. Soc.* 6:435-464, 2004.

[3] J.W. Barrett, L. Prigozhin. Dual formulations in critical-state problems. *In preparation.*

[4] A. Badía-Majós, C. López. Electric field in hard superconductors with arbitrary cross section and general critical current law. *J. Appl. Phys.* 95:8035-8040, 2004.

MATHEMATICAL MODELLING OF AN INNOVATIVE UNMANNED AIRSHIP FOR ITS CONTROL LAW DESIGN

M. Battipede,[1] M. Lando,[1] and P. Gili[1]
[1]*Aeronautical and Space Department, Politecnico di Torino, ITALY,*
{*manuela.battipede, marco.lando,piero.gili*} *@polito.it**

Abstract The paper is concerned with the dynamic modelling of the unconventional remotely-piloted Lighter-Than-Air vehicle patented by Nautilus S.p.A. and the Polytechnic of Turin. The airship mathematical model is based on a 6 degree-of-freedom non-linear model referring to the basic Newtonian mechanics. Emphasis is placed on those innovative and peculiar aspects of the dynamic modelling, such as aerodynamics, buoyancy and inertial features.

keywords: airship, dynamic modelling, flight simulation

1. Introduction

An innovative remotely-piloted airship has been designed and patented by Nautilus S.p.A. and the Polytechnic of Turin, and equipped with high precision sensors and communication devices by Galileo Avionica and Selex Communication. The object of this project is a low-speed, low-altitude advanced unmanned platform, named *Elettra Twin Flyers*, which should be employed for reconnaissance, monitoring and telecommunication purposes both in military and civil area. The development of a refined Flight Simulator [1] is essential to support the whole design process of this innovative unmanned airship and its subsequent marketing, as well as to provide a valid platform for the pilot training. In particular, the main task of the Flight Simulator is to assist the airship design process from the early stages, in which it is necessary to evaluate the global dynamic behavior of the vehicle, up to the more advanced phases, in which the single components and subsystems have to be correctly analyzed, dimensioned and integrated in the final product. To obtain a reliable Flight

*Paper written with financial support of Nautilus S.p.A.

Please use the following format when citing this chapter:

Battipede, M., Lando, M., and Gili, P., 2006, in IFIP International Federation for Information Processing, Volume 202, Systems, Control, Modeling and Optimization, eds. Ceragioli, F., Dontchev, A., Furuta, H., Marti, K., Pandolfi, L., (Boston: Springer), pp. 31-42.

Simulator it is essential to rely upon a mathematical model, which describes all the peculiar features of the airship from the flight mechanics point of view.

2. Airship characteristics

The new concept airship presented in Figure 1 features a double-hull architecture with a central plane housing structure, propellers, on-board energetic system and payload. This unconventional airship does not use aerodynamic control surfaces, therefore, the primary command system is based on six propellers, moved by electrical motors, suitably set in order to produce the desired forces and moments, necessary to control and maneuver the airship both in hovering and forward flight. In particular, two vertical propellers provide the vertical thrust for climbing, descent and pitching maneuvers, while four thrust-vectoring propellers mounted on rotating arms allow to control the lateral-directional attitude of the airship. The lift is basically generated by a hybrid system consisting of aerostatic lift, the *buoyancy*, provided by the helium inside the hulls, and the vertical thrust given by the vertical propellers. In forward flight, the buoyancy is boosted by the aerodynamic lift developed by the double fuse-shaped body of the airship. In addition, this airship is equipped with two *ballonets*, one for each hull, which can be blown up with air and deflated, respectively during the descent and climb operations, in order to handle altitude variations without losing helium from the hulls and avoid any significant change in the hull shape [2].

3. Mathematical model: general assumptions

The airship mathematical model is based on a six-DOF non linear dynamic model [3], in which the airship is treated as a rigid body without aeroelastic effects and symmetric with respect to the center-line vertical plane XZ. In particular, the model is described by the Newtonian non linear equations of motion, which are expressed through 12 ordinary differential equations. This 12-state formulation basically follows the standard dynamic modelling of the conventional aircraft [4] and assumes that the Earth is fixed in space and its curvature is neglected. However, the presence of aerostatic lift provided by a huge gas volume and the large volume of air displaced by the airship motion give rise to significant additions to the familiar aircraft equations of motion, such as the buoyancy force B and the apparent mass and inertia terms M_{app}.

These equations of motion are referred to the XYZ body-axes reference frame fixed in the Center of Gravity (CG) of the airship. Due to the airship symmetry, both the CG and the Center of Buoyancy (CB) lie in the XZ plane and their coordinates are evaluated in the body-fixed reference frame O_{XYZ}, as shown in Figure 2. Specifically, the CG position (a_x, a_y, a_z) is referred to the geometric point O fixed on the central plane in between the two vertical propellers, while the CB coordinates (b_x, b_y, b_z) are related to the CG position by means of the following expression:

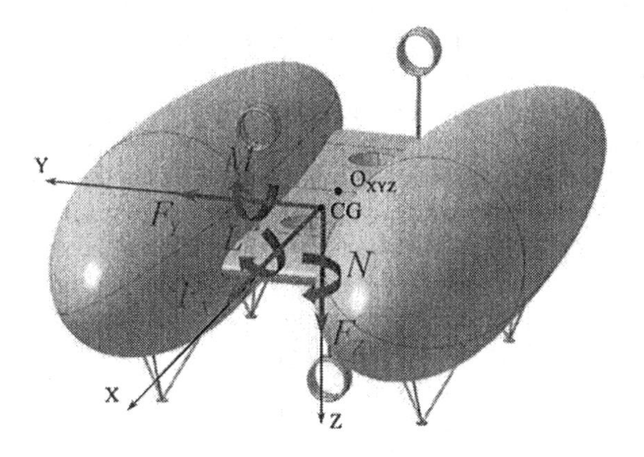

Figure 1. Force & moment system for the Nautilus unmanned airship

$$\bar{b} = \begin{bmatrix} x_{CB} - a_x \\ y_{CB} - a_y \\ z_{CB} - a_z \end{bmatrix} \equiv \begin{bmatrix} b_x \\ b_y \\ b_z \end{bmatrix} \quad (1)$$

where x_{CB}, y_{CB} and z_{CB} are the CB coordinates with respect to O_{XYZ}, which depend on some flight parameters and are gathered into look-up tables for different flight conditions. In this way, they can be evaluated at each time step of the numerical integration of the equations of motion.

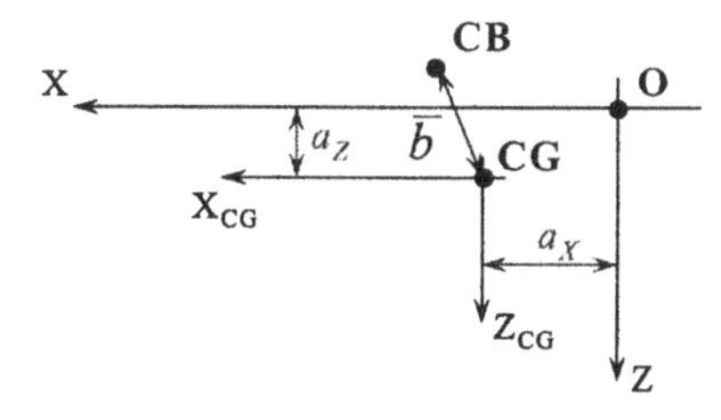

Figure 2. Scheme of the airship reference points

The general scheme of the airship model is illustrated in Figure 3. The complete state vector $\bar{x}$ consists of twelve elements:

$$\bar{x} = [u \ v \ w \ p \ q \ r \ \varphi \ \theta \ \psi \ N \ E \ H]^T \tag{2}$$

where (u, v, w) are the linear velocities, (p, q, r) are the angular velocities, (φ, θ, ψ) are the Euler angles which define the attitude of the airship relatively to the Earth, while (N, E, H) are the coordinates defining the *North-East-Up* position of the airship relatively to the Earth. These twelve state variables can be obtained by integrating their time-derivatives with respect to time through the *Runge-Kutta* numerical integration method. Moreover, the state variables need to be coupled back to all the force and moment equations, as well as to the equations of motion themselves in order to compute their time-derivatives. The inputs of the dynamic model consist of the rotational speeds n_{pr} of all the six propellers and the orientation angles δ_{pr} of the four thrust-vectoring propellers. These inputs feed the propulsion system and generate the desired propulsive forces and moments needed to maneuver the airship. In particular, these signals are generated by the pilot acting on a joystick and two throttles. Successively, the pilot commands are pre-processed and re-allocated by the Control Allocation System [5] modelled in the Flight Control Computer and, finally, are filtered by first order transfer functions that account for the actuator dynamics.

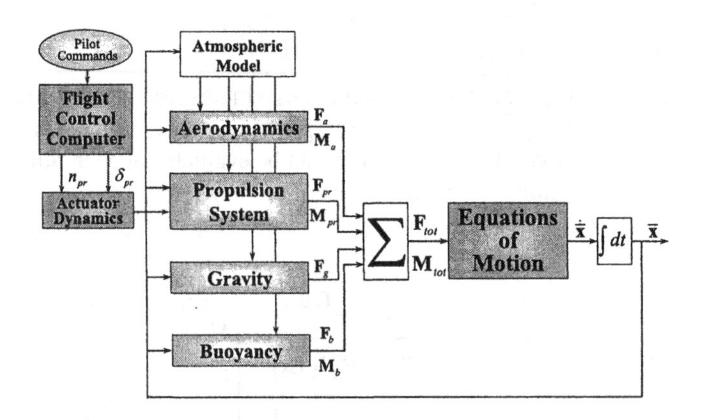

Figure 3. General scheme of the airship model

4. Equations of motion

The airship equations of motion are expressed in the non linear state-space format and consist of six force and moment equations, three kinematic equations and three navigation equations. The axial, side and normal force equations and the rolling, pitching and yawing moments equations with respect to the CG reference frame are computed by implementing the Newton's second law of motion for each degree of freedom. In a matricial form:

$$M\dot{\bar{x}}_v = \bar{F}_d + \bar{F} \quad \Rightarrow \quad \dot{\bar{x}}_v = M^{-1}\left(\bar{F}_d + \bar{F}\right) \tag{3}$$

where $\dot{\bar{x}}_v = [\dot{u} \;\; \dot{v} \;\; \dot{w} \;\; \dot{p} \;\; \dot{q} \;\; \dot{r}]^T$ is the partial state vector of linear and angular accelerations, M is the total mass matrix, $\bar{F}_d$ and $\bar{F}$ represent respectively the dynamic contributions and the external contributions depending on *a*erodynamics, static *b*uoyancy, *pr*opulsion system and *g*ravitational force:

$$\bar{F}_d = \begin{bmatrix} -m_z w\, q + m_y r\, v \\ -m_x u\, r + m_z p\, w \\ -m_y v\, p + m_x q\, u \\ -\left(J_z - J_y\right) rq + J_{yz}\left(q^2 - r^2\right) + J_{zx} p\, q - J_{xy} p\, r \\ -\left(J_x - J_z\right) pr - J_{yz} p\, q + J_{zx}\left(r^2 - p^2\right) + J_{xy} q\, r \\ -\left(J_y - J_x\right) qp + J_{yz} p\, r - J_{zx} q\, r + J_{xy}\left(p^2 - q^2\right) \end{bmatrix} \tag{4}$$

$$\bar{F} = \begin{bmatrix} F_{X_a} + F_{X_b} + F_{X_{pr}} + F_{X_g} \\ F_{Y_a} + F_{Y_b} + F_{Y_{pr}} + F_{Y_g} \\ F_{Z_a} + F_{Z_b} + F_{Z_{pr}} + F_{Z_g} \\ L_a + L_b + L_{pr} \\ M_a + M_b + M_{pr} \\ N_a + N_b + N_{pr} \end{bmatrix} \tag{5}$$

$F_{X,Y,Z}$ are the generic force components in the CG reference frame, while L, M, N are the generic moments around the $X-$, $Y-$ and $Z-$body axes respectively, as shown in Figure 1. The terms $m_{x,\,y,\,z}$ and $J_{x,\,y,\,z,\,xy,\,zx,\,yz}$ represent the airship inertial elements of the total mass matrix M:

$$M = M_i + M_g - M_{app} \tag{6}$$

The airship inertial data, in fact, have been computed on the assumption that M has to account for three contributions [6]: 1) the mass and inertia terms M_i of all the airship structure components; 2) the inertial properties M_g of the gas inside hulls and ballonets; 3) the apparent mass and inertia terms M_{app} arising from the large volumes of air displaced by the airship motion, especially in non stationary conditions. The inertial characteristics of airship structure components and gas are expressed as follows:

$$M_i + M_g = \begin{bmatrix} m & 0 & 0 & 0 & 0 & 0 \\ 0 & m & 0 & 0 & 0 & 0 \\ 0 & 0 & m & 0 & 0 & 0 \\ 0 & 0 & 0 & I_{xx} & -I_{xy} & -I_{xz} \\ 0 & 0 & 0 & -I_{xy} & I_{yy} & -I_{yz} \\ 0 & 0 & 0 & -I_{xz} & -I_{yz} & I_{zz} \end{bmatrix} + \begin{bmatrix} m_g & 0 & 0 & 0 & 0 & 0 \\ 0 & m_g & 0 & 0 & 0 & 0 \\ 0 & 0 & m_g & 0 & 0 & 0 \\ 0 & 0 & 0 & I_{xx_g} & -I_{xy_g} & -I_{xz_g} \\ 0 & 0 & 0 & -I_{xy_g} & I_{yy_g} & -I_{yz_g} \\ 0 & 0 & 0 & -I_{xz_g} & -I_{yz_g} & I_{zz_g} \end{bmatrix} \tag{7}$$

The apparent mass and inertial terms are defined by the following matrix:

$$M_{app} = \begin{bmatrix} X_{\dot{u}} & X_{\dot{v}} & X_{\dot{w}} & X_{\dot{p}} & X_{\dot{q}} & X_{\dot{r}} \\ Y_{\dot{u}} & Y_{\dot{v}} & Y_{\dot{w}} & Y_{\dot{p}} & Y_{\dot{q}} & Y_{\dot{r}} \\ Z_{\dot{u}} & Z_{\dot{v}} & Z_{\dot{w}} & Z_{\dot{p}} & Z_{\dot{q}} & Z_{\dot{r}} \\ L_{\dot{u}} & L_{\dot{v}} & L_{\dot{w}} & L_{\dot{p}} & L_{\dot{q}} & L_{\dot{r}} \\ M_{\dot{u}} & M_{\dot{v}} & M_{\dot{w}} & M_{\dot{p}} & M_{\dot{q}} & M_{\dot{r}} \\ N_{\dot{u}} & N_{\dot{v}} & N_{\dot{w}} & N_{\dot{p}} & N_{\dot{q}} & N_{\dot{r}} \end{bmatrix} \tag{8}$$

The effects of the gas M_g are estimated by modelling the gas/air mass inside hulls and ballonets through the *Catia* CAD code [7]. In this way, the gas inertial characteristics, such as mass m_g, gas center-of-gravity position (x_g, y_g, z_g) and inertia moments $(J_{xx_g}, J_{yy_g}, J_{zz_g})$ with respect to O_{XYZ}, can be evaluated and gathered into look-up tables in order to compute the gas inertial contributions with respect to the airship CG reference frame:

$$\begin{cases} I_{xx_g} &= J_{xx_g} - m_g \left[(z_g - a_z)^2 + (y_g - a_y)^2 \right] \\ I_{yy_g} &= J_{yy_g} - m_g \left[(x_g - a_x)^2 + (z_g - a_z)^2 \right] \\ I_{zz_g} &= J_{zz_g} - m_g \left[(x_g - a_x)^2 + (y_g - a_y)^2 \right] \end{cases} \tag{9}$$

On the contrary, the products of inertia I_{xy_g}, I_{xz_g}, I_{yz_g} in the matrix M_g are assumed to be equal to zero.

The apparent mass and inertial effects M_{app} may be considered as added forces and moments, therefore, they can be described by the dimensional derivatives of aerodynamic forces and moments with respect to linear and angular acceleration perturbations, i.e. $X_{\dot{u}} = \partial F_{X_a} / \partial \dot{u}$ or $L_{\dot{p}} = \partial L / \partial \dot{p}$. The non dimensional coefficients of these derivatives have been calculated with respect to the O_{XYZ} reference frame by using NSAERO [8], a multi-block computational fluid dynamics code, which solves the Navier-Stokes equations including also the viscous effects. Before being included into Eq. (6), however, the matrix M_{app} must be rewritten with respect to the CG reference frame. This translation can be done on the assumptions that the fluid kinetic energy does not depend on the reference frame:

$$\frac{1}{2}\bar{x}_{vCG}^T M_{appCG} \bar{x}_{vCG} = \frac{1}{2}\bar{x}_{vO}^T M_{appO} \bar{x}_{vO} \tag{10}$$

and the velocity vector $\bar{x}_{vO}$ of the body axis origin O can be related to the velocity vector $\bar{x}_{vCG}$ of the airship CG:

$$\bar{x}_{vO} = \left[\begin{array}{c|c} I & A \\ \hline 0 & I \end{array}\right]^{-1} \bar{x}_{vCG} \equiv \left[\begin{array}{c|c} I & -A \\ \hline 0 & I \end{array}\right] \bar{x}_{vCG} \tag{11}$$

where A is the rotational matrix from the O_{XYZ} to the CG reference frame and it is made up by the airship CG coordinates (a_x, a_y, a_z):

$$A = \begin{bmatrix} 0 & a_z & -a_y \\ -a_z & 0 & a_x \\ a_y & -a_x & 0 \end{bmatrix} \tag{12}$$

Finally, the apparent mass matrix in the CG reference frame will be:

$$M_{appCG} = \left[\begin{array}{c|c} I & 0 \\ \hline A & I \end{array}\right] M_{appO} \left[\begin{array}{c|c} I & -A \\ \hline 0 & I \end{array}\right] \tag{13}$$

In order to solve the equations of motion of Eq. (3), it is also necessary to know the attitude, defined by the Euler angles (φ, θ, ψ), and the altitude H of the airship, because some contributions to the external forces and moments $\bar{F}$ of Eq. (5) depend upon these variables. Moreover, it is useful to evaluate the coordinates of the vehicle with respect to the Earth-fixed reference frame to be able to simulate navigational tasks. Firstly, the three kinematic equations defining the airship attitude rates are respectively:

$$\begin{cases} \dot{\varphi} = p + \dot{\psi}\sin\theta \\ \dot{\theta} = q\cos\varphi - r\sin\varphi \\ \dot{\psi} = \dfrac{q\sin\varphi + r\cos\varphi}{\cos\theta} \end{cases} \tag{14}$$

Secondly, the three navigation equations computing the Earth-relative velocities are defined by the following expressions:

$$\begin{cases} \dot{N} = u\cos\theta\cos\psi + v\left(\sin\varphi\sin\theta\cos\psi - \cos\varphi\sin\psi\right) + \\ \qquad + w\left(\sin\varphi\sin\psi + \cos\varphi\sin\theta\cos\psi\right) \\ \dot{E} = u\cos\theta\sin\psi + v\left(\sin\varphi\sin\theta\sin\psi + \cos\varphi\cos\psi\right) + \\ \qquad + w\left(\cos\varphi\sin\theta\sin\psi - \sin\varphi\cos\psi\right) \\ \dot{H} = u\sin\theta - \left(v\sin\varphi + w\cos\varphi\right)\cos\theta \end{cases} \tag{15}$$

5. Aerodynamics and buoyancy

The airship aerodynamic modelling is based on the six aerodynamic coefficients $(C_{X_a}, C_{Y_a}, C_{Z_a}, C_{l_a}, C_{m_a}, C_{n_a})$ and the eighteen damping adimensional derivatives $(C_{X_{p,q,r}}, C_{Y_{p,q,r}}, C_{Z_{p,q,r}}, C_{l_{p,q,r}}, C_{m_{p,q,r}}, C_{n_{p,q,r}})$, which have been computed in the O_{XYZ} reference frame by using the NSAERO code. These coefficients depend on some flight parameters, such as the airspeed, the angle of attack α and the sideslip angle β. In particular, they are estimated on a rough grid for $0 \leq \alpha \leq 90^o$ and $0 \leq \beta \leq 180^o$, at different airspeeds and at zero altitude. These data are successively processed, interpolated and extrapolated into thicker grids and more extended ranges of the airspeed, the angle of attack $(-90^o \leq \alpha \leq +90^o)$, and the sideslip angle $(-180^o \leq \beta \leq +180^o)$, in order to obtain suitable 3-D look-up tables, as shown in Figure 4, in which the coefficient signs are opportunely changed according to the axis conventions. This 360-degree aerodynamic modelling allows the airship dynamic model to handle either hovering or any other forward flight condition and avoid discontinuities in passing from hovering to forward flight.

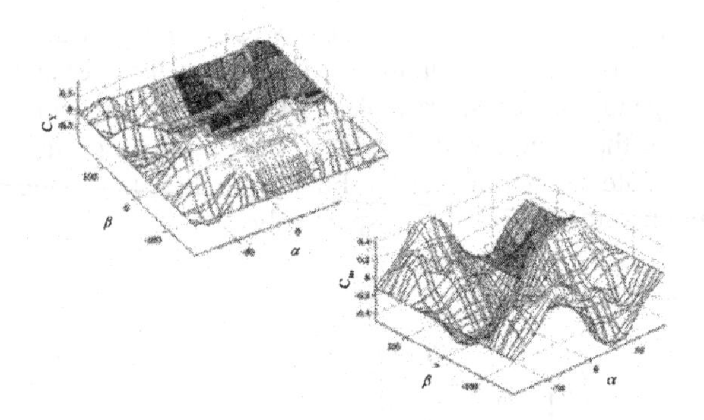

Figure 4. Aerodynamic modelling through Look-Up Tables

Finally, the aerodynamic forces and moments can be expressed in the standard notation as follows:

$$
\begin{cases}
F_{X_a} & = & (C_{X_a} + C_{X_p}\hat{p} + C_{X_q}\hat{q} + C_{X_r}\hat{r}) \cdot 0.5\rho U^2 V^{2/3} \\
F_{Y_a} & = & (C_{Y_a} + C_{Y_p}\hat{p} + C_{Y_q}\hat{q} + C_{Y_r}\hat{r}) \cdot 0.5\rho U^2 V^{2/3} \\
F_{Z_a} & = & (C_{Z_a} + C_{Z_p}\hat{p} + C_{Z_q}\hat{q} + C_{Z_r}\hat{r}) \cdot 0.5\rho U^2 V^{2/3} \\
L_a & = & \left(C_{l_a} + C_{l_p}\hat{p} + C_{l_q}\hat{q} + C_{l_r}\hat{r}\right) \cdot 0.5\rho U^2 V^{2/3} D \\
M_a & = & (C_{m_a} + C_{m_p}\hat{p} + C_{m_q}\hat{q} + C_{m_r}\hat{r}) \cdot 0.5\rho U^2 V^{2/3} l \\
N_a & = & (C_{n_a} + C_{n_p}\hat{p} + C_{n_q}\hat{q} + C_{n_r}\hat{r}) \cdot 0.5\rho U^2 V^{2/3} D
\end{cases}
\tag{16}
$$

where $\hat{p}=bp/2U$, $\hat{q}=cq/2U$, $\hat{r}=br/2U$ are the adimensional angular velocities, ρ is the air density, U is the free-stream airspeed, V is the airship volume, D is the hull diameter and l, c, b are reference lengths. Obviously, these aerodynamic forces need to be properly translated in the CG reference frame through the rotational matrix A, before being integrated in Eq. (5):

$$\left\{ \begin{array}{c} \bar{F}_a \\ \bar{M}_a \end{array} \right\}_{CG} = \left[\begin{array}{c|c} I & 0 \\ \hline A & I \end{array} \right] \left\{ \begin{array}{c} \bar{F}_a \\ \bar{M}_a \end{array} \right\}_{O} \tag{17}$$

The main contribution to the airship lift is supplied by the aerostatic buoyancy provided by the helium inside the hulls. The two inner *ballonets* can be blown up with air and deflated to handle altitude variations. In particular, air is initially released from ballonets during climb up to the *plenitude altitude* H_{max}, that is the altitude defined before each mission according to the amount of helium contained in the hulls and to which the gas is completely expanded filling the hulls themselves. Beyond H_{max}, helium has to be released from hulls causing the reduction of buoyancy B. Differently, during descent ballonets are blown up by the on-board pneumatic system [2]. Assuming constant pressure and temperature inside the hulls, B does not vary from the ground to the plenitude altitude. Beyond H_{max}, B decreases proportionally to the air specific weight according to the second relation of Eq. (14):

$$\begin{cases} B = V_b \, \Lambda_{air_0} \left(1 - \frac{1}{\varepsilon_{He}}\right) & H \leq H_{max} \\ B = V_b \, \Lambda_{air_0} \frac{p_a T_{K_0}}{p_{a_0} T_K} \left(1 - \frac{1}{\varepsilon_{He}}\right) & H > H_{max} \end{cases} \tag{18}$$

where V_b is the helium volume, Λ_{air_0} is the specific weight of air at zero altitude, ε_{He} is the air/helium specific weight ratio, p_a and T_K are respectively the pressure and the temperature of the atmosphere.

The buoyancy action is applied in CB, whose location, defined in Eq. (1), changes with the altitude depending on the amount of air contained in the ballonets. The presence of ballonets and their functioning are mathematically modelled through the variations Δx and Δz of the CB position with respect to the O_{XYZ} reference frame. The lateral shift Δy of the buoyancy is assumed to be negligible, while Δx and Δz have been estimated through the *Catia* CAD code by considering an initial air-helium subdivision of the hulls and, successively, varying the ballonet volume to simulate the altitude variation [9]. Finally, the Δx and Δz variations, needed to evaluate the CB coordinates at each time step of the simulation, are gathered into look-up tables as a function of the altitude H and the plenitude altitude H_{max}. Analogously to the gravitational force mg, the buoyancy B acts along the Z-Earth axis (NED inertial reference frame), thus, it needs to be transferred in the CG body reference frame through the Euler rotational matrix $\{E\}$:

$$\bar{F}_{b_{CG}} = \{E\}\{\ 0 \quad 0 \quad -B\ \}^T_{NED} \tag{19}$$

$$\{E\} = \begin{bmatrix} \cos\theta\cos\psi & \cos\theta\sin\psi & -\sin\theta \\ -\cos\varphi\sin\psi+\sin\varphi\sin\theta\cos\psi & \cos\varphi\cos\psi+\sin\varphi\sin\theta\sin\psi & \sin\varphi\cos\theta \\ \sin\varphi\sin\psi+\cos\varphi\sin\theta\cos\psi & -\sin\varphi\cos\psi+\cos\varphi\sin\theta\sin\psi & \cos\varphi\cos\theta \end{bmatrix} \tag{20}$$

The buoyancy moments $\bar{M}_{b_{CG}}$ are then computed by means of the CB coordinates (b_x, b_y, b_z): $\bar{M}_{b_{CG}} = \bar{b} \wedge \bar{F}_{b_{CG}}$.

6. Propulsive actions

The thrust T of all the six propellers is modelled through the first *Renard* formula $T = \tau\,\rho\omega^2 R^4$, where ρ is the air density, R is the radius and ω is the angular rate of the propellers. τ represents the thrust coefficient, which is a function of the propeller working point $\gamma = U_{ax}/\omega R$, where U_{ax} is the axial airspeed component at the propeller disc along its rotational axis. The first command input n_{pr} acts on the propeller angular rate ω. Figure 5 shows the disposition of the six propellers in the airship central plane. In particular, there are four thrust-vectoring propellers in asymmetrical positions: front up (FU), front down (FD), rear up (RU), rear down (RD); while the two vertical ducted propellers are in the fore (VF) and aft (VA) part of the central plane. The orientation δ of each of the four thrust-vectoring propellers is generated by the second command input δ_{pr}.

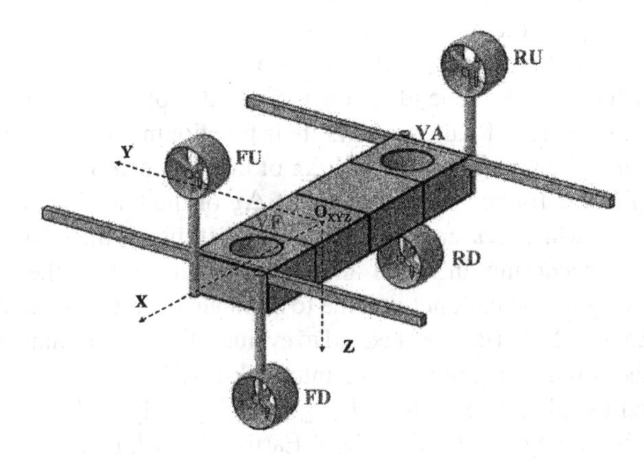

Figure 5. Scheme of the primary control system

The total propulsive forces and moments in the O_{XYZ} reference frame are carried out by adding up the contributions of all the propellers:

$$\begin{cases} F_{X_{pr}} = T_{FU}\cos\delta_{FU} + T_{RD}\cos\delta_{RD} + T_{RU}\cos\delta_{RU} + T_{FD}\cos\delta_{FD} \\ F_{Y_{pr}} = T_{FU}\sin\delta_{FU} + T_{RD}\sin\delta_{RD} + T_{RU}\sin\delta_{RU} + T_{FD}\sin\delta_{FD} \\ F_{Z_{pr}} = -T_{VT} - T_{VA} \end{cases} \tag{21}$$

$$\bar{M}_{pr} = \bar{F}_{pr} \wedge \bar{l}_{pr} \equiv \begin{bmatrix} 0 & F_{Z_{pr}} & -F_{Y_{pr}} \\ -F_{Z_{pr}} & 0 & F_{X_{pr}} \\ F_{Y_{pr}} & F_{X_{pr}} & 0 \end{bmatrix} \begin{Bmatrix} x_{pro} \\ y_{pro} \\ z_{pro} \end{Bmatrix} \tag{22}$$

where $\bar{l}_{pr}$ represents the position vector of each propeller with respect to O_{XYZ}. Successively, $\bar{F}_{pr}$ and $\bar{M}_{pr}$ have to be transferred in the CG reference frame by using an expression analogous to Eq. (17).

7. Conclusions

The airship mathematical model previously described was implemented in the *Matlab/Simulink* environment and, undoubtedly, represented the first step in the development of a refined Flight Simulator, whose availability is definitely essential for the design, test and implementation of the most suitable flight control laws for the innovative Nautilus airship. The general scheme of this Flight Simulator consists basically of two entities, the pilot station and the remote station, which are clearly illustrated in Figure 6.

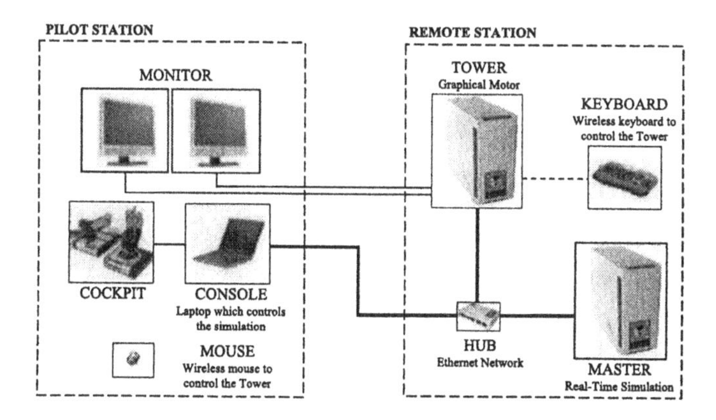

Figure 6. General scheme of the Flight Simulator

References

[1] M. Battipede, P.A. Gili, M. Lando. Ground Station and Flight Simulator for a Remotely-Piloted Non Conventional Airship. In *AIAA Guidance, Navigation and Control Conference.* Reston, VA, 2005.

[2] M. Battipede, M. Lando, P.A. Gili, P. Vercesi. Peculiar Performance of a New Lighter-Than-Air Platform for Monitoring. In *AIAA Aviation Technology , Integration and Operation Forum.* Reston, VA, 2004.

[3] G.A. Khoury, J.D. Gillett. *Airship Technology.* Cambridge University Press, 1999.

[4] B.L. Stevens, F.L. Lewis. *Aircraft Control and Simulation.* John Wiley & Sons, Inc., New York, 1992.

[5] M. Battipede, P.A. Gili, M. Lando. Control Allocation System for an Innovative Remotely-Piloted Airship. In *AIAA Atmospheric Flight Mechanics Conference.* Reston, VA, 2004.

[6] P.G. Thomasson. Equations of Motion of a Vehicle in a Moving Fluid. *Journal of Aircraft.* Vol. 37, 4:630-639, 2000.

[7] CATIA V5R12. *3-D CAD Software.* Dassault Systèmes S.A., 2003.

[8] NSAERO. *Advanced Fluid Dynamics Software.* Analytical Methods, Inc., Redmond, WA.

[9] M. Battipede, P.A. Gili, M. Lando, L. Massotti. Flight Simulator for the Control Law Design of an Innovative Remotely-Piloted Airship. In *AIAA Modeling and Simulation Technologies Conference.* Reston, VA, 2004.

SHAPE OPTIMAL CONCEPTION OF ANTENNA ARRAYS

L. Blanchard [1] and J.P. Zolésio[2]
[1] *INRIA Sophia-Antipolis, France, Louis.Blanchard@sophia.inria.fr,* [2] *CNRS and INRIA Sophia-Antipolis, France, Jean-Paul.Zolesio@sophia.inria.fr*

Abstract The synthesis of an array antenna is an inverse problem. The solution of this problem is the optimization of the complex excitation law of the antenna's radiating elements and the shape of the antenna which provides a specified radiation pattern.

The general character of this method permits to find numerous application in civil and military areas (satellites, telecommunication for earth mobiles, radar antennas, etc ...).

keywords: Shape optimisation, array antenna, Newton method.

1. Introduction

An array antenna is made of many elementary antennas located on a surface in R^3 so as to direct the radiated power towards a desired angular sector. As a general rule, we can consider many different shapes for the array geometry like cylindrical, spherical or conformal surface. However, in order to reduce the difficulty of the problem we consider in this paper planar array antennas. Consequently we note $\zeta \in R^{2M}$ the geometry parameter where M is the number of elements. Finally we call "weight" the amplitude and the phase of an element that we note $\omega \in C^M$.

On one hand, the basic property of an array is that the relative displacements of the antenna elements with respect to each other introduce relative phase shifts in the radiation vectors, which can then add constructively in some directions or destructively in others.

On the other hand, once an array has been designed to focus towards a particular direction, it becomes a simple matter to steer it towards some other direction by changing the relative phases of the array elements. Moreover it is also possible to modify the size of a desired angular sector by changing the amplitudes of the array elements.

The difficult problem is the dependence of the weights of the array elements towards the array's geometry, which we note henceforth $\zeta(\omega)$. On that account,

Please use the following format when citing this chapter:

Blanchard, L., and Zolésio, J.P., 2006, in IFIP International Federation for Information Processing, Volume 202, Systems, Control, Modeling and Optimization, eds. Ceragioli, F., Dontchev, A., Furuta, H., Marti, K., Pandolfi, L., (Boston: Springer), pp. 43-54.

the method which consists in optimizing both elements positions and weights is divided into sub-problems owing to the fact that each array's geometry is associated with a weight's optimization procedure.

We should note that for the general problem joining together the optimization of elements positions and weights, only stochastic methods such as genetics algorithms (GA) (see [8],[6]), simulated annealing (see [7]) or Integer linear programming (see [8]) were used in the literature. However, the problem which consists in optimizing elements weights for a fixed geometrical arrangement, is commonly solved by analytical methods. For example, the least squares technique which minimize the difference between desired and synthesized radiation pattern.

In this framework, this paper proposes a new analytical model , based on a *bicriterion* optimization method, which takes advantage of the derivative of functional expressed in term of min or max, see [4],[1],[3],[5].

2. Mathematical Formulation

We consider a planar array antenna Ω shown in Figure (1), whose M elements are located on the (x, y) plane according to the notation:

$$\zeta_m = (x_m, y_m) \ , \ \forall m \in [1..M].$$

In this paper, we are interested in fields that have radiated to large distances from the antenna. The *far field* approximation assumes that the field point r is very far from the antenna ($r >> \zeta_m$, $\forall m \in [1..M]$). In this case, the electromagnetic wave radiation in the far field is a spherical wave and the radiation pattern is independent of the field point distance $\|r\|$. Consequently we consider the field point $\hat{r}$ on the unit sphere in R^3 noted by S^2, where $\hat{r}$ gives the angular position in the field point.

2.1 Radiation Pattern and Directive Gain:

The radiation pattern factor of the array antenna is defined as follows:

$$|\, E(\zeta, \omega, \hat{r})\, |^2 = \Big| \sum_{m=1}^{M} \omega_m \; g_m(\hat{r}) \; e^{j \frac{2\pi}{\lambda} \hat{r}.\zeta_m} \Big|^2$$

where λ is the wavelength, ω_m is the weight coefficient of element m and $g_m(\hat{r})$ is the radiation pattern of element m.

The *Directive Gain* of an antenna towards a given direction $\hat{r}$ is the radiation pattern nomalized by the corresponding *isotropic radiation intensity*:

$$D(\zeta,\omega,\hat{r}) = \frac{|\, E(\zeta,\omega,\hat{r}) \,|^2}{\frac{1}{4\pi}\int_{S^2} |\, E(\zeta,\, \omega, u) \,|^2 \, ds(u)} \tag{1}$$

We note that the *Directive Gain* is often expressed in dB, that is $D_{dB} = 10\log_{10} D$.

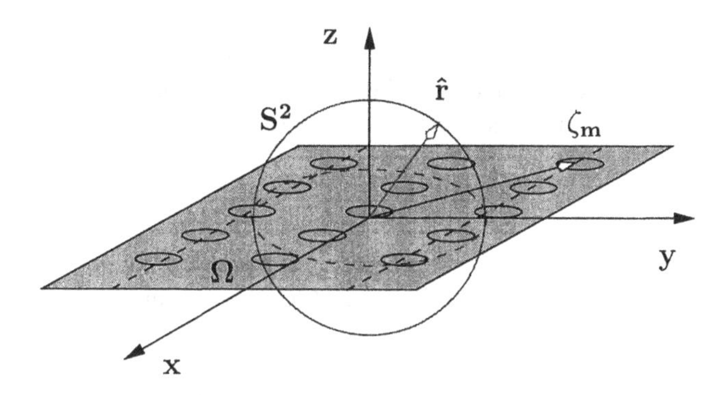

Figure 1. Representation of the planar array antenna Ω

2.2 Spatial Sampling and Grating Lobes :

If we consider a uniformly spaced array antenna, we can apply the *Nyquist-Shannon* sampling theorem (used in signal processing) to spatial sampling. This criterion implies that in order to avoid spatial aliasing the spacial sampling interval must be $d \leq \frac{\lambda}{2}$.

Consequently, the array elements should have interelement spacing of $\frac{\lambda}{2}$ or less in order to remove the grating lobes in the visible region.
The shape optimization method will create a non-periodical geometry in order to reduce the grating lobes. Indeed the presence of grating lobes has an adverse effect on the peak of the main lobe. When a grating lobe appears, energy is taken away from the main lobe and placed into the grating lobe resulting in a loss of *Directive Gain*.

3. Shape Optimization Method

The design problem consists in optimizing geometrical arrangement (ζ) and relative weights (ω) of the array elements so as to focus towards an angular pattern centered at the particular direction $u_p \in S^2$ noted $\mathcal{V}_\lambda(u_p)$.

The optimization criteria have to create beampatterns with main lobe conforms to the desired angular sector $\mathcal{V}_\lambda(u_p)$ and low sidelobe elsewhere in the spherical area $S^2/\mathcal{V}_\lambda(u_p)$. Thus in a mathematical formulation this *bicriterion* becomes:

- **Maximize the Directive Gain** in the angular pattern $\mathcal{V}_\lambda(u_p)$: using the minimisation of the functional

$$R_{in}(\zeta,\omega) = ||D(\zeta,\omega)||^{-1}_{L^\infty(\mathcal{V}_\lambda(u_p))}$$

- **Minimize the Directive Gain** in the angular pattern $S^2/\mathcal{V}_\lambda(u_p)$: using the minimisation of the functional

$$R_{out}(\zeta,\omega) = ||D(\zeta,\omega)||_{L^\infty(S^2/\mathcal{V}_\lambda(u_p))}$$

The Shape Optimization method should find an optimal geometrical arrangement (ζ^*) and its optimal weigth (ω^*)(i.e. the shortand notation for $\omega^*(\zeta^*)$), which satisfy:

$$(\zeta^*,\omega^*) = arg\Big[\min_{(\zeta,\omega)\in\mathcal{U}} [R_{in}(\zeta,\omega)\,,\, R_{out}(\zeta,\omega)]^T\Big] \tag{2}$$

where the feasible set is defined as follows:

$$\mathcal{U} = R^{2M}\times\mathcal{K} \quad \text{and} \quad \mathcal{K} = \{\omega\in C^M \mid ||\omega||_{C^M} = 1\}$$

ASSUMPTION 1 *A vector* $(\zeta^*,\omega^*)\in\mathcal{U}$ *is Edgeworth-Pareto optimal for problem (2) if and only if there exists no vector* $(\zeta,\omega)\in\mathcal{U}$ *such that:*

$$R_{in}(\zeta,\omega) \le R_{in}(\zeta^*,\omega^*) \quad and \quad R_{out}(\zeta,\omega) \le R_{out}(\zeta^*,\omega^*) \tag{3}$$

where at least one of them is strict.

A commonly used procedure to optimize several criteria simultaneously is to combine them. We choose the functional which is the product of the two criteria: $J(\zeta,\omega) = R_{in}(\zeta,\omega)\,R_{out}(\zeta,\omega)$. The global optimization problem is

$$(\mathcal{P}) \quad \min_{\zeta\in R^{2M}} \min_{\omega\in\mathcal{K}} J(\zeta,\omega) \tag{4}$$

For a fixed geometry $\zeta\in R^{2M}$, the weight optimization problem is

$$(\mathcal{P}_1) \quad \min_{\omega\in\mathcal{K}} J(\zeta,\omega) = \min_{\substack{\omega\in C^M\\ ||\omega||_{C^M}=1}} J(\zeta,\omega) \tag{5}$$

3.1 Tangential Newton Method to Solve the Problem $(\mathcal{P}_1)$

A necessary condition to solve the problem $(\mathcal{P}_1)$ is to solve the vector equation:

$$\nabla_{\mathcal{K}} J(\zeta, \omega) = 0. \quad (6)$$

where $\nabla_{\mathcal{K}} J(\omega)$ is the tangential gradient on the manifold $\mathcal{K}$. Tus, this zero finding problem (6) can by solve with the *Newton-Raphson* method, which consists in solving the Newton equation through the algorithm:

1 Given ω such that $\omega^t \omega = 1$.

2 Compute Δ solution of the newton equation:

$$\nabla_{\mathcal{K}} J(\zeta, \omega) + D_{\mathcal{K}} [\nabla_{\mathcal{K}} J(\zeta, \omega)] \Delta = 0 \quad (7)$$

(where $D_{\mathcal{K}} [\nabla_{\mathcal{K}} J(\zeta, \omega)]\Delta$ denotes the directional tangential derivate of $\nabla_{\mathcal{K}} J(\zeta, \omega)$ in the direction of Δ.)

3 Compute the update $\omega_+ = \omega + \Delta$.

ASSUMPTION 2 *Let* $\Pi_\omega = (Id_{2M} - \omega \otimes \omega)$ *be the orthogonal projection onto the tangent space at* $\omega \in \mathcal{K}$. *We have :*

$$\nabla_{\mathcal{K}} J(\zeta, \omega) = Pi_\omega \nabla J(\zeta, \omega) \quad \textit{and} \quad D_{\mathcal{K}} [\nabla_{\mathcal{K}} J(\zeta, \omega)] = D [\nabla_{\mathcal{K}} J(\zeta, \omega)] \Pi_\omega. \quad (8)$$

Remarks:

i) The function $J(\zeta, \omega)$ is homogeneous of degree zero:

$$\forall \omega \in C^M \quad \forall \lambda \in C \qquad J(\zeta, \lambda \omega) = \lambda J(\zeta, \omega).$$

On that account the solutions of (6) are not isolated in C^M, namely, if ω is a solution, then all the elements of the equivalence class $\{\lambda\omega \mid \forall\lambda \in C\}$ are solutions too, therefore the feasible set of the weigth is the unit sphere in C^M:

$$\mathcal{K} = \{\omega \in C^M \quad | \quad \|\omega\|_{C^M} = 1 \}$$

ii) The solution of the Newton equation (7), when unique, is $\Delta = -\omega$. So any point ω is mapped to $\omega_+ = 0$. This is clearly a solution of the equation (6) but it is a trivial solution. A way to avoid it consists in constraining Δ to belong to the *horizontal space* orthogonal to ω defined as follows:

$$H_\omega = \{y \in C^M \quad | y^t \omega = 0\} \quad (9)$$

With this constraint on Δ, the solution Δ of $\nabla_{\mathcal{K}} J(\zeta, \omega + \Delta) = 0$ become isolated. Moreover, the Newton equation (7) has in general no solution in H_ω, so the Newton equation must be relaxed. Therefore we project the Newton equation onto H_ω and replace it by the system:

$$\begin{cases} \Pi_\omega \left(\nabla_{\mathcal{K}} J(\zeta,\omega) \; + \; D_{\mathcal{K}} \left[\nabla_{\mathcal{K}} J(\zeta,\omega) \right] \Delta \right) = 0 \\ \omega^t \, \Delta = 0 \end{cases} \tag{10}$$

Implementation: Numerically the computation of the optimal weight is done by a *Newton-Raphson* under constraints method preceded by a projected conjugate gradient method. The goal of this *Newton-Raphson* method is of course to find the optimal weight $\omega^*(\zeta)$, solution of the problem $(\mathcal{P}_1)$ but especially to solve the vector equation (6) as we can see in the next paragraph.

3.2 Method to Solve the Problem $(\mathcal{P})$:

Let us consider $J(\zeta, \omega^*(\zeta)) = \min_{\omega \in \mathcal{K}} \; J(\zeta, \omega)$, then we can rewrite the problem $(\mathcal{P})$ as follows: $\min_{\zeta \in R^{2M}} \; J(\zeta, \omega^*(\zeta))$. A necessary condition to solve the problem $(\mathcal{P})$ is to solve the vector equation $\frac{d}{d\zeta} J(\zeta, \omega(\zeta)) = 0$, which by a chain rule becomes:

$$\nabla_\zeta J(\zeta, \omega(\zeta)) + \nabla_\omega J(\zeta, \, \omega(\zeta)) . \frac{\partial}{\partial \zeta} \omega(\zeta) \;=\; 0 \tag{11}$$

However, since we solved previously the problem $(\mathcal{P}_1)$ by the *Newton-Raphson* method, we have obtain the optimal weight $\omega^*(\zeta)$. In this way we avoid the fastidious calculation of the term $\frac{\partial}{\partial \zeta}\omega(\zeta)$. Indeed, we have:

$$\begin{aligned} &\omega^*(\zeta) \text{ solution of } (\mathcal{P}_1) \;\Rightarrow\; \nabla_\omega J(\zeta, \, \omega^*(\zeta)) = 0 \\ &\Rightarrow \; \frac{d}{d\,\zeta} J(\zeta, \omega^*(\zeta)) \;=\; \nabla_\zeta J(\zeta, \omega^*(\zeta)). \end{aligned}$$

3.3 Problem $(\mathcal{P})$ with Geometrical Constraints:

In the shape optimization of an antenna, we often need to constrain the antenna geometry. As a consequence we replace the problem $(\mathcal{P})$ by the problem under constraints :

$$(\mathcal{P}') \;\; \min_{\zeta \in C} \;\; J(\zeta, \omega^*(\zeta)) \qquad \text{where } C \text{ is the feasible set} \tag{12}$$

In this article, we consider two different kinds of constraints concerning the geometry of the antenna in order to control the expansion or the contraction of its surface. Thus in order to manage the feasible geometry we ought to:

- constrain the maximum size of the antenna by including the realizable geometries in a circle of radius R, therefore we have for the feasible set :

$$\mathcal{C} = \Big\{\zeta \in R^{2M} \quad | \quad \forall p \in [1..M] \quad g_p(\zeta) = ||\zeta_p||_2 - R \leq 0 \quad \Big\}.$$

- constrain the overlap of each array element not to have a superposition of elements between them. Thus for each element we defined its neighbors as follows:

$$\forall p \in [1..M] \qquad F_p = \Big\{q \in [1..M] \quad | \quad ||\zeta_p - \zeta_q||_2 \leq d_{min}\Big\}.$$

and the feasible set becomes:

$$C = \Big\{\zeta \in R^{2M} \quad | \quad \forall p \in [1..M], \forall q \in F_p$$
$$g_{p,q}(\zeta) = d_{min} - ||\zeta_p - \zeta_q||_2 \leq 0\Big\}.$$

4. Numerical Results

To assess the effectiveness of the proposed approach, we consider two different planar antenna arrays, where we approximate the single element pattern by : $EP(\hat{r}) = (\cos\theta)^2$. In the first example, we consider a sparse array with a fixed maximum size of the antenna. In the second example, we consider a planar antenna array constituted of a great number of elements. However, in order to reduce the size of the optimization parameter in the problem, we use the well-known idea which consists in dividing the array into groups of subarrays. Thus, each subarray m is fed through a single weight (ω_m) and (ζ_m) represents its center. In this example we use the shape optimization method under geometrical constraints to manage the overlap of the subarrays and obtain a feasible geometry.

4.1 Example 1 : Shape Optimization for a Sparse Array

We consider a sparse hexagonal array where the interelement spacing is $d = 2\lambda$, according to the *Nyquist-Shannon* criterion (2.2) this geometry creates grating lobes in the visible region as we can see in the Figure (3). We apply the shape optimization method under geometrical constraint in order to solve the problem $(\mathcal{P}')$. We compare the result of the optimal array to the hexagonal array in Figure (2).

4.2 Example 2 : Shape Optimization Antenna Using Subarrays

In this example we assume that all the M subarrays are identical and refer to each one as a *primary array*. Each subarray is made of a uniformly square

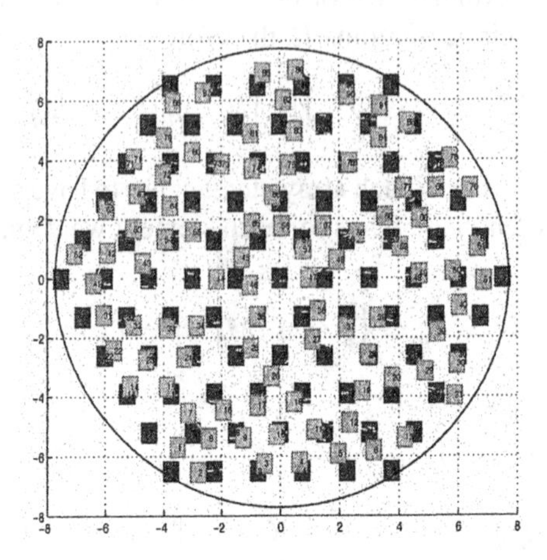

Figure 2. Geometrical representation of the hexagonal antenna array (dark) and the optimal antenna array (clear).

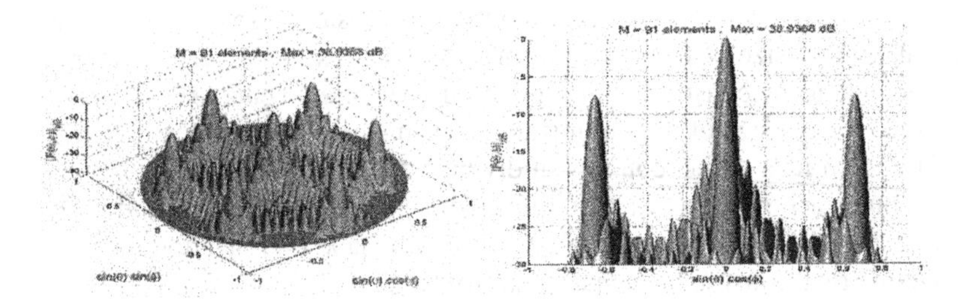

Figure 3. *Directive Gain* in dB for the antenna array with an hexagonal geometry.

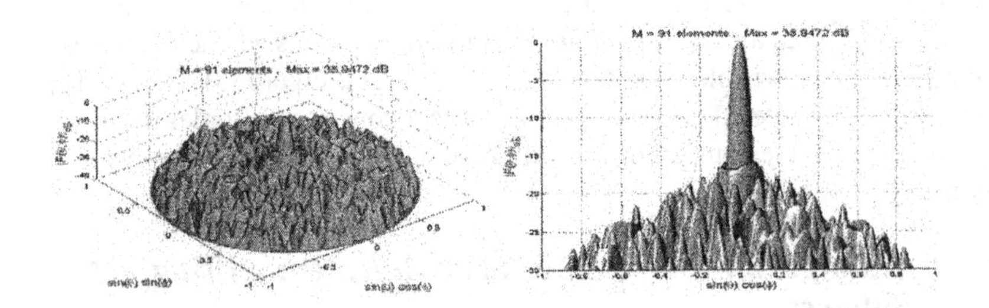

Figure 4. *Directive Gain* in dB for the antenna array with an optimal geometry.

array with N^2 elements where the interelement spacing is $d = \frac{\lambda}{2}$. The radiation pattern of each subarray is defined as follows: $|\, g_m(\hat{r})\,|^2 \;=\; |\, g(\hat{r})\,|^2$ is

$$EP(\hat{r})\,|^2 \left| \frac{\sin\left[\frac{N\pi}{2}(\sin\theta\cos\varphi)\right]}{N\sin\left[\frac{\pi}{2}(\sin\theta\cos\varphi)\right]} \frac{\sin\left[\frac{N\pi}{2}(\sin\theta\sin\varphi)\right]}{N\sin\left[\frac{\pi}{2}(\sin\theta\sin\varphi)\right]} \right|^2 \qquad \forall m \in [1..M]\,,$$

where (θ, φ) is the parametrization of the field point $\hat{r}$ on the unit sphere S^2 and $EP(\hat{r})$ is the element pattern. The array of the primary array is called the *secondary array*. The secondary array is a sparse square array where the interelement spacing is $d = \frac{N\lambda}{2}$, according to the *Nyquist-Shannon* criterion (2.2) this geometry creates grating lobes in the visible region as we can see in the Figure (1). The combined array factor will be equal to the product of these two array factors such that

$$|\, d(\varsigma, \omega, \hat{r})\,|^2 \;=\; |\, g(\hat{r})\,|^2 \left| \sum_{m=1}^{M} \omega_m \; e^{j\,\frac{2\pi}{\lambda}\,\hat{r}.\varsigma_m} \right|^2$$

We desire to center the angular pattern $\mathcal{V}_\lambda(u_p)$ at the particular direction $u_p = (\theta_p, \varphi_p) \;=\; (9^o, 9^o)$ and we choose $M \;=\; 100$, $N \;=\; 4$. We compare the result between the *Weights Optimization method* from the problem $(\mathcal{P}_1)$ and the *Shape Optimization under constraints method* from the problem $(\mathcal{P}')$ which optimizes both elements positions and weights. The comparison between the optimal array and the initial array is shown in Figure (5).

4.3 Weights Optimization versus Shape Optimization under constraints

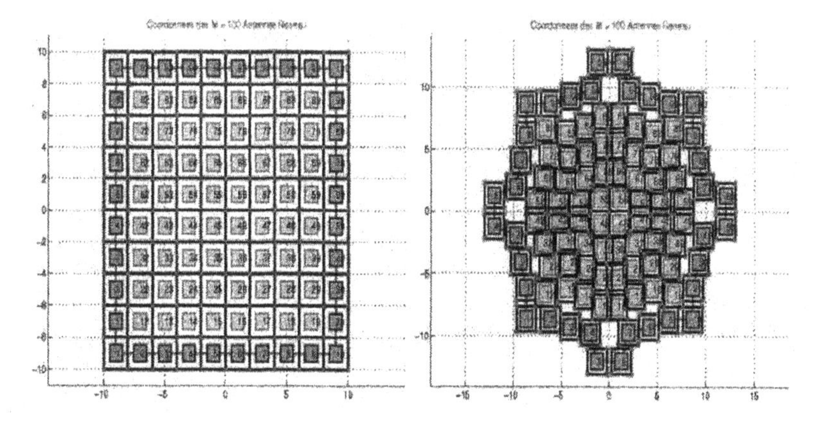

Figure 5. Geometrical representation of the initial antenna (on the left) and the optimal antenna (on the right).

From now on, all the figures on the left side will result from the **Weights Optimization** method (with the initial antenna) , and all those on the right side will result from the **Shape Optimization under constraints** method.

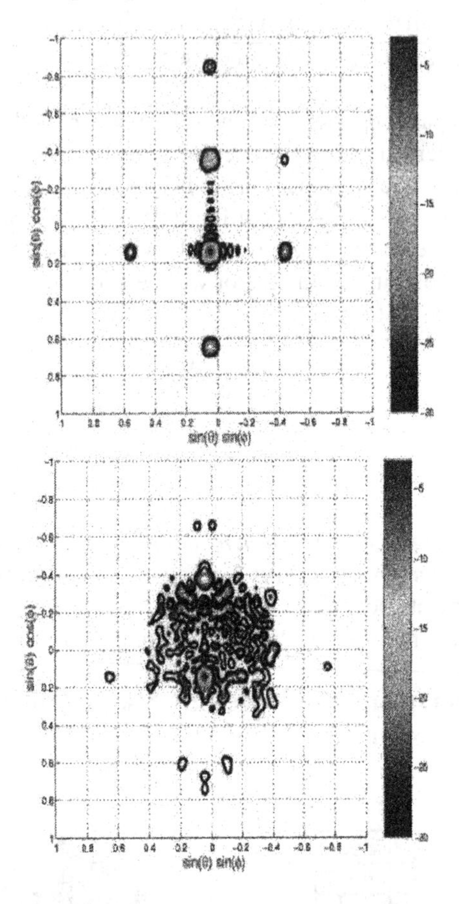

Figure 6. Isovalues of *Directive Gain* in dB.

Results from the initial antenna and the Weights Optimization method.

	Directive Gain mean	*Directive Gain* maximum
in the zone $\mathcal{V}_\lambda(u_p)$	**29.78 dB**	**35.33 dB**
in the zone $S^2/\mathcal{V}_\lambda(u_p)$	**3.27 dB**	**27.02 dB**

difference between the main lobe and the highest grating lobe : $8.31\,dB$.

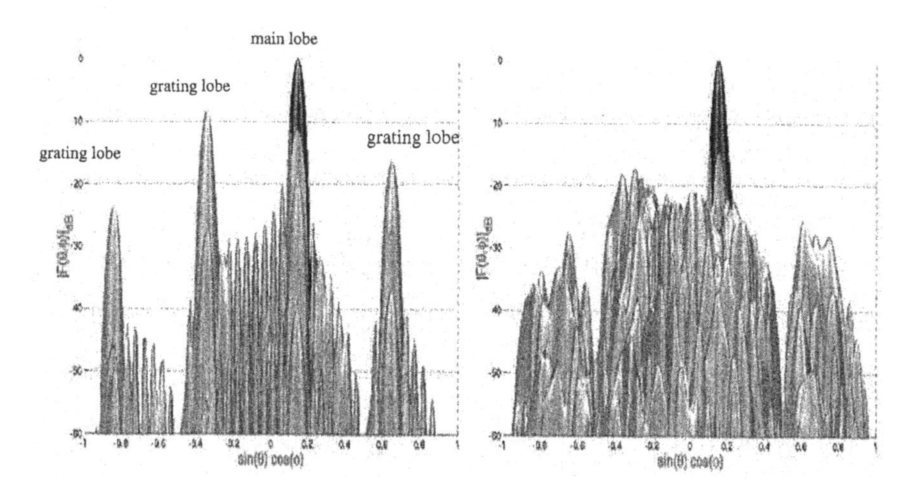

Figure 7. *Directive Gain* in dB in the areas $\mathcal{V}_\lambda(u_p)$ (dark) and $S^2/\mathcal{V}_\lambda(u_p)$ (clear).

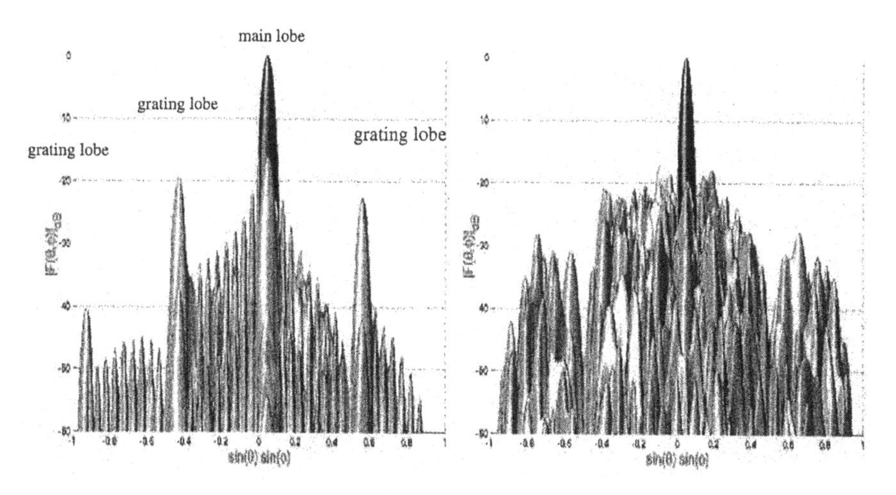

Figure 8. *Directive Gain* in dB in the areas $\mathcal{V}_\lambda(u_p)$ (dark) and $S^2/\mathcal{V}_\lambda(u_p)$ (clear).

Results from the optimal antenna and the Shape Optimization method.

	Directive Gain mean	*Directive Gain* maximum
in the zone $\mathcal{V}_\lambda(u_p)$	**29.05 dB**	**35.48 dB**
in the zone $S^2/\mathcal{V}_\lambda(u_p)$	**3.60 dB**	**18.28 dB**

difference between the main lobe and the highest grating lobe : $17.20\,dB$.

5. Conclusion

A computational method has been successfully designed for the shape optimization of an array antenna. Intuitively the non-periodicity of the array antenna is a necessary condition to avoid the grating lobes because the hypotheses of the Nyquyst's theorem are not satisfied any longer. Nevertheless, the optimization method is essential in order to definitively prevents the grating lobes and reduces the sidelobes. This analysis can be generalized for array antenna distributed on a surface by making use of intrinsic geometry approach induced by the oriented distance function [2].

References

[1] M. Delfour and J.P. Zolésio "Shape and Geometry" Advances in Design and Control,04, SIAM, 2001

[2] Delfour, Michel C. and Zolésio, Jean-Paul "Oriented distance function and its evolution equation for initial sets with thin boundary". SIAM J. Control Optim. 42 (2004), no. 6, 2286–2304

[3] Delfour, M. C. and Zolésio, J.-P., *Shape sensitivity analysis via min max differentiability*, SIAM J. Control Optim. , 1988, 26, 4, 834-862.

[4] J.P. Zolésio "In *Optimization of Distributed Parameter structures*", vol.II, (E. Haug and J. Céa eds.), Adv. Study Inst. Ser. E: Appl. Sci., 50, Sijthoff and Nordhoff, Alphen aan den Rijn, 1981 :

i) The speed method for Shape Optimization, 1089-1151 . ii) Domain Variational Formulation for Free Boundary Problems, 1152-1194. iii) Semiderivative of reapeted eigenvalues , 1457-1473.

[5] M. Cuer and J.P. Zolésio "Control of singular problem via differentiation of a min-max". Systems Control Lett. 11 (1988), no. 2, 151–158.

[6] Donelli, M., Caorsi, S., De Natale F., Pastorino M. and Massa A. "Linear antenna synthesis with a hybrid genetic algorithm" Progress In Electromagnetics Research, PIER 49, 1-22,2004

[7] Murino, V.,A. Trucco, and C.S. Regazzoni "Synthesis of unequally spaced arrays by simulated annealing" IEEE Trans.Signal Processing, Vol. 44,119-123, 1996

[8] S. Holm, A. Austeng, K. Iranpour and J. F. Hopperstad, "Sparse sampling in array processing", chapter 19 "in Sampling theory and practice" (F. Marvasti Ed.) Plenum, New York, 2000.

A SEMIGROUP APPROACH TO STOCHASTIC DYNAMICAL BOUNDARY VALUE PROBLEMS

S. Bonaccorsi [1] and G. Ziglio [1]

[1] *University of Trento, Department of Mathematics, Trento, Italy, stefano.bonaccorsi@unitn.it, ziglio@science.unitn.it*

Abstract In many physical applications, the evolution of the system is endowed with *dynamical boundary conditions*, i.e., with boundary operators containing time derivatives. In this paper we discuss a generalization of such systems, where stochastic perturbations affect the way the system evolves in the interior of the domain as well as on the boundary.

keywords: Stochastic differential equations, boundary noise, semigroup theory, dynamical boundary conditions.

1. Introduction

In this paper we apply the technique of product spaces and operator matrices to solve stochastic evolution equations with randomly perturbed dynamic boundary conditions. Similar results for deterministic problems were recently reached for instance by [1, 10]; in their approach, the starting point is to convert an equation with inhomogeneous boundary conditions into an abstract Cauchy problem, and therefore use semigroup theory. Here, we combine these techniques with stochastic analysis to solve some "model" problems.

In the following lines, we present an appropriate abstract setting for our program. Let X and ∂X be two Hilbert spaces, called the *state space* and *boundary space*, respectively, and $\mathcal{X} = X \times \partial X$ their product space.

We consider the following linear operators:

- $A_m : D(A_m) \subset X \to X$, called *maximal operator*, and $B : D(B) \subset \partial X \to \partial X$;
- $L : D(A_m) \to \partial X$, called *boundary operator*;
- $\Phi : D(\Phi) \subset X \to \partial X$, called *feedback operator*; we assume that $D(A_m) \subset D(\Phi)$.

Please use the following format when citing this chapter:

Bonaccorsi, S., and Ziglio, G., 2006, in IFIP International Federation for Information Processing, Volume 202, Systems, Control, Modeling and Optimization, eds. Ceragioli, F., Dontchev, A., Furuta, H., Marti, K., Pandolfi, L., (Boston: Springer), pp. 55-65.

With these operators, we consider the *stochastic dynamic boundary value problem*

$$\begin{cases} \mathrm{d}u(t) = [A_m u(t) + F(u(t))]\,\mathrm{d}t + G(u(t))\,\mathrm{d}W(t), & t > 0 \\ x(t) := Lu(t), & \\ \mathrm{d}x(t) = [Bx(t) + \Phi u(t)]\,\mathrm{d}t + \Gamma(x(t))\,\mathrm{d}V(t), & t > 0 \\ u(0) = u_0, \quad x(0) = x_0. & \end{cases} \tag{1}$$

In the next section, we introduce the abstract setting, and discuss sufficient conditions in order to solve (1); then in section 3 we discuss some examples that fit into our framework.

In general, we apply the abstract theory of section 2 by checking, first, that the leading operators of the internal and boundary dynamics both generate a strongly continuous (or even analytic) semigroup, while the stochastic perturbation is defined by a Lipschitz mapping. Then, we need to control the feedback operator Φ; in several cases, for instance if it is a bounded operator or if the boundary space is finite dimensional (compare example 3.1), the generation results (Lemma 7 and 8) hold. Then the existence and uniqueness result for the solution of equation (1) are given by Lemma 6.

2. Abstract setting for dynamic boundary value problems

Given Hilbert spaces U, V, we shall denote $\mathcal{L}(U; V)$ (resp. $\mathcal{L}(U)$) the space of linear bounded operators from U into V (resp. into U itself) and $\mathcal{L}_2(U; V)$ the space of Hilbert-Schmidt operators from U into V.

In order to consider the evolution of the system with dynamic boundary conditions, we start by introducing the operator A_0, defined by

$$\begin{cases} D(A_0) = \{f \in D(A_m) \mid Lf = 0\} \\ A_0 f = A_m f \text{ for all } f \in D(A_0). \end{cases}$$

We are in the position to formulate the main set of assumptions on the deterministic dynamic of the system.

Assumption 1

1 *A_0 is the generator of a strongly continuous semigroup $(T_0(t))$, $t \geq 0$, on the space X;*

2 *B is the generator of a strongly continuous semigroup $(S(t))_{t \geq 0}$ on the space ∂X;*

3 *$L : D(A_m) \subset X \to \partial X$ is a surjective mapping;*

4 *the operator $\binom{A_m}{L} : D(A_m) \subset X \to \mathcal{X} = X \times \partial X$ is closed.*

2.1 No boundary feedback

In order to separate difficulties, in this section we consider the case $\Phi \equiv 0$. In order to treat (1) using semigroup theory, we consider the operator matrix $\mathcal{A}$ on $\mathcal{X}$ given by

$$\mathcal{A} = \begin{pmatrix} A_m & 0 \\ 0 & B \end{pmatrix}, \qquad D(\mathcal{A}) = \left\{ \begin{pmatrix} u \\ x \end{pmatrix} \in D(A_m) \times D(B) \mid Lu = x \right\}$$

Our first step is to introduce the *Dirichlet operator* D_μ. This construction is justified by [9, Lemma 1.2]. For given $\mu \in \rho(A_0)$, assume that the stationary boundary value problem

$$\mu w - A_m w = 0, \qquad Lw = x$$

has a unique solution $D_\mu x := w \in D(A_m)$ for arbitrary $x \in \partial X$. Then D_μ is the Green (or Dirichlet) mapping associated with A_m and L. For $\mu \in \rho(A_0)$ we define the operator matrix

$$\mathcal{D}_\mu = \begin{pmatrix} I_X & -D_\mu \\ 0 & I_{\partial X} \end{pmatrix};$$

from [12, Lemma 2] we obtain the representation

$$(\mu - \mathcal{A}) = (\mu - \mathcal{A}_0)\mathcal{D}_\mu$$

where $\mathcal{A}_0$ is the diagonal operator matrix

$$\mathcal{A}_0 = \begin{pmatrix} A_0 & 0 \\ 0 & B \end{pmatrix}$$

on $D(\mathcal{A}_0) = D(A_0) \times D(B)$.

Using [1, Theorem 2.7 and Corollary 2.8], we are in the position to characterize the generation property of $(\mathcal{A}, D(\mathcal{A}))$.

LEMMA 2 *Assume that A_0 is invertible. Then $\mathcal{A}$ generates a C_0-semigroup $\mathcal{T}(t)$ on $\mathcal{X}$ if and only if the operator $Q_0(t) : D(B) \subset \partial X \to X$,*

$$Q_0(t)y := -A_0 \int_0^t T_0(t-s) D_0 S(s) y \, ds, \tag{2}$$

has an extension to a bounded operator on ∂X, satisfying

$$\limsup_{t \searrow 0} \|Q_0(t)\| < +\infty. \tag{3}$$

Moreover, in this case we can also give a representation of $\mathcal{T}(t)$:

$$\mathcal{T}(t) = \begin{pmatrix} T_0(t) & Q_0(t) \\ 0 & S(t) \end{pmatrix}. \tag{4}$$

COROLLARY 3 *Assume that A_0 and B generate analytic semigroups on X and ∂X, respectively; then $\mathcal{A}$ generates an analytic semigroup $(\mathcal{T}(t))_{t\geq 0}$ on $\mathcal{X}$.*

COROLLARY 4 *If $B \in \mathcal{L}(\partial X)$ is bounded, then $\mathcal{A}$ generates a C_0 semigroup on $\mathcal{X}$; in particular, if A_0 is invertible and $B = 0$, then $Q_0(t) = (I - T_0(t))D_0$.*

We are now in position to write the original problem (1) in an equivalent problem as a *stochastic abstract Cauchy problem*

$$\begin{cases} \mathrm{d}\mathbf{x}(t) = [\mathcal{A}\mathbf{x}(t) + \mathcal{F}(\mathbf{x}(t))]\,\mathrm{d}t + \mathcal{G}(\mathbf{x}(t))\,\mathrm{d}\mathcal{W}(t), \\ \mathbf{x}(0) = \mathbf{x}_0 \end{cases} \tag{5}$$

In order to solve this stochastic problem, we introduce some assumptions on the nonlinear and stochastic terms which appear in (1); these assumptions, in turn, will be reflected to the operators $\mathcal{F}$ and $\mathcal{G}$ in (5).

ASSUMPTION 5

We are given a stochastic basis $(\Omega, \mathcal{F}, \{\mathcal{F}_t\}, \mathbf{P})$;

1. *W and V are Wiener noises on X and ∂X, respectively, and $\mathcal{W} = (W, V)$ is a Q-Wiener process on $\mathcal{X}$, with trace class covariance operator Q;*
2. *the mappings $F : X \to X$ and $G : X \to \mathcal{L}_2(X, X)$ are Lipschitz continuous*

$$|F(x) - F(y)| + \|G(x) - G(y)\| \leq C|x - y|,$$

with linear growth bound

$$|F(x)|^2 + \|G(x)\|^2 \leq C(1 + |x|^2);$$

3. *the mapping $\Gamma : \partial X \to \mathcal{L}_2(\partial X, \partial X)$ is Lipschitz continuous with linear growth bound:*

$$\|\Gamma(x) - \Gamma(y)\| \leq C|x - y|, \qquad \|\Gamma(x)\|^2 \leq C(1 + |x|^2).$$

In order to make this paper self-contained, let us recall the relevant result from [3].

LEMMA 6 *Assume that $\mathcal{A}$ is the generator of a C_0 semigroup on $\mathcal{X}$ and Assumption 5 holds. Then for every $\mathbf{x}_0 \in \mathcal{X}$, there exists a unique mild solution to (5); moreover, it has a continuous modification.*

2.2 Boundary feedback

We are now in the position to include the feedback operator Φ into our discussion. In order to simplify the exposition, and in view of the examples below, we choose to concentrate on two cases, which are far from being general.

We shall prove some generation results for the operator matrix

$$\mathcal{A} = \begin{pmatrix} A_m & 0 \\ \Phi & B \end{pmatrix}, \qquad D(\mathcal{A}) = \left\{ \begin{pmatrix} u \\ x \end{pmatrix} \in D(A_m) \times D(B) \mid Lu = x \right\},$$

where we assume that A_0 is the generator of a strongly continuous semigroup with $0 \in \rho(A_0)$.

As in Section 2.1 we write

$$\mathcal{A} = \mathcal{A}_0 \mathcal{D}_0^{\Phi},$$

where the operator matrix $\mathcal{D}_0^{\Phi}$ is given by

$$\mathcal{D}_0^{\Phi} = \begin{pmatrix} I_X & -D_0 \\ B^{-1}\Phi & I_{\partial X} \end{pmatrix} = I_{\mathcal{X}} + \begin{pmatrix} 0 & -D_0 \\ B^{-1}\Phi & 0 \end{pmatrix}.$$

The first result can be proved as in [1, Section 4].

LEMMA 7 *Assume that the feedback operator $\Phi : X \to \partial X$ is bounded. Then the matrix operator $\mathcal{A}$ is the generator of a C_0 semigroup.*

Next, we consider a generation result in case Φ is unbounded, which is the case in several applications (see for instance [2]). This case may be treated using the techniques of *one-sided coupled operators*, compare [8, Theorem 3.13 and Corollary 3.17].

LEMMA 8 *Assume that A_0 is the generator of an analytic semigroup, that $B \in \mathcal{L}(\partial X)$ and $D_0\Phi$ is a compact operator; then the matrix operator $\mathcal{A}$ is the generator of an analytic semigroup.*

Assume that the boundary space ∂X is finite dimensional. Then $B \in \mathcal{L}(\partial X)$ is bounded and $D_0\Phi$ is a finite rank operator, hence it is compact, so that $\mathcal{A}$ is the generator of an analytic semigroup thanks to previous lemma.

3. Motivating examples

We are concerned with the following examples. The first two are also considered in the paper [2]; notice that the first one has some applications in mathematical biology (for instance, to study impulse propagation along a neuron). The third example was considered in the paper [4] and (in the special case discussed here) in the paper [6].

3.1 Impulse propagation with boundary feedback

A widely accepted model for a dendritic spine with passive electric activity can be described by means of the following equation for the potential

$$\frac{\partial u}{\partial t}(t,\xi) = \frac{\partial^2 u}{\partial \xi^2}(t,\xi) + f(\xi, u(t,\xi)), \qquad t > 0, \quad \xi > 0;$$

the extremal point $\xi = 0$ denotes the cellular *soma*, where the potential evolves with a different dynamic; setting $x(t) = u(t,0)$, the following equation is a possible model for this dynamic

$$\mathrm{d}x(t) = [-bx(t) + cu'(t,0)]\,\mathrm{d}t + \sigma(x(t))\,\mathrm{d}W(t),$$

where $W(t)$ is a real standard brownian motion.

In order to set the problem in an abstract setting, we consider the spaces $X = L^2(\mathbf{R}_+)$ and $\partial X = \mathbf{R}$; the matrix operator $\mathcal{A}$ is given by

$$\mathcal{A} = \begin{pmatrix} \frac{\partial^2}{\partial x^2} & 0 \\ c\frac{\partial}{\partial x}|_{x=0} & -b \end{pmatrix}.$$

Since the boundary space ∂X is finite dimensional and the leading operator $\frac{\partial^2}{\partial x^2}$ on $\mathbf{R}_+$ with Dirichlet boundary condition generates an analytic semigroup, then so does $\mathcal{A}$ on $\mathcal{X} = X \times \partial X$. Therefore, we write our problem in the equivalent form

$$\mathrm{d}\mathbf{x}(t) = [\mathcal{A}\mathbf{x}(t) + \mathcal{F}(\mathbf{x}(t))]\,\mathrm{d}t + \mathcal{G}(\mathbf{x}(t))\,\mathrm{d}\mathcal{W}(t)$$

and we obtain existence and uniqueness of the solution thanks to Lemma 6.

3.2 Dynamic on a domain with mixed boundary conditions

In previous example, the boundary space was finite dimensional. Here, we shall be concerned with a dynamical system which evolves in a bounded region $\mathcal{O} \subset \mathbf{R}^d$, with smooth boundary $\Gamma = \partial\mathcal{O}$. We assume that $\Gamma = \bar{\Gamma}_1 \cup \bar{\Gamma}_2$, where Γ_i are open subsets of Γ with $\Gamma_1 \cap \Gamma_2 = \emptyset$.

Let $\mathcal{O}$ represent a solid body; suppose that a classical heat diffusion process occurs inside $\mathcal{O}$, so if $u = u(t,x)$ represents the temperature at point x and time t, the process can be modelled by the classical heat equation

$$\frac{\partial u}{\partial t}(t,x) = \Delta u(t,x)$$

where the thermal conductivity $\rho > 0$ is taken to be 1 for simplicity. This equation needs to be completed by boundary conditions and initial data. Inspired by physical considerations, different sorts of boundary conditions exist in the literature. In this paper, we consider dynamical ones, that is

$$\mathrm{d}x_1(t) = Bx_1(t)\,\mathrm{d}t + \Gamma(x_1(t))\,\mathrm{d}V(t),$$

where x_1 is the valuation of u on the (portion of) boundary Γ_1 and a stochastic perturbation acts on the boundary behavior. This kind of boundary conditions appears when the boundary material has a large thermal conductivity and sufficiently small thickness. Hence, the boundary material is regarded as the boundary of the domain. For instance, one considers an iron ball in which water and ice coexists. Notice that we can cover the case where the boundary conditions are dynamical only on a part of the boundary.

We are concerned with the Sobolev spaces $H^s(\mathcal{O})$, $s > 0$ (see for instance [11] for the definition). The construction of the Sobolev spaces $H^s(\Gamma)$ for functions defined on the boundary $\Gamma = \partial\mathcal{O}$ is given in terms of the Laplace-Beltrami operator $B := \Delta_\Gamma$ on Γ; indeed we have

$$H^s(\Gamma) = \text{ domain of } (-\Delta_\Gamma)^s.$$

Denote $B^s(\Gamma) = H^{s-\frac{1}{2}}(\Gamma)$, for $s > \frac{1}{2}$, and similarly for $B^s(\Gamma_i)$. Then the trace mapping γ (and similarly for γ_i) is continuous from $H^s(\mathcal{O})$ into $B^s(\Gamma)$, for $s > \frac{1}{2}$.

In order to state the equation in an abstract setting, we introduce, on the Hilbert space $X = L^2(\mathcal{O})$, the operator

$$\begin{aligned}&A_m u := \Delta u(x), \qquad \text{with domain}\\ &D(A_m) = \{\varphi \in H^{1/2}(\mathcal{O}) \cap H^2_{loc}(\mathcal{O}) \mid A_m\varphi \in X\}.\end{aligned}$$

We also consider the normal boundary derivative

$$\mathcal{B}_2(x, \partial) = \sum_{k,j=1}^{d} a_{kj}(x)\nu_k\gamma_2\frac{\partial}{\partial x_j}, \qquad x \in \Gamma_2,$$

where $\nu = (\nu_1, \ldots, \nu_d)$ is the outward normal vector field to Γ. Then we consider the following linear equation

$$\begin{cases} \mathrm{d}u(t) = A_m u(t)\,\mathrm{d}t, & \\ x_1 = \gamma_1 u, & \mathrm{d}x_1(t) = Bx_1(t)\,\mathrm{d}t + \Gamma(x_1(t))\,\mathrm{d}V(t), \\ x_2 = \mathcal{B}_2 u, & \mathrm{d}x_2(t) = 0, \end{cases} \tag{6}$$

with the initial conditions

$$u(0) = u_0, \qquad x_1(0) = \gamma_1 u_0, \qquad x_2(0) = 0.$$

We shall transform our problem in an abstract Cauchy problem in a larger space. We define the Hilbert space $\mathcal{X} = L^2(\mathcal{O}) \times L^2(\Gamma_1) \times L^2(\Gamma_2)$, and we denote $\mathbf{x} \in \mathcal{X}$ the column vector with components (u, x_1, x_2). On the product space $\mathcal{X}$ we introduce the matrix operator $\mathcal{A}$, defined as

$$\mathcal{A} := \begin{pmatrix} A_m & 0 & 0 \\ 0 & B & 0 \\ 0 & 0 & 0 \end{pmatrix}$$

on

$$D(\mathcal{A}) = \Big\{ \mathbf{x} = (u, x_1, x_2) \mid u \in H^2(\mathcal{O}),\ x_1 = \gamma_1 u,\ x_1 \in D(B), \\ x_2 = \mathcal{B}_2 u,\ x_2 = 0 \Big\}.$$

Then $\mathcal{A}$ satisfies the assumptions of Corollary 3, hence it is the generator of an analytic semigroup; we solve problem (6) in the equivalent form

$$\begin{cases} \mathrm{d}\mathbf{x}(t) = \mathcal{A}\mathbf{x}(t)\,\mathrm{d}t + \mathcal{G}(\mathbf{x}(t))\,\mathrm{d}\mathcal{W}(t) \\ \mathbf{x}(0) = \mathbf{x}_0 \end{cases}$$

using Assumption 5 and Lemma 6.

3.3 Inhomogeneous boundary conditions

Let us consider, with the notation of previous example, the case when the boundary conditions on Γ_2 are given by $x_2(t) = f(t)$ for a function $f : \mathbf{R}_+ \to \Gamma_2$ that is continuously differentiable in time. The boundary space ∂X is given by the product $\partial X_1 \times \partial X_2$; denote Π_1, resp. Π_2, the immersions on ∂X_1 (resp. ∂X_2) into ∂X. We define the boundary operator $L : D(A_m) \to \partial X$ as $Lu = \binom{\gamma_1 u}{\mathcal{B}_2 u}$; the operator $\underline{B}$ is the operator matrix $\begin{pmatrix} B & 0 \\ 0 & 0 \end{pmatrix}$ on ∂X. Let R be a bounded operator from U into ∂X_1; in order to separate the difficulties we consider the following form of (1)

$$\begin{cases} du(t) = A_m u(t)\,\mathrm{d}t, & t > 0, \\ x(t) = Lu(t), & t \geq 0, \\ dx(t) = [\underline{B}x(t) + \Pi_2 f'(t)]\,\mathrm{d}t + \Pi_1 R\,\mathrm{d}V(t), & t > 0, \\ u(0) = u_0, \quad x(0) = x_0, \quad f(0) = f_0. \end{cases} \tag{7}$$

We define the abstract problem

$$\begin{cases} \mathrm{d}\mathbf{x}(t) = [\mathcal{A}\mathbf{x}(t) + \underline{f}'(t)]\,\mathrm{d}t + \mathcal{R}\,\mathrm{d}V(t), \\ \mathbf{x}(0) = \zeta, \end{cases} \tag{8}$$

where $\mathcal{R} \in L(U, \mathcal{X})$ is defined by $\mathcal{R} \cdot h := \binom{0}{\Pi_1 R \cdot h}$ for all h in the Hilbert space U, $\underline{f}(t) = \binom{0}{\Pi_2 f(t)}$ and $\zeta = (u_0, x_0, f_0)^*$.

The solution in *mild form* is given by the formula

$$\mathbf{x}(t) = \mathbf{x}(t,\zeta) = \mathcal{T}(t)\zeta + \int_0^t \mathcal{T}(t-s)\begin{pmatrix} 0 \\ \Pi_2 f'(s) \end{pmatrix} \mathrm{d}s \\ + \int_0^t \mathcal{T}(t-s)\mathcal{R}\,\mathrm{d}V(s). \tag{9}$$

We consider first the middle integral in (9); using the representation of the semigroup $\mathcal{T}(t)$ given by formula (4) we obtain

$$\int_0^t \mathcal{T}(t-s)\begin{pmatrix} 0 \\ \Pi_2 f'(s) \end{pmatrix} \mathrm{d}s = -\mathcal{T}(t)\underline{f}(0) + \underline{f}(t) \\ - \begin{pmatrix} A_0 \int_0^t T_0(t-\sigma)D_0\Pi_2 f(\sigma)\,\mathrm{d}\sigma \\ 0 \end{pmatrix}.$$

We then write (9) in the form

$$\mathbf{x}(t) = \mathcal{T}(t)\begin{pmatrix} u_0 \\ \Pi_1 x_0 \end{pmatrix} + \begin{pmatrix} 0 \\ \Pi_2 f(t) \end{pmatrix} - \begin{pmatrix} A_0 \int_0^t T_0(t-\sigma)D_0\Pi_2 f(\sigma)\,\mathrm{d}\sigma \\ 0 \end{pmatrix} \\ + \int_0^t \mathcal{T}(t-s)\mathcal{R}\,\mathrm{d}V(s);$$

notice that we do not need anymore the differentiability condition on f.

In the next statement, we are concerned with the properties of the stochastic convolution process

$$W_{\mathcal{A}}^{\mathcal{R}}(t) = \int_0^t \mathcal{T}(t-s)\mathcal{R}\,\mathrm{d}V(s). \tag{10}$$

COROLLARY 9 *Under the assumptions of Proposition 3, assume*

$$\int_0^t \|\mathcal{T}(s)\mathcal{R}\|_{HS}^2 \,\mathrm{d}s < +\infty \qquad \forall\, t \in [0,T]. \tag{11}$$

Then $W_{\mathcal{A}}^{\mathcal{R}}$ is a gaussian process, centered, with covariance operator defined by

$$\mathrm{Cov}\; W_{\mathcal{A}}^{\mathcal{R}}(t) = Q_t := \int_0^t [\mathcal{T}(s)\mathcal{R}\mathcal{R}^*\mathcal{T}^*(s)]\mathrm{d}s \tag{12}$$

and $Q_t \in \mathcal{L}_2(H)$ for every $t \in [0,T]$.

REMARK 10 *Condition (11) is verified whenever $R \in \mathcal{L}_2(U, \partial X)$ and, in particular, in case $R \in \mathcal{L}(U, \partial X)$ and U is finite dimensional.*

3.4 Stochastic boundary conditions

In several papers, the case of a white-noise perturbation on the boundary $f(t) = \dot{V}(t)$ is considered; oure interest is motivated by the results in [4–6].

Following [4], we shall define as mild solution of (9) in the interior of the domain the process $u(t)$ given by

$$u(t) = T_0(t)u_0 + Q_0(t)x_0 - A_0 \int_0^t T_0(t-\sigma)D_0(\Pi_2 1)\,\mathrm{d}V(\sigma) \tag{13}$$

We study this example with state space $\mathcal{O} = [0,1]$ and mixed boundary condition on $\partial\mathcal{O}$, that is, $X = L^2(\mathcal{O})$, $D(A_0) = \{u \in X : u(0) = 0,\ u'(1) = 0\}$, $A_0 u(\xi) = u''(\xi)$, $U = \mathbf{R}$, $Q = I$, $R = I$ and we choose

$$B = \begin{pmatrix} -1 & 0 \\ 0 & 0 \end{pmatrix} \qquad \Longrightarrow \qquad S(t) = \begin{pmatrix} e^{-t} & 0 \\ 0 & 1 \end{pmatrix},$$

which determine the boundary condition at 0, while there exists a continuous function $f(t)$ which determines the boundary condition $\frac{\mathrm{d}}{\mathrm{d}x}u(t,1) = f(t)$.

In this case we can explicitly work out the solution. At first, notice that $(D_0\Pi_2\alpha)(x) = \alpha x$; next, we construct the orthogonal basis $\{g_k : k \in \mathbf{N}\}$, setting $g_k(x) = \sin((\pi/2 + k\pi)x)$, to which it correspond the eigenvalues $\lambda_k = -(\pi/2 + k\pi)^2$. Since

$$T_0(t)(x) = \sum_{k=1}^{\infty} \langle x, g_k(x)\rangle e^{\lambda_k t} g_k(x)$$

and

$$\int_0^1 x \sin((\pi/2 + k\pi)x)\,\mathrm{d}x = \frac{(-1)^k}{|\lambda_k|},$$

we obtain

$$-A_0 \int_0^t T_0(t-\sigma)D_0(\Pi_2 1)\,\mathrm{d}V(\sigma) = -\sum_{k=1}^{\infty}(-1)^{k+1} g_k(x) \int_0^t e^{\lambda_k(t-\sigma)}\,\mathrm{d}V(\sigma).$$

In this setting estimate (11) is verified due to the choice of $\partial X = \mathbf{R}^2$, see Remark 10. Also, the new stochastic term is well defined for every $t \geq 0$, since

$$\mathbf{E}\left| A_0 \int_0^t T_0(t-\sigma)D_0(\Pi_2 1)\,\mathrm{d}V(\sigma)\right|^2 \leq C \sum_{k=1}^{\infty} \frac{1}{\lambda_k} < +\infty, \tag{14}$$

with a constant C independent from t.

References

[1] V. Casarino, K.-J. Engel, R. Nagel and G. Nickel, 2003. A semigroup approach to boundary feedback systems. *Integral Equations Operator Theory* **47**(3): 289–306.

[2] I. Chueshov and B. Schmalfuss, 2004. Parabolic stochastic partial differential equations with dynamical boundary conditions, *Differential Integral Equations*, **17**(7-8): 751–780.

[3] G. Da Prato and J. Zabczyk, 1992. *Stochastic equations in infinite dimensions*. Cambridge University Press.

[4] G. Da Prato and J. Zabczyk, 1993. Evolution equations with white-noise boundary conditions. *Stochastics Stochastics Rep.* **42**(3-4): 167–182.

[5] G. Da Prato and J. Zabczyk, 1996. *Ergodicity for infinite-dimensional systems*. Cambridge University Press.

[6] Debussche, A. and Fuhrman, M. and Tessitore, G., *Optimal control of a stochastic heat equation with boundary noise and boundary control*, preprint 2004.

[7] T. E. Duncan, B. Maslowski and B. Pasik-Duncan, Ergodic boundary/point control of stochastic semilinear systems, *SIAM J. Control Optim.* **36** (1998), no. 3, 1020–1047 (electronic).

[8] K.-J. Engel, 1999. Spectral theory and generator property for one-sided coupled operator matrices. *Semigroup Forum* **58** (2): 267–295.

[9] G. Greiner, 1987. Perturbing the boundary conditions of a generator. *Houston J. Math.* **13**: 213-229.

[10] M. Kumpf and G. Nickel, 2004. Dynamic boundary conditions and boundary control for the one-dimensional heat equation. *J. Dynam. Control Systems* **10**(2): 213–225.

[11] Lions, J.-L. and Magenes, E., 1972. *Non-homogeneous boundary value problems and applications. Vol. I*, Springer-Verlag, New York.

[12] R. Nagel, 1990. The spectrum of unbounded operator matrices with nondiagonal domain. *J. Funct. Anal.* **89**(2): 291–302.

A UNIQUENESS THEOREM FOR A CLASSICAL NONLINEAR SHALLOW SHELL MODEL

J. Cagnol,[1] C.G. Lebiedzik,[2] and R.J. Marchand[3]

[1] *Pôle Universitaire Léonard de Vinci, ESILV, DER CS 92916 Paris La Défense Cedex, France, John.Cagnol@devinci.fr* *, [2] *Wayne State University, 656 W. Kirby, Room 1150, Detroit, MI 48202, USA, kate@math.wayne.edu*†, [3] *Slippery Rock University, Slippery Rock, PA 16057, USA, richard.marchand@sru.edu*

Abstract The main goal of this paper is to establish the uniqueness of solutions of finite energy for a classical dynamic nonlinear thin shallow shell model with clamped boundary conditions. The static representation of the model is an extension of a *Koiter* shallow shell model. Until now, this has been an open problem in the literature. The primary difficulty is due to a lack of regularity in the nonlinear terms. Indeed the nonlinear terms are not *locally Lipshitz* with respect to the energy norm. The proof of the theorem relies on sharp PDE estimates that are used to prove uniqueness in a lower topology than the space of finite energy.

keywords: Nonlinear shells, weak solutions, uniqueness

1. Introduction

The model considered governs the vibrations of a thin shallow shell structure. The importance of the model stems from a variety of engineering applications such as helicopter rotor blades, propellors, acoustic chambers (for noise suppression) with curved walls such as aircraft cabins, aircraft control surfaces, curved surfaces on modern turbo-jet engines, etc. The primary goals in applications are often to achieve uniform stability or control of the vibrations as described, for example, in [9]. These control theoretic applications coupled with necessary numerical methods for implementation require a thorough knowledge of existence and uniqueness of finite energy (weak) solutions. In the next subsection we will describe classical shell theory notation and present the variational form of the model followed by a brief review of related literature. An overview of the primary results and relevant proofs will comprise the remaining two sections.

*Research supported by NSF-INRIA grant number NSF-Int-0226961.
†Supported by NSF grant number DMS-0408565.

Please use the following format when citing this chapter:

Cagnol, J., Lebiedzik, C.G., and Marchand, R.J., 2006, in IFIP International Federation for Information Processing, Volume 202, Systems, Control, Modeling and Optimization, eds. Ceragioli, F., Dontchev, A., Furuta, H., Marti, K., Pandolfi, L., (Boston: Springer), pp. 67-78.

1.1 The Model

The model to be considered is an extension of *Koiter's* nonlinear model for a thin shallow shell considered in various forms in [7, 4–1, 6, 5]. The shallowness assumption as given in [7] is represented by the expression $L/R \ll 1$ where R is the minimum principal radius of curvature, and L is the wave length of the deformation pattern on the middle surface. We will utilize much of the notation used in [4] and principally developed in [7]. Throughout the paper, we will assume that Greek letters belong to the set {1,2} and Latin letters belong to the set {1,2,3}. We define the middle surface of the shell S as the image of a connected bounded open set $\Omega \subset \mathcal{E}^2$ with boundary Γ under the mapping $\mathbf{\Phi} : (\xi^1, \xi^2) \in \overline{\Omega} \to \mathcal{E}^3$ where $\Phi \in [C^3(\overline{\Omega})]^3$ and $\mathcal{E}^n$ is the n-dimensional Euclidean Space. It is assumed that the two tangent vectors given by $\mathbf{a}_\alpha = \partial\mathbf{\Phi}/\partial\xi^\alpha$ are linearly independent at any point on the surface of the shell, and combining these two vectors with the normal vector, $\mathbf{a}_3 = \frac{\mathbf{a}_1 \times \mathbf{a}_2}{|\mathbf{a}_1 \times \mathbf{a}_2|}$ defines a covariant basis for a local reference frame on the surface of the shell. We will denote partial derivatives with the notation $\mathbf{\Phi}_{,\alpha} = \partial\mathbf{\Phi}/\partial\xi^\alpha$ for any point $(\xi^1, \xi^2) \in \overline{\Omega}$.

In order to simplify the notation, the summation convention will be used in which letters repeated in the roof and cellar of a product will be summed over. For example, the contravariant basis for the tangent plane at any point on the surface of the shell is given by the two vectors $\mathbf{a}^\beta$ defined by the relation $\mathbf{a}_\alpha \cdot \mathbf{a}^\beta = \delta_\alpha^\beta$ where δ_α^β denotes the usual Kronecker delta symbol. The matrix $a_{\alpha\beta} = \mathbf{a}^\alpha \cdot \mathbf{a}^\beta$ is the so-called *first fundamental form* of the surface with its inverse given by the matrix $(a^{\alpha\beta})$. The *second fundamental form*, denoted by $(b_{\alpha\beta})$ measures the normal curvatures of the middle surface of the shell. It is defined by $b_{\alpha\beta} = b_{\beta\alpha} = -\mathbf{a}_\alpha \cdot \mathbf{a}_{3,\beta} = \mathbf{a}_3 \cdot \mathbf{a}_{\alpha,\beta}$.

We represent as $\mathbf{u} = (u_1, u_2, u_3)$ the vector function of all three displacements, while in-surface displacements are given by $\overrightarrow{u} = (u_1, u_2)$ and the transverse displacement is denoted by u_3. The *Christoffel symbols* given by $\Gamma^\alpha_{\beta\lambda} = \mathbf{a}^\alpha \cdot \mathbf{a}_{\beta,\lambda}$ are used to define the covariant derivatives for the displacement vector of the middle surface $\mathbf{u}(\xi^1, \xi^2) = u_i\mathbf{a}^i$ in a fixed reference frame. In particular, $u_{\alpha|\beta} = u_{\alpha,\beta} - \Gamma^\lambda_{\alpha\beta}u_\lambda$ and $u_{3|\alpha\beta} = u_{3,\alpha\beta} - \Gamma^\lambda_{\alpha\beta}u_{3,\lambda}$. In addition, the linear strain tensor is denoted by $\epsilon_{\alpha\beta}(\overrightarrow{u}) = \frac{1}{2}(u_{\alpha|\beta} + u_{\beta|\alpha})$. The model assumes that the tangential (in surface) displacements are small relative to the transverse displacement. It is also assumed that all deflections are small and finite in the sense that for any parameter $-1 \leq t \leq 1$ the displacements $t\mathbf{u}$ shall be less than or equal to the middle surface strains in order of magnitude as described in [7]. In this case, suitable expressions for the *middle surface strain tensor* $\gamma_{\alpha\beta}$ and the *change of curvature tensor* $\rho_{\alpha\beta}$ are given respectively by $\gamma_{\alpha\beta}(\mathbf{u}) = \epsilon_{\alpha\beta}(\overrightarrow{u}) - b_{\alpha\beta}u_3 + \frac{1}{2}u_{3,\alpha}u_{3,\beta}$ and $\rho_{\alpha\beta}(\mathbf{u}) = u_{3|\alpha\beta}$. We note that the nonlinearity of the model arises from the third term in $\gamma_{\alpha\beta}$.

These strain measures, along with an appropriate constituitve law, are used to derive corresponding stress measures. The exact form of the stress measures depends on the shell material. Thus, we assume the simplest possible scenario, as done originally by Koiter in [7] and duplicated in [4] and [1]: that the shell consists of elastic, homogeneous, and isotropic material and the strains are small everywhere. Additionally, we assume that all nonzero stress components are imposed on surfaces which are parallel to the middle surface of the shell. Then the tensor of elastic moduli is given by

$$E^{\alpha\beta\lambda\mu} = \frac{E}{2(1+\nu)}\left[a^{\alpha\lambda}a^{\beta\mu} + a^{\alpha\mu}a^{\beta\lambda} + \frac{2\nu}{1-\nu}a^{\alpha\beta}a^{\lambda\mu}\right].$$

Here, E is Young's modulus and ν is Poisson's ratio for the material. See [2] for a proof of the positivity of the tensor of elastic moduli. Throughout the exposition, we let $\dot{u}$ represent differentiation with respect to time.

Since our primary interest is weak solutions, we state the model in the following variational form: we look for solutions $u_\alpha(t) \in \left[H_0^1(\Omega)\right]^2, u_3(t) \in H_0^2(\Omega)$, $\dot{u}_\alpha(t) \in \left[L^2(\Omega)\right]^2$, $\dot{u}_3(t) \in H_0^1(\Omega)$ satisfying clamped boundary conditions, $\mathbf{u}|_\Gamma = \left.\frac{\partial u_3}{\partial \nu}\right|_\Gamma = 0$, such that for test functions $\mathbf{v} = (v_1, v_2, v_3) = (\overrightarrow{v}, v_3) \in \left[H_0^1(\Omega)\right]^2 \cap H_0^2(\Omega)$ we have

$$m(\ddot{\mathbf{u}}, \mathbf{v}) + a(\mathbf{u}, \mathbf{v}) - n(\mathbf{u}, \mathbf{v}) = 0 \tag{1}$$

where

$$\begin{aligned}
m(\ddot{\mathbf{u}}, v) &= \rho e\left\{\left(a^{\alpha\beta}\ddot{u}_\alpha, v_\beta\right) + \gamma\left(a^{\alpha\beta}\left[b_1^1 b_2^2 - b_1^2 b_2^1\right]\ddot{u}_\alpha, v_\beta\right)\right\} \\
&+ \rho e\gamma\left\{\left(a^{\alpha\beta}\left[\ddot{u}_{3|\alpha} + b_\alpha^\lambda \ddot{u}_\lambda\right], \left[v_{3|\beta} + b_\beta^\mu v_\mu\right]\right)\right\} \\
&+ \rho e\gamma\left\{\left(a^{\alpha\beta}\ddot{u}_\alpha, v_{3|\beta}b_\eta^\eta\right) + \left(a^{\alpha\beta}\left[\ddot{u}_{3|\alpha} + 2b_\alpha^\lambda \ddot{u}_\lambda\right], v_\beta b_\eta^\eta\right)\right\} \\
&+ \rho e\left\{(\ddot{u}_3, v_3) + \gamma\left(\left[b_1^1 b_2^2 - b_1^2 b_2^1\right]\ddot{u}_3, v_3\right)\right\}
\end{aligned}$$

$$\begin{aligned}
a(\mathbf{u}, \mathbf{v}) &= e\left(E^{\alpha\beta\lambda\mu}\left[\epsilon_{\alpha\beta}(\overrightarrow{u}) - b_{\alpha\beta}u_3\right], \epsilon_{\lambda\mu}(\overrightarrow{v})\right) \\
&+ e\gamma\left(E^{\alpha\beta\lambda\mu}u_{3|\alpha\beta}, v_{3|\lambda\mu}\right) - e\left(E^{\alpha\beta\lambda\mu}\left[\epsilon_{\alpha\beta}(\overrightarrow{u}) - b_{\alpha\beta}u_3\right], b_{\lambda\mu}v_3\right)
\end{aligned}$$

$$\begin{aligned}
n(\mathbf{u}, \mathbf{v}) &= \frac{e}{2}\left(E^{\alpha\beta\lambda\mu}u_{3,\alpha}u_{3,\beta}, \epsilon_{\lambda\mu}(\overrightarrow{v})\right) - \frac{e}{2}\left(E^{\alpha\beta\lambda\mu}\left[u_{3,\alpha}u_{3,\beta}\right], b_{\lambda\mu}v_3\right) \\
&+ e\left(E^{\alpha\beta\lambda\mu}\left[\epsilon_{\alpha\beta}(\overrightarrow{u}) - b_{\alpha\beta}u_3 + \frac{1}{2}u_{3,\alpha}u_{3,\beta}\right], u_{3,\lambda}v_{3,\mu}\right)
\end{aligned}$$

where $\gamma = e^2/12$ and initial conditions $u_\alpha(0) = u_\alpha^0 \in \left[H_0^1(\Omega)\right]^2$, $\dot{u}_\alpha(0) = u_\alpha^1 \in \left[L^2(\Omega)\right]^2$, $u_3(0) = u_3^0 \in H_0^2(\Omega)$, $\dot{u}_3(0) = u_3^1 \in H_0^1(\Omega)$ Here we make use of the notation $(u, v) = \int_\Omega uv\sqrt{a}\, d\xi_1 d\xi_2$ with $a = det(a_{\alpha\beta})$ and $a \neq 0$. A linear version of this model is considered in [3] to model turbine blades. An associated static (nonlinear) model is given in [4] as an extension of Koiter's model. It accounts for tangential surface loads in order to obtain, as a special case, the nonlinear model for thin elastic plates when there is no curvature, $b_{\alpha\beta} = 0$. The energy of the system (1) is $E(t) = E_p(t) + E_k(t)$ where $E_k(t) = m(\dot{\mathbf{u}}, \dot{\mathbf{u}})$ and $E_p(t) = a(\mathbf{u}, \mathbf{u})$. It is well-known that $E_p(t)$ is topologically equivalent to $\left[H_0^1(\Omega)\right]^2 \times H^2(\Omega)$.

2. Main Results

LEMMA 1 **(Regular Solutions)** *For arbitrary $T > 0$ and initial data $(\vec{u}^0, \vec{u}^1)$ in $\left[H_0^2(\Omega)\right]^2 \times \left[H_0^1(\Omega)\right]^2$, $(u_3^0, u_3^1) \in H_0^3(\Omega) \times H_0^2(\Omega)$, there exists a unique, global solution to the variational form (1)*

$$\mathbf{u} \in C([0,T], [H_0^2(\Omega)]^2 \times H_0^3(\Omega)) \cap C^1([0,T], [H_0^1(\Omega)]^2 \times H_0^2(\Omega)).$$

Proof: The proof of this lemma follows by means of a rather standard nonlinear Galerkin argument. As such, we omit the details.

The existence of finite energy (weak) solutions follows similarly from a standard nonlinear Galerkin argument. However, to our knowledge, uniqueness has until now been an open problem.

THEOREM 2 **(Weak Solutions)** *For arbitrary $T > 0$ and initial data*

$$(\vec{u}^0, \vec{u}^1) \in \left[H_0^1(\Omega)\right]^2 \times \left[L^2(\Omega)\right]^2, \qquad (u_3^0, u_3^1) \in H_0^2(\Omega) \times H_0^1(\Omega),$$

there exists a **unique** *solution to the variational form (1) such that*

$$(\vec{u}, u_3) \in C_w\left([0,T]; \left[H_0^1(\Omega)\right]^2 \times H_0^2(\Omega)\right)$$
$$\left(\dot{\vec{u}}, \dot{u}_3\right) \in C_w\left([0,T]; \left[L^2(\Omega)\right]^2 \times H_0^1(\Omega)\right)$$

The primary mathematical difficulty is due to a lack of regularity in the nonlinear terms. Indeed, the nonlinear terms are not bounded in the space of finite energy. The proof is an adaptation of the method used by Sedenko in [10] and by Lasiecka in [8]. One of the key estimates used in the proof below was proved in [8] and is expressed as the following Lemma.

LEMMA 3 *Given $\epsilon, r > 0$ and $\mathbf{w}^1 \in \left[H^{\epsilon+r}\right]^3$ and $\mathbf{w}^2 \in \left[H^1\right]^3$ then*

$$\|(w_i^1, w_i^2)\|_{L^2}^2 \leq C\left[\ln(1+\lambda_n)\|w_i^1\|_{L^2}^2 + \frac{1}{\lambda_n^r}\|w_i^1\|_{H^{\epsilon+r}}^2\right]\|w_i^2\|_{H^1}^2 \quad (2)$$

$$\|w_i^1 w_j^2\|_{L^2}^2 \leq C\left[\ln(1+\lambda_n)\|w_i^1\|_{L^2}^2 + \frac{1}{\lambda_n^r}\|w_i^1\|_{H^{\epsilon+r}}^2\right]\|w_j^2\|_{H^1}^2 \quad (3)$$

where λ_n denotes the eigenvalues of the Laplace operator. Note the sequence $\{\lambda_n\}$ is increasing with $\lim \lambda_n = \infty$.

3. Proof of Theorem 2

Let $\tilde{\mathbf{u}} = \mathbf{u}^2 - \mathbf{u}^1$ where $\mathbf{u}^1$ and $\mathbf{u}^2$ are any two weak solutions of (1) satisfying the same initial conditions. The first step is to derive an abstract formulation for the model. To this end, we make the following definitions.

DEFINITION 4 *Define linear operators $\mathcal{M}$ and $\mathcal{A}$ and space $\mathcal{H}$ as*

$$(\mathcal{M}\mathbf{u}, \tilde{\mathbf{u}}) = m(\mathbf{u}, \tilde{\mathbf{u}})\ ; \quad (\mathcal{A}\mathbf{u}, \tilde{\mathbf{u}}) = a(\mathbf{u}, \tilde{\mathbf{u}}). \tag{4}$$

$$\mathcal{H} = \mathcal{D}(\mathcal{A}^{1/2}) \times \mathcal{D}(\mathcal{M}^{1/2}) = [H_0^1(\Omega)]^2 \times H_0^2(\Omega) \times [(L^2(\Omega)]^2 \times H_0^1(\Omega) \tag{5}$$

and let $\mathbf{A} : \mathcal{H} \to \mathcal{H}$ be defined by

$$\mathbf{A} = \begin{pmatrix} 0 & -I \\ \mathcal{M}^{-1}\mathcal{A} & 0 \end{pmatrix} \tag{6}$$

The coercivity of $m(\mathbf{u}, \tilde{\mathbf{u}})$ and $a(\mathbf{u}, \tilde{\mathbf{u}})$ were proved in [3] and [4] respectively, so the operator $\mathbf{A}$ is well-defined. Also, if we define the nonlinear operator N such that $(N(\mathbf{u}), \mathbf{v}) = n(\mathbf{u}, \mathbf{v})$ with $N(\mathbf{u}) = N_\alpha(u_3) + N_3(\vec{u}, u_3)$ where

$$\begin{aligned} N_\alpha(u_3) &= \frac{e}{2}\left(E^{\alpha\beta\lambda\mu}u_{3,\lambda}, u_{3,\mu}\right)\Big|_\beta \\ N_3(\vec{u}, u_3) &= -\frac{e}{2}\left(b_{\alpha\beta}E^{\alpha\beta\lambda\mu}u_{3,\lambda}u_{3,\mu}\right) \\ &\quad -e\left(u_{3|\alpha}E^{\alpha\beta\lambda\mu}\left[\epsilon_{\lambda\mu}(\vec{u}) - b_{\lambda\mu}u_3 + \frac{1}{2}u_{3,\lambda}u_{3,\mu}\right]\right)\Big|_\beta \end{aligned}$$

then the variational form in (1) implies

$$\partial_t \begin{pmatrix} \mathbf{u} \\ \partial_t \mathbf{u} \end{pmatrix} + \mathbf{A}\begin{pmatrix} \mathbf{u} \\ \partial_t \mathbf{u} \end{pmatrix} = \begin{pmatrix} 0 \\ \mathcal{M}^{-1}N(\mathbf{u}) \end{pmatrix}. \tag{7}$$

This will be used to prove the following lemma.

LEMMA 5 *Using Definition 4 and (7), the following holds.*

$$\left\|\mathbf{A}^{-1}\begin{pmatrix} \tilde{\mathbf{u}} \\ \partial_t \tilde{\mathbf{u}} \end{pmatrix}\right\|_{\mathcal{H}} \leq \int_0^t \left\|\mathbf{A}^{-1}\begin{pmatrix} 0 \\ \mathcal{M}^{-1}[N(\mathbf{u}^2) - N(\mathbf{u}^1)] \end{pmatrix}\right\|_{\mathcal{H}} ds \tag{8}$$

Proof: Acting with $\mathbf{A}^{-1}$ on equation (7) and multiplying by $X(s)$ yields

$$\left(\partial_t \mathbf{A}^{-1}\begin{pmatrix} \tilde{\mathbf{u}} \\ \partial_t \tilde{\mathbf{u}} \end{pmatrix}, X(s)\right)_{\mathcal{H}} + \left(\mathbf{A}\mathbf{A}^{-1}\begin{pmatrix} \tilde{\mathbf{u}} \\ \partial_t \tilde{\mathbf{u}} \end{pmatrix}, X(s)\right)_{\mathcal{H}} = \left(\mathbf{A}^{-1}\begin{pmatrix} 0 \\ \mathcal{M}_\gamma^{-1}[N(\mathbf{u^2}) - N(\mathbf{u^1})] \end{pmatrix}, X(s)\right)_{\mathcal{H}} \tag{9}$$

We define

$$X(s) = \mathbf{A}^{-1}\begin{pmatrix} \tilde{\mathbf{u}}(s) \\ \partial_t \tilde{\mathbf{u}}(s) \end{pmatrix} \quad \text{and} \quad F(s) = \mathbf{A}^{-1}\begin{pmatrix} 0 \\ \mathcal{M}^{-1}[N(\mathbf{u^2}) - N(\mathbf{u^1})] \end{pmatrix}$$

Using $(\mathbf{A}X, X)_{\mathcal{H}} = 0$ on (9) yields $(\partial_t X(s), X(s))_{\mathcal{H}} = (X(s), F(s))_{\mathcal{H}}$ and

$$\frac{1}{2}\int_0^t \partial_s \|X(s)\|^2_{\mathcal{H}} ds = \int_0^t (X(s), F(s))_{\mathcal{H}}\, ds$$

it follows

$$\frac{1}{2}\|X(t)\|^2_{\mathcal{H}} \leq \frac{1}{2}\|X(0)\|^2_{\mathcal{H}} + \int_0^t \|X(s)\|_{\mathcal{H}} \|F(s)\|_{\mathcal{H}} ds$$

Since $\tilde{\mathbf{u}}(0) = \partial_t \tilde{\mathbf{u}}(0) = 0$ and $\mathbf{A}^{-1}$ is linear, we have $X(0) = 0$, so

$$\frac{1}{2}\|X(t)\|^2_{\mathcal{H}} \leq \left(\sup_{s\in[0,t]} \|X(s)\|_{\mathcal{H}}\right)\left(\int_0^t \|F(s)\|_{\mathcal{H}} ds\right) \tag{10}$$

thus, for all $\bar{t} \leq t$ we have

$$\begin{cases} \frac{1}{2}\|X(\bar{t})\|^2_{\mathcal{H}} \leq \left(\sup_{s\in[0,\bar{t}]} \|X(s)\|_{\mathcal{H}}\right)\left(\int_0^{\bar{t}} \|F(s)\|_{\mathcal{H}} ds\right) \\ \sup_{s\in[0,\bar{t}]} \|X(s)\|^2_{\mathcal{H}} \leq \sup_{s\in[0,t]} \|X(s)\|^2_{\mathcal{H}} \\ \int_0^{\bar{t}} \|F(s)\|_{\mathcal{H}} ds \leq \int_0^t \|F(s)\|_{\mathcal{H}} ds \end{cases} \tag{11}$$

Combining (10) and (11) gives

$$\frac{1}{2}\sup_{\bar{t}\in[0,t]} \|X(\bar{t})\|^2_{\mathcal{H}} \leq \left(\sup_{s\in[0,t]} \|X(s)\|^2_{\mathcal{H}}\right)\left(\int_0^t \|F(s)\|_{\mathcal{H}} ds\right)$$

consequently

$$\sup_{s\in[0,t]} \|X(s)\|_{\mathcal{H}} \leq \int_0^t \|F(s)\|_{\mathcal{H}} ds$$

and then using $\|X(t)\|_{\mathcal{H}} \leq \sup_{s\in[0,t]} \|X(s)\|_{\mathcal{H}}$, we get (8).

LEMMA 6 *The following inequality holds.*

$$\|\mathcal{M}^{1/2}\tilde{\mathbf{u}}\|_{L^2} \leq \left\| \mathbf{A}^{-1} \begin{pmatrix} \tilde{\mathbf{u}} \\ \partial_t \tilde{\mathbf{u}} \end{pmatrix} \right\|_{\mathcal{H}} \tag{12}$$

Proof: We have

$$\mathbf{A}^{-1} \begin{pmatrix} \tilde{\mathbf{u}} \\ \partial_t \tilde{\mathbf{u}} \end{pmatrix} = \begin{pmatrix} \mathcal{A}^{-1}\mathcal{M}\partial_t \tilde{\mathbf{u}} \\ -\tilde{\mathbf{u}} \end{pmatrix}$$

therefore

$$\left\| \mathbf{A}^{-1} \begin{pmatrix} \tilde{\mathbf{u}} \\ \partial_t \tilde{\mathbf{u}} \end{pmatrix} \right\|_{\mathcal{H}}^2 = \|\mathcal{A}^{-1/2}\mathcal{M}\partial_t \tilde{\mathbf{u}}\|_{L^2}^2 + \|\mathcal{M}^{1/2}\tilde{\mathbf{u}}\|_{L^2}^2$$

LEMMA 7 *The following inequality holds.*

$$\|\mathcal{M}^{1/2}\tilde{\mathbf{u}}\|_{L^2} \leq \int_0^t \|\mathcal{A}^{-1/2}[N(\mathbf{u^2}(s)) - N(\mathbf{u^1})(s)]\|_{L^2} ds \tag{13}$$

Proof: We have

$$\mathbf{A}^{-1} \begin{pmatrix} 0 \\ \mathcal{M}^{-1}[N(\mathbf{u^2}) - N(\mathbf{u^1})] \end{pmatrix} = \begin{pmatrix} \mathcal{A}^{-1}[N(\mathbf{u^2}) - N(\mathbf{u^1})] \\ 0 \end{pmatrix}$$

therefore

$$\left\| \mathbf{A}^{-1} \begin{pmatrix} 0 \\ \mathcal{M}^{-1}[N(\mathbf{u^2}) - N(\mathbf{u^1})] \end{pmatrix} \right\|_{\mathcal{H}} = \|\mathcal{A}^{-1/2}[N(\mathbf{u^2}) - N(\mathbf{u^1})]\|_{L^2}$$

From Lemma 5, we get

$$\left\| \mathbf{A}^{-1} \begin{pmatrix} \tilde{\mathbf{u}} \\ \partial_t \tilde{\mathbf{u}} \end{pmatrix} \right\|_{\mathcal{H}} \leq \int_0^t \|\mathcal{A}^{-1/2}[N(\mathbf{u^2}) - N(\mathbf{u^1})]\|_{L^2} ds$$

Applying Lemma 6 yields (13).

LEMMA 8 *Following Lemma 7, we obtain*

$$\begin{aligned} \|\tilde{u}_\alpha\|_{L^2}^2 + \|\tilde{u_3}\|_{H^1}^2 \quad \leq \quad & t \int_0^t (\|N_\alpha(u_3^2) - N_\alpha(u^1)\|_{H^{-1}}^2 \\ & + \|N_3(\overrightarrow{u}^2, u_3^2) - N_3(\overrightarrow{u}^1, u_3^1)\|_{H^{-2}}^2) ds \end{aligned} \tag{14}$$

Proof: The lemma follows from the use of $(\int_0^t f(t)dt)^2 \leq t \int_0^t f(t)^2 dt$ on (13).

LEMMA 9 *Given $\epsilon > 0$ and $r > 0$ with $\epsilon + r \leq 1$, there exists a constant C depending on λ, n, u_3^1, and u_3^2 such that*

$$||N_\alpha(u_3^2) - N_\alpha(u_3^1)||_{H^{-1}} \leq C \ln(1+\lambda_n)||\tilde{u}_3||_{H^1} + \frac{C}{\lambda_n^r}$$

Proof: The proof uses the positivity of the tensor of elastic moduli given in [2] and the fact that the divergence operator, $(\cdot)|_\beta$, is a bounded operator on $H^{-1}(\Omega)$. Recalling the definition of $N_\alpha(u_3)$, it can be shown that

$$\begin{aligned} N_\alpha(u_3^2) - N_\alpha(u_3^1) &= \frac{e}{2}\left[E^{\alpha\beta\lambda\mu}\left(u_{3,\lambda}^2 u_{3,\mu}^2 - u_{3,\lambda}^1 u_{3,\mu}^1\right)\right]\Big|_\beta \\ &= \frac{e}{2}\left[E^{\alpha\beta\lambda\mu}\tilde{u}_{3,\lambda}\left(u_{3,\mu}^2 + u_{3,\mu}^1\right)\right]\Big|_\beta \end{aligned}$$

This gives

$$\begin{aligned} ||N_\alpha(u_3^2) - N_\alpha(u_3^1)||_{H^{-1}} &= \left\|\frac{e}{2}\left[E^{\alpha\beta\lambda\mu}\tilde{u}_{3,\lambda}\left(u_{3,\mu}^2 + u_{3,\mu}^1\right)\right]\Big|_\beta\right\|_{H^{-1}} \\ &\leq C\left\|\tilde{u}_{3,\lambda}\left(u_{3,\mu}^2 + u_{3,\mu}^1\right)\right\|_{L^2} \end{aligned} \quad (15)$$

Now apply Lemma 3 to obtain

$$\left\|\tilde{u}_{3,\lambda}u_{3,\lambda}^2\right\| \leq C\left(\ln(1+\lambda_n)\,||\tilde{u}_3||_{H^1} + \frac{1}{\lambda_n^r}\left\|u_3^2 - u_3^1\right\|_{H^{1+\epsilon+r}}\right)\left\|u_3^2\right\|_{H^2}$$

Combining this result with an analogous result for $||\tilde{u}_{3,\lambda}u_{3,\lambda}^2||$ and (15) and using the fact that u_3^1 amd u_3^2 are H^2 functions leads to the conclusion of the lemma. We now prove a similar lemma for $N_3(\overrightarrow{u}, u_3)$.

LEMMA 10 *Given $\epsilon > 0$ and $r > 0$ with $\epsilon + r \leq 1$, there exists a constant C depending on $\lambda, n, \mathbf{u}^1$, and $\mathbf{u}_3^2$ such that*

$$\begin{aligned} &\left\|N_3(\overrightarrow{u}^2, u_3^2) - N_3(\overrightarrow{u}^1, u_3^1)\right\|_{H^{-2}} \\ &\quad \leq C\ln(1+\lambda_n)\,||\tilde{u}_3||_{H^1}^2 + C\ln(1+\lambda_n)\,||\tilde{u}_\alpha||_{L^2}^2 + \frac{C}{\lambda_n^r} \end{aligned}$$

Proof. Using the definition of $N_3(\overrightarrow{u}, u_3)$ it can be shown that

$$N_3(u_\alpha^2, u_3^2) - N_3(u_\alpha^1, u_3^1) = X^1\left(\tilde{u}_3, u_3^1, u_3^2\right) + X^2\left(\tilde{\mathbf{u}}, \mathbf{u}^1, \mathbf{u}^2\right)$$

with

$$X^1\left(\tilde{u}_3, u_3^1, u_3^2\right) = -\frac{e}{2} b_{\alpha\beta} E^{\alpha\beta\lambda\mu} \tilde{u}_{3,\lambda}\left(u_{3,\mu}^2 + u_{3,\mu}^1\right)$$
$$+e\left(u_{3|\alpha}^2 E^{\alpha\beta\lambda\mu} b_{\lambda\mu}\tilde{u}_3 + \tilde{u}_{3|\alpha} E^{\alpha\beta\lambda\mu} b_{\lambda\mu} u_3^1\right)\Big|_\beta$$
$$-\frac{e}{2}\left(u_{3|\alpha}^2 E^{\alpha\beta\lambda\mu} u_{3,\lambda}^2 \tilde{u}_{3,\mu} + u_{3|\alpha}^2 E^{\alpha\beta\lambda\mu} \tilde{u}_{3,\lambda} u_{3,\mu}^1 + \tilde{u}_{3|\alpha} E^{\alpha\beta\lambda\mu} u_{3,\lambda}^1 u_{3,\mu}^1\right)\Big|_\beta$$
$$X^2\left(\tilde{\mathbf{u}}, \mathbf{u}^1, \mathbf{u}^2\right) = e\left(u_{3|\alpha}^2 E^{\alpha\beta\lambda\mu}\epsilon_{\lambda\mu}(\vec{\tilde{u}}) + \tilde{u}_{3|\alpha} E^{\alpha\beta\lambda\mu}\epsilon_{\lambda\mu}(\vec{u}^1)\right)\Big|_\beta$$

Consequently,

$$\left\|N_3(\vec{u}^2, u_3^2) - N_3(\vec{u}^1, u_3^1)\right\|_{H^{-2}}$$
$$\leq \left\|X^1\left(\tilde{u}_3, u_3^1, u_3^2\right)\right\|_{H^{-2}} + \left\|X^2\left(\tilde{\mathbf{u}}, \mathbf{u}^1, \mathbf{u}^2\right)\right\|_{H^{-2}} \quad (16)$$

It suffices to estimate each term on the right-hand side of (16). In particular,

$$\left\|X^1\left(\tilde{u}_3, u_3^1, u_3^2\right)\right\|_{H^{-2}} \leq C\left\{\left\|b_{\alpha\beta} E^{\alpha\beta\lambda\mu}\tilde{u}_{3,\lambda}\left(u_{3,\mu}^2 + u_{3,\mu}^1\right)\right\|_{H^{-2}}\right.$$
$$+\left\|u_{3|\alpha}^2 E^{\alpha\beta\lambda\mu} b_{\lambda\mu}\tilde{u}_3 + \tilde{u}_{3|\alpha} E^{\alpha\beta\lambda\mu} b_{\lambda\mu} u_3^1\right\|_{H^{-1}} \quad (17)$$
$$\left.-\left\|u_{3|\alpha}^2 E^{\alpha\beta\lambda\mu} u_{3,\lambda}^2\tilde{u}_{3,\mu} + u_{3|\alpha}^2 E^{\alpha\beta\lambda\mu}\tilde{u}_{3,\lambda} u_{3,\mu}^1 + \tilde{u}_{3|\alpha} E^{\alpha\beta\lambda\mu} u_{3,\lambda}^1 u_{3,\mu}^1\right\|_{H^{-1}}\right\},$$

Using classical embedding theory and the fact that the second fundamental form and the tensor of elastic moduli are bounded yields the following.

$$\left\|b_{\alpha\beta} E^{\alpha\beta\lambda\mu}\tilde{u}_{3,\lambda}\left(u_{3,\mu}^2 + u_{3,\mu}^1\right)\right\|_{H^{-2}} \leq C\left\|\tilde{u}_{3,\lambda}\left(u_{3,\mu}^2 + u_{3,\mu}^1\right)\right\|_{L^2}$$
$$\leq C\left\|\tilde{u}_3\right\|_{H^1}\left\|u_3^2 + u_3^1\right\|_{H^1} \leq C\left\|\tilde{u}_3\right\|_{H^1}$$

Applying similar arguments to the remaining terms in (17) leads to

$$\left\|X^1\left(\tilde{u}_3, u_3^1, u_3^2\right)\right\|_{H^{-2}} \leq C\left\|\tilde{u}_3\right\|_{H^1} \quad (18)$$

The higher order terms in X^2 make it much more difficult to estimate. First,

$$\left\|X^2\left(\tilde{\mathbf{u}}, \mathbf{u}^1, \mathbf{u}^2\right)\right\|_{H^{-2}}$$
$$\leq C\left\|\tilde{u}_{3|\alpha} E^{\alpha\beta\lambda\mu}\epsilon_{\lambda\mu}(\vec{u}^1)\right\|_{H^{-1}} + C\left\|u_{3|\alpha}^2 E^{\alpha\beta\lambda\mu}\epsilon_{\lambda\mu}(\vec{\tilde{u}})\right\|_{H^{-1}} \quad (19)$$

Applying Lemma 3 to the first term in (19) gives

$$\left\|\tilde{u}_{3|\alpha} E^{\alpha\beta\lambda\mu}\epsilon_{\lambda\mu}(\vec{u}^1)\right\|_{H^{-1}} = \sup_{\{\|\varphi\|_{H_0^1}=1\}} \int_\Omega \tilde{u}_{3|\alpha} E^{\alpha\beta\lambda\mu}\epsilon_{\lambda\mu}(\vec{u}^1)\varphi_\beta\sqrt{a}\,d\xi^1 d\xi^2$$

$$\leq \sup_{\{\varphi,\, \|\varphi\|_{H_0^1}=1\}} C \left\|\epsilon_{\lambda\mu}(\vec{u}^1)\right\|_{L^2} \|\tilde{u}_{3|\alpha}\varphi_\beta\|_{L^2} \tag{20}$$

$$\leq C\left(\|u^1\|_{H^1}(\ln(1+\lambda_n)\|\tilde{u}_3\|_{H^1} + \frac{1}{\lambda_n^r}\|\tilde{u}_3\|_{H^{1+\epsilon+r}}\right)$$

after applying the supremum. Similarly,

$$\left\|u^2_{3|\alpha}E^{\alpha\beta\lambda\mu}\epsilon_{\lambda\mu}(\vec{\tilde{u}})\right\|_{H^{-1}} = \sup_{\{\|\varphi\|_{H_0^1}=1\}} \int_\Omega E^{\alpha\beta\lambda\mu}u^2_{3|\alpha}\varphi_\beta\epsilon_{\lambda\mu}(\vec{\tilde{u}})\sqrt{a}d\xi^1 d\xi^2$$

$$= \sup_{\{\varphi,\, \|\varphi\|_{H_0^1}=1\}} \int_\Omega E^{\alpha\beta\lambda\mu}u^2_{3|\alpha}\varphi_\beta\left(\frac{1}{2}\left(\tilde{u}_{\lambda|\mu}+\tilde{u}_{\mu|\lambda}\right)\right)\sqrt{a}d\xi^1 d\xi^2 \tag{21}$$

$$= \sup_{\{\varphi,\, \|\varphi\|_{H_0^1}=1\}} \frac{1}{2}\int_\Omega E^{\alpha\beta\lambda\mu}u^2_{3|\alpha}\varphi_\beta$$

$$\left(\tilde{u}_{\lambda,\mu} - \Gamma^\delta_{\lambda\mu}\tilde{u}_\delta + \tilde{u}_{\mu,\lambda} - \Gamma^\delta_{\mu\lambda}\tilde{u}_\delta\right)\sqrt{a}d\xi^1 d\xi^2 \tag{22}$$

where the definitions of the linear strain ($\epsilon_{\lambda\mu}$) and covariant differentiation were used in (21) and (22). Using the clamped boundary conditions,

$$\int_\Omega E^{\alpha\beta\lambda\mu}u^2_{3|\alpha}\varphi_\beta\tilde{u}_{\lambda,\mu}\sqrt{a}d\xi^1 d\xi^2 = -\int_\Omega E^{\alpha\beta\lambda\mu}\left(u^2_{3|\alpha}\varphi_\beta\right)\Big|_\mu \tilde{u}_\lambda\sqrt{a}d\xi^1 d\xi^2 \tag{23}$$

which applied to (22) yields

$$\left\|u^2_{3|\alpha}E^{\alpha\beta\lambda\mu}\epsilon_{\lambda\mu}(\vec{\tilde{u}})\right\|_{H^{-1}}$$

$$= \sup_{\{\varphi,\, \|\varphi\|_{H_0^1}=1\}} \frac{1}{2}\left\{\int_\Omega E^{\alpha\beta\lambda\mu}\left(\left(u^2_{3|\alpha}\varphi_\beta\right)\Big|_\mu\tilde{u}_\lambda + \left(u^2_{3|\alpha}\varphi_\beta\right)\Big|_\lambda\tilde{u}_\mu\right)\sqrt{a}d\xi^1 d\xi^2\right.$$

$$\left.-\int_\Omega E^{\alpha\beta\lambda\mu}\left(\Gamma^\delta_{\lambda\mu}\tilde{u}_\delta + \Gamma^\delta_{\mu\lambda}\tilde{u}_\delta\right)u^2_{3|\alpha}\varphi_\beta\sqrt{a}d\xi^1 d\xi^2\right\}$$

$$\leq \sup_{\{\varphi,\, \|\varphi\|_{H_0^1}=1\}} C\left\|\left(u^2_{3|\alpha}\varphi_\beta\right)\Big|_\mu\tilde{u}_\lambda\right\|_{L^2} + C\left\|\left(u^2_{3|\alpha}\varphi_\beta\right)\Big|_\lambda\tilde{u}_\mu\right\|_{L^2}$$

$$+C\left\|\Gamma^\delta_{\lambda\mu}\tilde{u}_\delta u^2_{3|\alpha}\varphi_\beta\right\|_{L^2} + C\left\|\Gamma^\delta_{\mu\lambda}\tilde{u}_\delta u^2_{3|\alpha}\varphi_\beta)\right\|_{L^2} \tag{24}$$

Expanding the divergence and using Cauchy-Schwarz gives

$$\left\|\left(u^2_{3|\alpha}\varphi_\beta\right)\Big|_\mu\tilde{u}_\lambda\right\|_{L^2} \leq \left\|u^2_{3|\alpha\mu}\right\|_{L^2}\|\varphi_\beta\tilde{u}_\lambda\|_{L^2} + \left\|\varphi_{\beta|\mu}\right\|_{L^2}\left\|u^2_{3|\alpha}\tilde{u}_\lambda\right\|_{L^2}$$

$$\leq \left\|u^2_3\right\|_{H^2}\|\varphi_\beta\tilde{u}_\lambda\|_{L^2} + \|\varphi_\beta\|_{H^1}\left\|u^2_{3|\alpha}\tilde{u}_\lambda\right\|_{L^2}$$

$$\leq C\left\|u^2_3\right\|_{H^2}\left[\ln(1+\lambda_n)\|\tilde{u}_\lambda\|_{L^2} + \frac{1}{\lambda_n^r}\|\tilde{u}_\lambda\|_{H^{r+\epsilon}}\right]\|\varphi_\beta\|_{H^1}$$

$$+C\|\varphi_\beta\|_{H^1}\left[\ln(1+\lambda_n)\|\tilde{u}_\lambda\|_{L^2}+\frac{1}{\lambda_n^r}\|\tilde{u}_\lambda\|_{H^{r+\epsilon}}\right]\|u_3^2\|_{H^2} \tag{25}$$

for the first term in (24). A similar argument for the third term on the right-hand side of (24) leads to

$$\left\|\Gamma^\delta_{\lambda\mu}\tilde{u}_\delta u^2_{3|\alpha}\varphi_\beta\right\|_{L^2}\leq C\left[\ln(1+\lambda_n)\|\tilde{u}_\delta\|_{L^2}+\frac{1}{\lambda_n^r}\|\tilde{u}_\lambda\|_{H^{r+\epsilon}}\right]\|\varphi_\beta\|_{H^1}\|u_3^2\|_{H^1} \tag{26}$$

Combining these with similar arguments for the remaining terms gives

$$\left\|u^2_{3|\alpha}E^{\alpha\beta\lambda\mu}\epsilon_{\lambda\mu}(\vec{\tilde{u}})\right\|_{H^{-1}}\leq C\left[\ln(1+\lambda_n)\|\tilde{u}_\alpha\|_{L^2}+\frac{1}{\lambda_n^r}\|\tilde{u}_\alpha\|_{H^{r+\epsilon}}\right]\|u_3^2\|_{H^2}. \tag{27}$$

Combining (16), (18), (19),(20), and (27) results in Lemma (10).

LEMMA 11 *Let $r=1-\epsilon$, then*

$$\|\tilde{u}_\alpha\|^2_{L^2}+\|\tilde{u}_3\|^2_{H^1}\leq Ct^2\lambda_n^{Ct^2-r/2}(1+\lambda_n^{-1})^{Ct^2} \tag{28}$$

Proof. Lemmas 7, 9, and 10 give

$$\|\tilde{u}_\alpha\|^2_{L^2}+\|\tilde{u}_3\|^2_{H^1}\leq Ct\int_0^t\left[\ln(1+\lambda_n)\|\tilde{u}_3\|^2_{H^1}+\ln(1+\lambda_n)\|\tilde{u}_\alpha\|^2_{L^2}+\frac{1}{\lambda_n^r}\right]ds$$

which implies

$$\|\tilde{u}_\alpha\|^2_{L^2}+\|\tilde{u}_3\|^2_{H^1}\leq\frac{Ct^2}{\lambda_n^{r/2}}+Ct\ln(1+\lambda_n)\int_0^t(\|\tilde{u}_3\|_{H^1}+\|\tilde{u}_\alpha\|_{L^2})\,ds \tag{29}$$

Applying Gronwall's inequality to (29) gives (28).

Since the sequence $\{\lambda_n\}$ tends to ∞, the right-hand side of (28)will go to zero as long as $t\in\left[0,\sqrt{\frac{r}{2C}}\right)$. Thus, $\tilde{\mathbf{u}}=0$ on $\left[0,\sqrt{\frac{r}{2C}}\right)$. The bootstrap argument completes the proof of the theorem.

References

[1] M. Bernadou. *Méthodes d'Eléments Finis pour les Problémes de Coques Minces.* Masson, Paris, 1994.

[2] M. Bernadou, P.G. Ciarlet and B. Miara. Existence theorems for two-dimensional linear shell theories. Technical Report 1771. Unité deRecherche INRIA-Rocquencourt, 1992.

[3] M. Bernadou, B. Lalanne. On the approximation of free vibration modes of a general thin shell application to turbine blades. In *The Third European Conference on Mathematics in Industry.* J. Manley, et al. Eds., Kluwer Academic Publishers and B. G. Teubner Stuttgart, 257-264, 1990.

[4] M. Bernadou and J.T. Oden. An existence theorem for a class of nonlinear shallow shell problems. *J. Math Pures Appl.* 60:285-308, 1981.

[5] P.G. Ciarlet. A two-dimensional non-linear shell model of Koiter's type. *Jean Leray '99 Conference Proceedings*. Kluwer Academic Publishers, 437-449, 2003.

[6] M.C. Delfour and J.P. Zolésio. Tangential differential equations for dynamical thin/shallow shells. *J. Differential Equations.* 128:125-167, 1995.

[7] W.T. Koiter. On the nonlinear theory of thin elastic shells. *Proc. Kon. Ned. Akad. Wetensch.* B:1-54, 1966.

[8] I. Lasiecka. Uniform stabilizability of a full von Karman system with nonlinear boundary feedback. *SIAM J. Control Optim.* 36: 1376-1422, 1998.

[9] I. Lasiecka, R. Marchand. Riccati equations arising in acoustic structure interactions with curved walls. *Journal of Dynamics and Control.* 8:269-292, 1998.

[10] V.I. Sedenko. On uniqueness of the generalized solutions of initial boundary value problems for Marguerre-Vlasov nonlinear oscillations of shallow shells. *Russian Izv.* 1-2, 1994.

RIEMANNIAN METRIC OF THE AVERAGED CONTROLLED KEPLER EQUATION

B. Bonnard,[1] J.-B. Caillau,[2] and R. Dujol[2]

[1] *Institut de mathématiques de Bourgogne, UMR CNRS 5584, 9 avenue Alain Savary, F-21078 Dijon, Bernard.Bonnard@univ-bourgogne.fr**, [2] *ENSEEIHT-IRIT, UMR CNRS 5505, 2 rue Camichel, F-31071 Toulouse, {caillau, dujol}@n7.fr*

Abstract A non-autonomous sub-Riemannian problem is considered: Since periodicity with respect to the independent variable is assumed, one can define the averaged problem. In the case of the minimization of the energy, the averaged Hamiltonian remains quadratic in the adjoint variable. When it is non-degenerate, a Riemannian problem and the corresponding metric can be uniquely associated to the averaged problem modulo the orthogonal group of the quadratic form. The analysis is applied to the controlled Kepler equation. Explicit computations provide the averaged Hamiltonian of the Kepler motion in the three-dimensional case. The Riemannian metric is given, and the curvature of a special subsytem is evaluated.

keywords: periodic sub-Riemannian problems, averaging, Riemannian metrics, minimum energy control, Kepler equation

Introduction

An elementary generalization of sub-Riemannian problems [6] is to take time-dependent vector fields: Instead of a linear in the control dynamics

$$\dot{x} = \sum_{i=1}^{m} u_i f_i(x)$$

where the f_i's generate a smooth distribution on the ambient manifold X of dimension n, $n \geq m$, one considers vector fields $f_i(\theta, x)$, periodic with respect to the additional variable θ. The dynamics of θ is known so that, up to a reparameterization of time, this amounts to dealing with non-autonomous vector fields. An approximation of the system is then provided by the averaged system:

*The three authors are supported in part by the French Space Agency, contract 02/CNES/0257/00, and by the HYCON Network of Excellence, contract number FP6-IST-511368.

Please use the following format when citing this chapter:

Bonnard, B., Caillau, J.-B., and Dujol, R., 2006, in IFIP International Federation for Information Processing, Volume 202, Systems, Control, Modeling and Optimization, eds. Ceragioli, F., Dontchev, A., Furuta, H., Marti, K., Pandolfi, L., (Boston: Springer), pp. 79-89.

Each vector field is averaged with respect to θ over a period. This approach is well known in optimal control, see [7] and [13], where the variable θ is regarded as the *fast angular variable*. Its application in celestial mechanics traces back to [12], and more recently to [8] in the case of orbit transfer problems with low thrust propulsion. Our aim is to analyze this problem, that is to contribute to the study of the controlled Kepler equation by means of averaging techniques initiated in [9]. The motivation is that the trajectories of the averaged system approximate those of the original problem which can only be computed numerically [5].

The first section is devoted to periodic sub-Riemannian problems in the previous sense. Since the averaging is performed on the Hamiltonian associated to the problem of minimization of the energy, we recall Pontryagin maximum principle before stating basic properties of the averaged Hamiltonian. The second section deals with the Riemannian structure defined by this Hamiltonian: When the quadratic form is non-degenerate on the cotangent bundle, a Riemannian metric can be canonically associated to the averaged problem modulo the action of the orthogonal group of the quadratic form. In the last section, we apply this approach to the controlled Kepler equation: The averaged Hamiltonian for minimum energy is explicitly computed in the three dimensional case, thus extending the results of [9]. Using an adapted change of coordinates, the metric corresponding to the two-dimensional problem is given in orthogonal form. As a first result in this Riemannian setting, the curvature of minimum energy transfers towards circular orbits is computed.

1. Periodic sub-Riemannian problems

Let X be an n-dimensional manifold, and let $f_i(\theta, x)$, $i = 1, m$, be smooth vector fields parameterized by θ in S^1,

$$f_i(\theta, x) \in T_x X,\ \theta \in S^1,\ x \in X.$$

The corresponding *periodic* sub-Riemannian dynamics is defined by

$$\dot{x} = \sum_{i=1}^{m} u_i f_i(\theta, x) \tag{1}$$

$$\dot{\theta} = g_0(\theta, x) + g_1(\theta, x, u). \tag{2}$$

The two functions g_0 and g_1 are smooth, g_0 positive, and g_1 is assumed to be linear in u. The broader class of *sub-Riemannian systems with drift* [5] may provide examples of such dynamics. Indeed, if $\dot{x} = f_0(x) + \sum_{i=1}^{m} u_i f_i(x)$ with $f_0 = g_0 \partial/\partial x_n$, the system falls into the previous class with $\theta = x_n$ provided periodicity of the f_i's with respect to x_n holds. As shall be stated in section 3, this is indeed the case for the controlled Kepler equation.

Standard performance indexes for (1-2) are minimum time, minimum length,

$$\int_0^{t_f} \sqrt{|u_1|^2 + \cdots + |u_m|^2}\, dt \to \min$$

or minimum energy :

$$\int_0^{t_f} \left(|u_1|^2 + \cdots + |u_m|^2\right)\, dt \to \min .$$

For the problem to make sense, one usually has to add a bound constraint on the control, $|u| \leq \varepsilon$. The choice of the finite- dimensional norm $|.|$ is of course crucial (the control set may not even be smooth). Since Maupertuis' principle does not hold here, minimizing the length—that is the L^1–norm of the control—may give rise to intricate optimal control problems (see for instance [10] in the case of Kepler equation). We focus here on the minimization of the energy, that is on the optimization of the L^2–norm of the control, the final time being fixed. In this particular case, we first relax the problem by dropping the bound on the control. Indeed, the underlying idea is that, for a given positive ε, the constraint will be automatically fulfilled for a big enough fixed final time. Hence, treating the control as a small quantity, it is natural to do the feedback $u = \varepsilon v$ and to reparameterize the trajectories by θ:

$$\frac{dx}{d\theta} = \frac{\varepsilon}{g_0(\theta, x) + \varepsilon g_1(\theta, x, v)} \sum_{i=1}^{m} v_i f_i(\theta, x).$$

The criterion becomes

$$\varepsilon^2 \int_{\theta_0}^{\theta_f} \left(|v_1|^2 + \cdots + |v_m|^2\right) \frac{d\theta}{g_0(\theta, x) + \varepsilon g_1(\theta, x, v)}$$

and Pontryagin maximum principle tells us that optimal trajectories are projection on X of the integral curves of the following Hamiltonian defined on the cotangent bundle T^*X:

$$H(\theta, x, p, v) = \frac{\varepsilon}{g_0(\theta, x) + \varepsilon g_1(\theta, x, v)} \left(p^0 \varepsilon |v|^2 + \sum_{i=1}^{m} v_i P_i(\theta, x, p)\right).$$

Here before, p^0 is a nonpositive constant, p is the adjoint state to x and belongs to the cotangent space T_x^*X, and the P_i's are the *Poincaré coordinates* or *Hamiltonian lifts* of the vector fields,

$$P_i(\theta, x, p) = \langle p, f_i(\theta, x)\rangle, \ i = 1, m.$$

We consider the so-called *normal case*, p^0 negative: For obvious homogeneity reasons, we use the normalization $p^0 = -1/2\varepsilon$. Consequently, up to first order in ε, we have the following approximation of H:

$$\begin{aligned} H(\theta,x,p,v) &= \frac{\varepsilon}{g_0}(1-\varepsilon g_1/g_0+\cdots)\left(-\frac{1}{2}|v|^2+\sum_{i=1}^m v_i P_i(\theta,x,p)\right) \\ &= \frac{\varepsilon}{g_0(x,\theta)}\left(-\frac{1}{2}|v|^2+\sum_{i=1}^m v_i P_i(\theta,x,p)\right)+o(\varepsilon). \end{aligned}$$

According to the maximum principle, the optimal control maximizes H so the maximized first order approximation of the Hamiltonian, which we still denote H, is the *true*[1] Hamiltonian function

$$H(\theta,x,p)=\frac{1}{2g_0(x,\theta)}\sum_{i=1}^m P_i^2(\theta,x,p) \tag{3}$$

where, for the sake of simplicity, we have dropped the multiplicative factor ε. This Hamiltonian is clearly invariant with respect to feedbacks $v=R(\theta,x)v'$, $R(\theta,x)$ in $\mathrm{SO}(m)$.

The *averaged Hamiltonian* is

$$\overline{H}(x,p)=\frac{1}{2\pi}\int_0^{2\pi} H(\theta,x,p)d\theta.$$

Under mild assumptions, the integral curves of the averaged system converge uniformly towards those of the original system, see for instance [1, chap. 10]. The function $H(\theta,x,p)$ is a non-negative quadratic form in p, possibly degenerate, with coefficients parameterized respectively by θ and x. We denote by $w(\theta,x)$ this form. Since the integral is positive and linear, the following holds.

LEMMA 1 *The averaged Hamiltonian also defines a non-negative quadratic form in p, denoted by $w(x)$. Moreover,*

$$\mathrm{Ker}\, w(x)=\bigcap_{\theta\in S^1}\mathrm{Ker}\, w(\theta,x).$$

According to this lemma, we can only expect the rank to increase. An interpretation is that the oscillations of the *fast variable* θ generate new control directions, namely brackets of the original vector fields. We will assume in the sequel that $w(x)$ is non- degenerate.

2. Riemannian structure of the averaged problem

Let w be a smooth function on X such that, for any point x, $w(x)$ defines a positive definite quadratic form on the fiber T_x^*X. Then, w can be represented by its polar form which is a two-times covariant symmetric tensor. In coordinates,

$$w(x)=\sum_{i,j=1}^m w_{ij}(x)\frac{\partial}{\partial x_i}\otimes\frac{\partial}{\partial x_j}.$$

The matrix $W(x) = (w_{ij}(x))_{ij}$ belongs to the cone of real positive definite symmetric matrices, $\mathrm{Sym}_+(n, \mathbf{R})$. This tensor defines a Riemannian structure [11] on the fiber space T^*X. The question is: Does it arise from a Riemannian problem on X? That is, is the Hamiltonian $H(x,p) = 1/2 \sum_{i,j=1}^m w_{ij}(x) p_i p_j$ associated with a control problem

$$\int_0^t \left(|u_1|^2 + \cdots + |u_n|^2\right) dt \to \min$$

and $\dot{x} = u_1 f_1(x) + \cdots + u_n f_n(x)$ (where there are $m = n$ vector fields and controls)? Now, the answer is obviously positive since one has just to write H as a sum of squares, using a change of adjoint variable of the type $p = q\,{}^tA(x)$,

$$H(x,p) = \frac{1}{2} \sum_{i=1}^{n} P_i^2(x,p)$$

with $P_i(x,p) = \langle p, f_i(x) \rangle$ (the last equalities defining so the f_i's). Indeed, the Hamiltonian is a quadratic form in p parameterized by x, so it can be written as a sum of independent linear forms in p, using for instance Gauss algorithm. To this end, let us recall the (right) action of the linear group $\mathrm{GL}(n, \mathbf{R})$ on $\mathrm{Sym}_+(n, \mathbf{R})$. For A in $\mathrm{GL}(n, \mathbf{R})$, one defines

$$A \cdot M = {}^tAMA, \ M \in \mathrm{Sym}_+(n, \mathbf{R}).$$

Given M in $\mathrm{Sym}_+(n, \mathbf{R})$, we look for A such that $A \cdot M = I$: The relevant set is hence the set of *M–orthonormal* matrices. The isotropy group of M is the *orthogonal group* $\mathrm{O}(M)$ of matrices O such that

$${}^tOMO = M.$$

Therefore, two matrices A and B are both M–orthogonal if and only if $B^{-1} \cdot (A \cdot M) = (AB^{-1}) \cdot M = M$, that is if AB^{-1} belongs to $\mathrm{O}(M)$. The set of M–orthonormal matrices is thus in one-to-one correspondance with the orthogonal group, and M is diagonalized to identity by a unique element of the quotient set $\mathrm{GL}(n, \mathbf{R})/\mathrm{O}(M)$.

Let now $\mathrm{O}(w(x))$ be the orthogonal group generated by the quadratic form $w(x)$, and let $A(x)$ be in $\mathrm{GL}(n, \mathbf{R})/\mathrm{O}(w(x))$: The Hamiltonian writes

$$H(x,p) = \frac{1}{2}|q|^2 = \frac{1}{2}|p\,{}^tA(x)^{-1}|^2$$

and comes from the Riemannian problem with dynamics $\dot{x} = {}^tA(x)^{-1}u$. Eventually, $|u|^2$ is equal to $|{}^tA(x)\dot{x}|^2$ and the Riemannian metric is $ds^2 =$

$\sum_{ij=1}^{n} g_{ij} dx_i dx_j$ where $G(x) = (g_{ij}(x))_{ij}$ is the positive definite symmetric matrix $(A^tA)(x)$. The following proposition summarizes the computations.

Proposition 1 *The Riemannian metric associated to the Riemannian structure on the cotangent bundle is unique modulo the action on positive definite symmetric matrices of the quadratic form defining the structure.*

Proof. Let M be a positive definite symmetric matrix. If both A and B are M–orthonormal, there is O in $\mathrm{O}(M)$ such that $A = OB$. Then $A^tA{=}O \cdot (B^tB)$ for the (left) action of $\mathrm{O}(M)$ on $\mathrm{Sym}_+(n, \mathbf{R})$, $O \cdot Y = OY{}^tO$, and A^tA and B^tB belong to the same orbit. □

We end this section by recalling the effect of a change of variables in X on the tensor w. Indeed, it is not realistic to look for coordinates that would trivialize w to $\sum_{i=1}^{n} (\partial/\partial y_i)^2$ since then, the associated metric would be flat. A less strong requirement may be to find such coordinates that diagonalize the quadratic form,

$$w(y) = \sum_{i=1}^{n} d_i(y) \left(\frac{\partial}{\partial y_i}\right)^2, \; d_i(y) > 0.$$

In this case, the associated metric on X is *orthogonal* [2],

$$ds^2 = \sum_{i=1}^{n} \frac{dy_i^2}{d_i(y)}.$$

Given M in $\mathrm{Sym}_+(n, \mathbf{R})$, this amounts to finding *M–orthogonal* matrices. As a consequence of Sylvester's law of inertia, two matrices A and B are M–orthogonal if and only if there is a *scaling*, that is an element of $(\mathbf{R}_+^*)^n$ (defining a diagonal matrix S with positive entries), such that $S \cdot (A \cdot M) = B \cdot M$. In other words, M–orthogonal matrices are in one-to-one correspondance with elements of the direct product $\mathrm{O}(M) \times (\mathbf{R}_+^*)^n$.

If $x = \varphi(y)$ is a change of coordinates on X, the new adjoint state q and tensor matrix U verify:

$$p = q d\varphi(y)^{-1}, \qquad W(x) = {}^t d\varphi(y) \cdot U(y). \tag{4}$$

Accordingly, one gets an orthogonal metric on X if and only if φ is such that $d\varphi$ is $(W \circ \varphi)^{-1}$–orthogonal. As we shall see now in the last section, it turns out that such a change of coordinates is available in the Kepler case.

3. Application to the controlled Kepler equation

We briefly state the control problem, see for instance [5] for a detailed exposition. The Kepler equation describes the motion of a body in a central field and can be normalized to

$$\ddot{q} = -\frac{q}{|q|^3} + \gamma .$$

where $|q|^2 = |q_1|^2 + |q_2|^2 + |q_3|^2$, q being the position vector in $\mathbf{R}^3$ and γ the control. Several sets of coordinates describing the geometry of the osculating conic are available. The dynamics also depends on the local frame chosen to express the control. In contrast with [8, 9] where the so-called *tangential-normal* frame is used, we use the feedback-invariance of the maximized Hamiltonian and prefer to write down the equations in the *radial-orthoradial* frame for reasons that will be made clear in the next paragraph: $\gamma = u_1 f_1 + u_2 f_2 + u_3 f_3$ with $f_1 = q/|q|$, $f_2 = f_3 \times f_1$ and $f_3 = q \times \dot{q}/|q \times \dot{q}|$. The state is described by five *equinoctial* elements, for instance $x = (P, e, h)$ where P is the *semi-latus rectum*, $e = (e_x, e_y)$ the *eccentricity vector*, $h = (h_x, h_y)$ the *inclination vector*, and by an angle, the *true longitude* l. We restrict the problem to the manifold X of elliptic trajectories,

$$X = \{(P, e, h) \mid P > 0 \text{ and } |e| < 1\}$$

so that the vector fields are smooth on $S^1 \times X$ and define a periodic sub-Riemannian problem as in section 34:

$$f_1 = \sqrt{P}\left(\sin l \frac{\partial}{\partial e_x} - \cos l \frac{\partial}{\partial e_y}\right)$$

$$f_2 = \sqrt{P}\left[\frac{2P}{W}\frac{\partial}{\partial P} + \left(\cos l + \frac{e_x + \cos l}{W}\right)\frac{\partial}{\partial e_x} + \left(\sin l + \frac{e_y + \sin l}{W}\right)\frac{\partial}{\partial e_y}\right]$$

$$f_3 = \frac{\sqrt{P}}{W}\left(-Ze_y\frac{\partial}{\partial e_x} + Ze_x\frac{\partial}{\partial e_y} + \frac{C\cos l}{2}\frac{\partial}{\partial h_x} + \frac{C\sin l}{2}\frac{\partial}{\partial h_y}\right)$$

where

$$\begin{aligned} W &= 1 + e_x \cos l + e_y \sin l \\ Z &= h_x \sin l - h_y \cos l \\ C &= 1 + |h|^2. \end{aligned}$$

The variation of the angle is $\dot{l} = g_0(l, x) + g_1(l, x, u)$ with

$$g_0 = \frac{W^2}{P^{3/2}} \qquad g_1 = \sqrt{P}\frac{Z}{W}u_3.$$

An important remark is that, because there is some decoupling between these vector fields, the two-dimensional Kepler problem is obtained by letting $h = 0$ in the previous equations. We start with the computation on this subproblem.

Applying the process described in section 34, one gets the maximized first order approximation of the Hamiltonian, $H = (P_1^2 + P_2^2)/2$ with

$$P_1 = \frac{P^{5/4}}{W}(p_{e_x} \sin l - p_{e_y} \cos l)$$

$$P_2 = \frac{P^{5/4}}{W}\left[p_P\frac{2P}{W} + p_{e_x}\left(\cos l + \frac{e_x + \cos l}{W}\right) + p_{e_y}\left(\sin l + \frac{e_y + \sin l}{W}\right)\right].$$

Hence, the computation of the averaged has the complexity of integrating terms of the form $P(\cos l, \sin l)/W^k$, where P is a polynomial and k and integer comprised between 2 and 4. Since

$$\int_0^{2\pi} \frac{P(\cos l, \sin l)}{W^k} dl = \int_{S^1} \frac{P(z/2 + 1/2z, z/2i - 1/2iz)}{W^k} \frac{dz}{iz}$$

the averaged is evaluated by computing the residues of the integrands. Since $W = (\bar{e}z^2 + 2z + e)/2z$ (we use the complex eccentricity $e = e_x + ie_y$), we have two poles,

$$z = \frac{-1 \pm \sqrt{1 - |e|^2}}{\bar{e}}.$$

The product of the poles is $e/\bar{e}$ so it has modulus one, and $z_1 = (-1 - \sqrt{1 - |e|^2})/\bar{e}$ clearly does not belong to the unit disk: The only pole to consider is $z_2 = (-1 + \sqrt{1 - |e|^2})/\bar{e}$. In contrast, if one uses the tangential-normal frame as in [9], W is replaced by $W(1 + 2e_x \cos l + 2e_y \sin l + |e|^2)$, and two poles among the four are to be taken into account.

An inspection of the Hamiltonian shows that the following averages have to be computed, for which we give the results:

$$\overline{1/W^2} = \delta^3$$

$$\overline{\cos l/W^3} = -3e_x\delta^5/2 \qquad \overline{\sin l/W^3} = -3e_y\delta^5/2$$
$$\overline{\cos^2 l/W^3} = \delta^3/2 + 3e_x^2\delta^5/2 \qquad \overline{\sin^2 l/W^3} = \delta^3/2 + 3e_y^2\delta^5/2$$
$$\overline{\cos l \sin l/W^3} = 3e_x e_y \delta^5/2$$

$$\overline{1/W^4} = (2 + 3|e|^2)\delta^7/2$$
$$\overline{\cos l/W^4} = -e_x(4 + |e|^2)\delta^7/2 \qquad \overline{\sin l/W^4} = -e_y(4 + |e|^2)\delta^7/2$$
$$\overline{\cos^2 l/W^4} = \delta^5/2 + 5e_x^2\delta^7/2 \qquad \overline{\sin^2 l/W^4} = \delta^5/2 + 5e_y^2\delta^7/2$$
$$\overline{\cos l \sin l/W^4} = 5e_x e_y \delta^7/2$$

with $\delta = 1/\sqrt{1 - |e|^2}$. Substituing these expressions, we get the averaged Hamiltonian

$$\begin{aligned} \overline{H}(x,p) &= \frac{P^{5/2}}{4(1-|e|^2)^{5/2}} \left[4p_P^2 P^2 \left(-3 + \frac{5}{1-|e|^2} \right) \right. \\ &+ p_{e_x}^2 \left(5(1-|e|^2) + e_y^2 \right) + p_{e_y}^2 \left(5(1-|e|^2) + e_x^2 \right) \\ &- \left. 20 p_P p_{e_x} P e_x - 20 p_P p_{e_y} P e_y - 2 p_{e_x} p_{e_y} e_x e_y \right]. \end{aligned}$$

At this point, we take advantage of the computation in [9] and make the following change of variables:

$$P = \frac{1 - \rho^2}{n^{2/3}}$$

$$e_x = \rho \cos\theta \qquad e_y = \rho \sin\theta$$

where n is the so-called *mean movement* [14]. Using 4), we obtain

$$\overline{H} = \frac{1}{4n^{5/3}} \left[18n^2 p_n^2 + 5(1-\rho^2)p_\rho^2 + (5-4\rho^2)\frac{p_\theta^2}{\rho^2} \right].$$

Up to a scalar, this is the result that was obtained by symbolic machine computation in [9]. However, the complexity of the computation did not allow the authors to tackle the three-dimensional problem whereas we shall be able to do so, here. Let us write before the Riemannian metric of the two-dimensional Kepler problem in orthogonal form:

$$ds^2 = \frac{1}{9n^{1/3}} dn^2 + \frac{2n^{5/3}}{5(1-\rho^2)} d\rho^2 + \frac{2n^{5/3}}{5-4\rho^2} \rho^2 d\theta^2.$$

The three-dimensional case has the same complexity: Indeed, the Hamiltonian is $H = (P_1^2 + P_2^2 + P_3^2)/2$ with

$$P_3 = \frac{P^{5/4}}{W^2} \left(-Zp_{e_x} e_y + Zp_{e_y} e_x + C/2\, p_{h_x} \cos l + C/2\, p_{h_y} \sin l \right)$$

and P_1, P_2 unchanged. Hence, the previous averaging computations allow us to conclude: Extending the change of coordinates to h according to $h_x = \sigma \cos \Omega$, $h_y = \sigma \sin \Omega$, we get

$$\begin{aligned} \overline{H} &= \frac{1}{4n^{5/3}} \left[18n^2 p_n^2 + 5(1-\rho^2)p_\rho^2 + (5-4\rho^2)\frac{p_\theta^2}{\rho^2} \right] \\ &+ \frac{1}{4n^{5/3}} \frac{(1+\sigma^2)^2}{4} \frac{1+4\rho^2}{1-\rho^2} \left(\cos\omega p_\sigma + \sin\omega \frac{p_{\theta\Omega}}{\sigma} \right)^2 \\ &+ \frac{1}{4n^{5/3}} \frac{(1+\sigma^2)^2}{4} \left(-\sin\omega p_\sigma + \cos\omega \frac{p_{\theta\Omega}}{\sigma} \right)^2 \end{aligned}$$

where $\omega = \theta - \Omega$ is the *angle of the pericenter* and where

$$p_{\theta\Omega} = \frac{2\sigma^2}{1+\sigma^2} p_\theta + p_\Omega.$$

We conclude the exposition by a preliminary computation of curvature. In the two-dimensional case, one can restrict the metric to $\{\theta = 0\}$. The associated trajectories are, for instance, those reaching a circular orbit. On this submanifold of dimension two, the metric is

$$ds^2 = \frac{1}{9n^{1/3}} dn^2 + \frac{2n^{5/3}}{5(1-\rho^2)} d\rho^2$$

and the following holds.

Proposition 2 *The curvature is zero.*

Proof. In orthogonal coordinates, the Gaussian curvature is

$$K = -\frac{1}{2\sqrt{g}}\left[\frac{\partial}{\partial n}\left(\frac{1}{\sqrt{g}}\frac{\partial g_{11}}{\partial n}\right) + \frac{\partial}{\partial \rho}\left(\frac{1}{\sqrt{g}}\frac{\partial g_{22}}{\partial \rho}\right)\right]$$

with $g = g_{11}g_{22}$, whence the result. □

As a result, the metric is locally isomorphic to $ds^2 = dx^2 + dy^2$. Actually, we prove in [3] that the result is global and that we can find coordinates in which the two-dimensional subsystem is flat. The insight on the control of the Kepler equation provided by this Riemannian point of view will be developed in forthcoming papers. See for instance [4] for a preliminary evaluation of Riemannian balls of the two-dimensional Kepler motion.

Notes

1. True in the sense that it is not parameterized by the control anymore.

References

[1] V. I. Arnold. *Mathematical Methods of Classical Mechanics.* Springer-Verlag, New-York, 1978.

[2] M. Berger, B. Gostiaux. *Géométrie différentielle.* Presses Universitaires de France, Paris, 1981.

[3] B. Bonnard, J.-B. Caillau. Riemannian metric of the averaged energy minimization problem in orbital transfer with low thrust. Submitted to *Annales de l'Institut Henri Poincaré.*

[4] B. Bonnard, J.-B. Caillau, R. Dujol. Averaging and optimal control of elliptic Keplerian orbits with low propulsion. Submitted to *13th IFAC Workshop on Control Applications of Optimisation*, Paris, April 2006.

[5] B. Bonnard, J.-B. Caillau, E. Trélat. Geometric optimal control of elliptic Keplerian orbits. *Disc. and Cont. Dyn. Syst. Series B* 5(4):929-956, 2005.

[6] B. Bonnard, M. Chyba. *Singular trajectories and their role in optimal control.* Math. and Applications 40, Springer Verlag, Paris, 2003.

[7] F. Chaplais. Averaging and deterministic optimal control. *SIAM J. Control and Opt.* 25(3):767-780, 1987.

[8] R. Epenoy, S. Geffroy. Optimal low-thrust transfers with constraints. Generalization of averaging techniques. *Acta Astronautica* 41(3):133-149, 1997.

[9] S. Geffroy. Les techniques de moyennisation en contrôle optimal. Application aux transferts orbitaux à poussée continue. Master thesis, ENSEEIHT, Institut National Polytechnique de Toulouse, 1994.

[10] J. Gergaud, T. Haberkorn. Homotopy method for minimum consumption orbit transfer problem. *ESAIM Control Opt. and Calc. of Var.*, to appear.

[11] C. Godbillon. *Géométrie différentielle et mécanique analytique.* Hermann, Paris, 1985.

[12] J. Kevorkian, J. D. Cole. *Perturbation methods in applied mathematics.* Springer Verlag, New-York, 1981.

[13] S. G. Peng. Analyse asymptotique et problème homogénéisé en contrôle optimal avec vibrations rapides. *SIAM J. Control and Opt.* 27(4):673-696, 1989.

[14] O. Zarrouati. *Trajectoires spatiales.* CNES–Cepadues, Toulouse, 1987.

ERROR ESTIMATES FOR THE NUMERICAL APPROXIMATION OF BOUNDARY SEMILINEAR ELLIPTIC CONTROL PROBLEMS. CONTINUOUS PIECEWISE LINEAR APPROXIMATIONS

E. Casas,[1] and M. Mateos[2]

[1]*Dpto. de Matemática Aplicada y Ciencias de la Computación, E.T.S.I. Industriales y de Telecomunicación, Universidad de Cantabria, 39071 Santander, Spain, eduardo.casas@unican.es**,

[2]*Dpto. de Matemáticas, E.P.S.I. de Gijón, Universidad de Oviedo, Campus de Viesques, 33203 Gijón, Spain, mmateos@uniovi.es*

Abstract We discuss error estimates for the numerical analysis of Neumann boundary control problems. We present some known results about piecewise constant approximations of the control and introduce some new results about continuous piecewise linear approximations. We obtain the rates of convergence in $L^2(\Gamma)$. Error estimates in the uniform norm are also obtained. We also discuss the semidiscretization approach as well as the improvement of the error estimates by making an extra assumption over the set of points corresponding to the active control constraints.

keywords: Boundary control, semilinear elliptic equation, numerical approximation, error estimates.

1. Introduction

This paper continues a series of works about error estimates for the numerical analysis of control problems governed by semilinear elliptic partial differential equations. In [1] a distributed problem approximated by piecewise constant controls was studied. In [7] the control appears in the boundary. This makes the task more difficult since the states are now less regular than in the distributed case. Piecewise constant approximations were used in that reference. The advantage of these is that we have a pointwise expression both for the control and its approximation, which we can compare to get uniform convergence. The reader is addressed to these papers for further references about error estimates for the approximation of linear-quadratic problems governed by partial differ-

*Paper written with financial support of Ministerio de Ciencia y Tecnología (Spain).

Please use the following format when citing this chapter:

Casas, E., and Mateos, M., 2006, in IFIP International Federation for Information Processing, Volume 202, Systems, Control, Modeling and Optimization, eds. Ceragioli, F., Dontchev, A., Furuta, H., Marti, K., Pandolfi, L., (Boston: Springer), pp. 91-101.

ential equations and for the approximation of control problems governed by ordinary differential equations.

In the case of continuous piecewise linear approximations of the control, there exists not such a pointwise formula in general. If the functional is quadratic with respecto to the control, recent results in [8] about the stability of L^2 projections in Sobolev $W^{s,p}(\Gamma)$ spaces allow us to obtain uniform convergence and adapt the proofs. The general case is more delicate. Results for distributed control problems can be found in [3]. The main purpose of this paper is to obtain similar results for Neumann boundary controls. This is done in Theorem 10.

We also refer to the works for distributed linear-quadratic problems about semidiscretization [9] and postprocessing [10]. The first proposes only discretizing the state, and not the control. The solution can nevertheless be expressed with a finite number of parameters via the adjoint-state and the problem can be solved with a computer with a slightly changed optimization code. The second one proposes solving a completely discretized problem with piecewise constant approximations of the control and finally construct a new control using the pointwise projection of the discrete adjoint state. We are able to reproduce the first scheme for Neumann boundary controls, a general functional and a semilinear equation.

The rest of the paper is as follows. In the next section, we define precisely the problem. In Section 3 we recall several results about this control problem. Section 4 contains the main results of this paper: we discretize the problem and obtain error estimates for the solutions.

2. Statement of the problem

Throughout the sequel, Ω denotes an open convex bounded polygonal set of $\mathbf{R}^2$ and Γ is the boundary of Ω. We will also take $p > 2$. In this domain we formulate the following control problem

$$
\text{(P)}\left\{\begin{array}{l}
\inf \ J(u) = \displaystyle\int_\Omega L(x, y_u(x))\,dx + \int_\Gamma l(x, y_u(x), u(x))\,d\sigma(x) \\
\\
\text{subject to } (y_u, u) \in H^1(\Omega) \times L^\infty(\Gamma), \\
u \in U^{ad} = \{u \in L^\infty(\Gamma) \mid \alpha \le u(x) \le \beta \text{ a.e. } x \in \Gamma\}, \\
(y_u, u) \text{ satisfying the state equation (1)}
\end{array}\right.
$$

$$
\left\{\begin{array}{rcll}
-\Delta y_u(x) & = & a_0(x, y_u(x)) & \text{in } \Omega \\
\partial_\nu y_u(x) & = & b_0(x, y_u(x)) + u(x) & \text{on } \Gamma,
\end{array}\right. \tag{1}
$$

where $-\infty < \alpha < \beta < +\infty$. Here u is the control while y_u is said to be the associated state. The following hypotheses are assumed about the functions involved in the control problem (P):

(A1) The function $L : \Omega \times \mathbf{R} \longrightarrow \mathbf{R}$ is measurable with respect to the first component, of class C^2 with respect to the second, $L(\cdot, 0) \in L^1(\Omega)$,

$\frac{\partial L}{\partial y}(\cdot,0) \in L^p(\Omega)$ $\frac{\partial^2 L}{\partial y^2}(\cdot,0) \in L^\infty(\Omega)$ and for all $M > 0$ there exists a constant $C_{L,M} > 0$ such that $\left|\frac{\partial^2 L}{\partial y^2}(x,y_2) - \frac{\partial^2 L}{\partial y^2}(x,y_1)\right| \le C_{L,M}|y_2 - y_1|$, for a.e. $x \in \Omega$ and $|y|, |y_i| \le M$, $i = 1, 2$.

(A2) The function $l : \Gamma \times \mathbf{R}^2 \longrightarrow \mathbf{R}$ is Lipschitz with respect to the first component, of class C^2 with respect to the second and third variables, $l(\cdot, 0, 0) \in L^1(\Gamma)$, $D^2_{(y,u)}l(\cdot,0,0) \in L^\infty(\Gamma)$ and for all $M > 0$ there exists a constant $C_{l,M} > 0$ such that

$$\left|\frac{\partial l}{\partial y}(x_2,y,u) - \frac{\partial l}{\partial y}(x_1,y,u)\right| + \left|\frac{\partial l}{\partial u}(x_2,y,u) - \frac{\partial l}{\partial u}(x_1,y,u)\right| \le C_{l,M}|x_2 - x_1|,$$

$$\|D^2_{(y,u)}l(x,y_2,u_2) - D^2_{(y,u)}l(x,y_1,u_1)\| \le C_{l,M}(|y_2 - y_1| + |u_2 - u_1|),$$

for a.e. $x, x_i \in \Gamma$ and $|y|, |y_i|, |u|, |u_i| \le M$, $i = 1, 2$, where $D^2_{(y,u)}l$ denotes the second derivative of l with respect to (y, u). Moreover we assume that there exists $\Lambda > 0$ such that

$$\frac{\partial^2 l}{\partial u^2}(x,y,u) \ge \Lambda, \quad \text{a.e. } x \in \Gamma \text{ and } (y,u) \in \mathbf{R}^2. \tag{2}$$

Let us remark that this inequality implies the strict convexity of l with respect to the third variable.

(A3) The function $a_0 : \Omega \times \mathbf{R} \longrightarrow \mathbf{R}$ is measurable with respect to the first variable and of class C^2 with respect to the second, $a_0(\cdot, 0) \in L^p(\Omega)$, $\frac{\partial a_0}{\partial y}(\cdot,0) \in L^\infty(\Omega)$, $\frac{\partial^2 a_0}{\partial y^2}(\cdot,0) \in L^\infty(\Omega)$, $\frac{\partial a_0}{\partial y}(x,y) \le 0$ a.e. $x \in \Omega$ and $y \in \mathbf{R}$ and for all $M > 0$ there exists a constant $C_{a_0,M} > 0$ such that

$$\left|\frac{\partial^2 a_0}{\partial y^2}(x,y_2) - \frac{\partial^2 a_0}{\partial y^2}(x,y_1)\right| < C_{a_0,M}|y_2 - y_1| \text{ a.e. } x \in \Omega \text{ and } |y_1|, |y_2| \le M.$$

(A4) The function $b_0 : \Gamma \times \mathbf{R} \longrightarrow \mathbf{R}$ is Lipschitz with respect to the first variable and of class C^2 with respect to the second, $b_0(\cdot, 0) \in W^{1-1/p,p}(\Gamma)$, $\frac{\partial^2 b_0}{\partial y^2}(\cdot,0) \in L^\infty(\Gamma)$, $\frac{\partial b_0}{\partial y}(x,y) \le 0$ and for all $M > 0$ there exists a constant $C_{b_0,M} > 0$ such that

$$\left|\frac{\partial b_0}{\partial y}(x_2,y) - \frac{\partial b_0}{\partial y}(x_1,y)\right| \le C_{b_0,M}|x_2 - x_1|,$$

$$\left|\frac{\partial^2 b_0}{\partial y^2}(x,y_2) - \frac{\partial^2 b_0}{\partial y^2}(x,y_1)\right| \leq C_{b_0,M}|y_2 - y_1|.$$

for a.e. $x, x_1, x_2 \in \Gamma$ and $|y|, |y_1|, |y_2| \leq M$.

(A5) At least one of the two conditions must hold: either $\frac{\partial a_0}{\partial y}(x,y) < 0$ in $E_\Omega \times \mathbf{R}$ with $E_\Omega \subset \Omega$ of positive n-dimensional measure or $\frac{\partial b_0}{\partial y}(x,y) < 0$ on $E_\Gamma \times \mathbf{R}$ with $E_\Gamma \subset \Gamma$ of positive $(n-1)$-dimensional measure.

3. Analysis of the control problem

Let us briefly state some useful results known for this control problem. The proofs can be found in [7].

THEOREM 1 *For every $u \in L^2(\Gamma)$ the state equation (1) has a unique solution $y_u \in H^{3/2}(\Omega)$, that depends continuously on u. Moreover, there exists $p_0 > 2$ depending on the measure of the angles in Γ such that if $u \in W^{1-1/p,p}(\Gamma)$ for some $2 \leq p \leq p_0$, then $y_u \in W^{2,p}(\Omega)$.*

Let us note that the inclusion $H^{3/2}(\Omega) \subset C(\bar{\Omega})$ holds for Lipschitz domains in $\mathbf{R}^2$. As a consequence of the theorem above, we know that the functional J is well defined in $L^2(\Gamma)$. Let us discuss the differentiability properties of J.

THEOREM 2 *Suppose that assumptions (A3)–(A4) are satisfied. Then the mapping $G : L^\infty(\Gamma) \longrightarrow H^{3/2}(\Omega)$ defined by $G(u) = y_u$ is of class C^2. Under the assumptions (A1)–(A4), the functional $J : L^\infty(\Gamma) \to \mathbf{R}$ is of class C^2. Moreover, for every $u, v \in L^\infty(\Gamma)$*

$$\int_\Gamma \left(\frac{\partial l}{\partial u}(x, y_u, u) + \varphi_u\right) v \, d\sigma.$$

where the adjoint state $\varphi_u \in H^{3/2}(\Omega)$ is the unique solution of the problem

$$\begin{cases} -\Delta\varphi = \dfrac{\partial a_0}{\partial y}(x, y_u)\varphi + \dfrac{\partial L}{\partial y}(x, y_u) & \text{in } \Omega \\ \partial_\nu\varphi = \dfrac{\partial b_0}{\partial y}(x, y_u)\varphi + \dfrac{\partial l}{\partial y}(x, y_u, u) & \text{on } \Gamma. \end{cases}$$

Expressions for the derivatives of G and the second derivative of J can be found in [7].

The existence of a solution for problem (P) follows easily from our assumptions (A1)–(A5). In particular, we underline the important fact that the function l is convex with respect to the third variable. See (2). The first order optimality conditions for Problem (P) follow readily from Theorem 2.

THEOREM 3 *Assume that $\bar{u}$ is a local solution of Problem (P). Then there exist $\bar{y}, \bar{\varphi} \in H^{3/2}(\Omega)$ such that*

$$\left\{ \begin{array}{rcll} -\Delta \bar{y}(x) & = & a_0(x, \bar{y}(x)) & \text{in} \quad \Omega \\ \partial_\nu \bar{y}(x) & = & b_0(x, \bar{y}(x)) + \bar{u}(x) & \text{on} \quad \Gamma, \end{array} \right. \tag{3}$$

$$\left\{ \begin{array}{rcll} -\Delta \bar{\varphi} & = & \dfrac{\partial a_0}{\partial y}(x, \bar{y})\bar{\varphi} + \dfrac{\partial L}{\partial y}(x, \bar{y}) & \text{in} \quad \Omega \\ \partial_\nu \bar{\varphi} & = & \dfrac{\partial b_0}{\partial y}(x, \bar{y})\bar{\varphi} + \dfrac{\partial l}{\partial y}(x, \bar{y}, \bar{u}) & \text{on} \quad \Gamma, \end{array} \right. \tag{4}$$

$$\int_\Gamma \left(\frac{\partial l}{\partial u}(x, \bar{y}, \bar{u}) + \bar{\varphi} \right) (u - \bar{u})\, d\sigma(x) \geq 0 \;\; \forall u \in U^{ad}. \tag{5}$$

First order optimality conditions allow us to deduce extra regularity for the optimal control.

THEOREM 4 *Suppose that $\bar{u}$ is a local solution of (P), then for all $x \in \Gamma$ the equation*

$$\bar{\varphi}(x) + \frac{\partial l}{\partial u}(x, \bar{y}(x), t) = 0$$

has a unique solution $\bar{t} = \bar{s}(x)$. The mapping $\bar{s} : \Gamma \longrightarrow \mathbf{R}$ is Lipschitz and it is related with $\bar{u}$ through the formula

$$\bar{u}(x) = Proj_{[\alpha,\beta]}(\bar{s}(x)) = \max\{\alpha, \min\{\beta, \bar{s}(x)\}\}. \tag{6}$$

Moreover $\bar{u} \in C^{0,1}(\Gamma)$ and $\bar{y}, \bar{\varphi} \in W^{2,p}(\Omega) \subset C^{0,1}(\bar{\Omega})$ for some $p > 2$.

In order to establish the second order optimality conditions we define the cone of critical directions. The derivative of J can be represented by the function in $L^2(\Gamma)$:

$$\bar{d}(x) = \frac{\partial l}{\partial u}(x, \bar{y}(x), \bar{u}(x)) + \bar{\varphi}(x).$$

The cone is:

$$C_{\bar{u}} = \{v \in L^2(\Gamma) \text{ satisfying (7) and } v(x) = 0 \text{ if } |\bar{d}(x)| > 0\},$$

$$v(x) = \left\{ \begin{array}{ll} \geq 0 & \text{for a.e. } x \in \Gamma \text{ where } \bar{u}(x) = \alpha, \\ \leq 0 & \text{for a.e. } x \in \Gamma \text{ where } \bar{u}(x) = \beta. \end{array} \right. \tag{7}$$

Now we formulate the second order necessary and sufficient optimality conditions. See Casas and Mateos [4]

THEOREM 5 *If $\bar{u}$ is a local solution of (P), then $J''(\bar{u})v^2 \geq 0$ holds for all $v \in C_{\bar{u}}$. Conversely, if $\bar{u} \in U^{ad}$ satisfies the first order optimality conditions (3)–(5) and the coercivity condition $J''(\bar{u})v^2 > 0$ holds for all $v \in C_{\bar{u}} \setminus \{0\}$, then there exist $\delta > 0$ and $\varepsilon > 0$ such that*

$$J(u) \geq J(\bar{u}) + \delta \|u - \bar{u}\|^2_{L^2(\Gamma)}$$

is satisfied for every $u \in U^{ad}$ such that $\|u - \bar{u}\|_{L^\infty(\Omega)} \leq \varepsilon$.

4. Discretization

Here, we define a finite-element based approximation of the optimal control problem (P). To this aim, we consider a regular family of triangulations $\{\mathcal{T}_h\}_{h>0}$ of $\bar{\Omega}$: $\bar{\Omega} = \cup_{T \in \mathcal{T}_h} T$.

For fixed $h > 0$, we denote by $\{T_j\}_{j=1}^{N(h)}$ the family of triangles of $\mathcal{T}_h$ with a side on the boundary of Γ. If the edges of $T_j \cap \Gamma$ are x_Γ^j and x_Γ^{j+1} then $[x_\Gamma^j, x_\Gamma^{j+1}] := T_j \cap \Gamma$, $1 \leq j \leq N(h)$, with $x_\Gamma^{N(h)+1} = x_\Gamma^1$.

4.1 Discretization of the state equation

Associated with this triangulation we set

$$Y_h = \{y_h \in C(\bar{\Omega}) \mid y_{h|T} \in \mathcal{P}_1, \text{ for all } T \in \mathcal{T}_h\},$$

where $\mathcal{P}_1$ is the space of polynomials of degree less than or equal to 1. For each $u \in L^\infty(\Gamma)$, we denote by $y_h(u)$ the unique element of Y_h that satisfies

$$a(y_h(u), z_h) = \int_\Omega a_0(x, y_h(u)) z_h \, dx + \int_\Gamma [b_0(x, y_h(u)) + u] z_h \, dx \;\; \forall z_h \in Y_h, \tag{8}$$

where $a : Y_h \times Y_h \longrightarrow \mathbf{R}$ is the bilinear form defined by

$$a(y_h, z_h) = \int_\Omega \nabla y_h(x) \nabla z_h(x) \, dx.$$

The existence and uniqueness of a solution of (8) follows in the standard way from the monotonicity of a_0 and b_0 (see [7]).

Let us now introduce the approximate adjoint state associated to a control. To every $u \in U_{ad}$ we relate $\varphi_h(u) \in Y_h$, the unique function satisfying

$$a(\varphi_h(u), z_h) = \int_\Omega \left(\frac{\partial a_0}{\partial y}(x, y_h(u)) \varphi_h(u) + \frac{\partial L}{\partial y}(x, y_h(u)) \right) z_h \, dx +$$

$$\int_\Gamma \left(\frac{\partial b_0}{\partial y}(x, y_h(u)) \varphi_h(u) + \frac{\partial l}{\partial y}(x, y_h(u), u) \right) z_h \, d\sigma(x) \;\; \forall z_h \in Y_h.$$

The following approximation properties are essential to study the approximation of the control problem. They follow from real interpolation. See Brenner and Scott [2, Section 12.3] and [5]. Proof of inequality (9) is more technical. It is done adapting the proof of Aubin-Nietsche Lemma to semilinear equations as in [5] and taking into account the $H^{3/2}(\Omega)$ regularity of the solution of the Neumann problem with data in $L^2(\Gamma)$. A full proof will appear in the forthcoming paper [6].

THEOREM 6 *(i) For every $u \in H^{1/2}(\Gamma)$ there exists $C > 0$, depending continuously on $\|u\|_{H^{1/2}(\Gamma)}$, such that*

$$\|y_u - y_h(u)\|_{H^s(\Omega)} + \|\varphi_u - \varphi_h(u)\|_{H^s(\Omega)} \leq Ch^{2-s} \text{ for all } 0 \leq s \leq 1,$$

and

$$\|y_u - y_h(u)\|_{L^2(\Gamma)} + \|\varphi_u - \varphi_h(u)\|_{L^2(\Gamma)} \leq Ch^{3/2}. \tag{9}$$

(ii) For every $u \in L^2(\Gamma)$ there exists $C_0 > 0$, depending continuously on $\|u\|_{L^2(\Gamma)}$, such that

$$\|y_u - y_h(u)\|_{H^s(\Omega)} + \|\varphi_u - \varphi_h(u)\|_{H^s(\Omega)} \leq C_0 h^{3/2-s} \text{ for all } 0 \leq s \leq 1.$$

(iii) For every $u_1, u_2 \in L^2(\Gamma)$ there exists a constant $C > 0$ such that

$$\|y_{u_1} - y_{u_2}\|_{H^1(\Omega)} + \|y_h(u_1) - y_h(u_2)\|_{H^1(\Omega)} +$$

$$\|\varphi_{u_1} - \varphi_{u_2}\|_{H^1(\Omega)} + \|\varphi_h(u_1) - \varphi_h(u_2)\|_{H^1(\Omega)} \leq C\|u_1 - u_2\|_{L^2(\Gamma)}.$$

(iv) Moreover, if $u_h \rightharpoonup u$ weakly in $L^2(\Gamma)$, then $y_h(u_h) \to y_u$ and $\varphi_h(u_h) \to \varphi_u$ strongly in $C(\bar{\Omega})$.

4.2 Discrete optimal control problem

We have several choices to write a discrete optimal control problem. Set

$$U_h^0 = \{u \in L^\infty(\Gamma) \mid u_{|(x_\Gamma^j, x_\Gamma^{j+1})} \in \mathcal{P}_0 \text{ for } 1 \leq j \leq N(h)\},$$

$$U_h^1 = \{u \in C(\Gamma) \mid u_{|(x_\Gamma^j, x_\Gamma^{j+1})} \in \mathcal{P}_1 \text{ for } 1 \leq j \leq N(h)\}.$$

and, following Hinze [9], we can semidiscretize the problem and take $U_h^2 = L^2(\Gamma)$. The corresponding approximated control problems are, for $i \in \{0, 1, 2\}$,

$$(\mathrm{P}_h^i) \begin{cases} \min J_h(u_h) = \displaystyle\int_\Omega L(x, y_h(u_h)(x))\, dx + \int_\Gamma l(x, y_h(u_h)(x), u_h(x))\, d\sigma(x), \\ \text{subject to } (y_h(u_h), u_h) \in Y_h \times U_h^{ad,i} \text{ satysfying (8)}, \end{cases}$$

where $U_h^{ad,i} = U_h^i \cap U^{ad}$.

The first order optimality conditions can be written as follows:

THEOREM 7 *Fix $i \in \{0, 1, 2\}$ and assume that $\bar{u}_h$ is a local optimal solution of (P_h^i). Then there exist $\bar{y}_h$ and $\bar{\varphi}_h$ in Y_h satisfying*

$$a(\bar{y}_h, z_h) = \int_\Omega a_0(x, \bar{y}_h) z_h \, dx + \int_\Gamma (b_0(x, \bar{y}_h) + \bar{u}_h) z_h \, dx \quad \forall z_h \in Y_h,$$

$$a(\bar{\varphi}_h, z_h) = \int_\Omega \left(\frac{\partial a_0}{\partial y}(x, \bar{y}_h)\bar{\varphi}_h + \frac{\partial L}{\partial y}(x, \bar{y}_h) \right) z_h \, dx +$$

$$\int_\Gamma \left(\frac{\partial b_0}{\partial y}(x, \bar{y}_h)\bar{\varphi}_h + \frac{\partial l}{\partial y}(x, \bar{y}_h, \bar{u}_h) \right) z_h \, d\sigma(x) \quad \forall z_h \in Y_h,$$

$$\int_\Gamma \left(\bar{\varphi}_h + \frac{\partial l}{\partial u}(x, \bar{y}_h, \bar{u}_h) \right) (u_h - \bar{u}_h) \, d\sigma(x) \geq 0 \ \forall u_h \in U_h^{ad,i}. \quad (10)$$

REMARK 8 *At this point, we can show the difficulty introduced by the fact that U_h^1 is formed by continuous piecewise linear functions instead of piecewise constant functions. To make a clear presentation, let us assume for a while that $l(x, y, u) = \ell(x, y) + \frac{\Lambda}{2}u^2$. In the case where U_h^0 is formed by piecewise constant functions, we get from (10) that*

$$\bar{u}_{h|(x_\Gamma^j, x_\Gamma^{j+1})} = Proj_{[\alpha,\beta]} \left(-\frac{1}{\Lambda} \int_{x_\Gamma^j}^{x_\Gamma^{j+1}} \bar{\varphi}_h(x) d\sigma(x) \right).$$

Comparing this representation of $\bar{u}_h$ with (6) we can prove that $\bar{u}_h \to \bar{u}$ strongly in $L^\infty(\Gamma)$; see [7].

Since we are considering piecewise linear controls in the present paper, no such pointwise projection formula can be deduced. We only can say that $\bar{u}_h$ is the convex projection of $-\frac{1}{\Lambda}\bar{\varphi}_h(x)$. More precisely, $\bar{u}_h$ is the solution of problem

$$\min_{v_h \in U_h} \|\bar{\varphi}_h + \Lambda v_h\|^2_{L^2(\Gamma)}. \quad (11)$$

This makes the analysis of the convergence more difficult than in [7]. In particular, we can prove that $\bar{u}_h \to \bar{u}$ strongly in $L^2(\Gamma)$, but this convergence cannot be obtained in $L^\infty(\Gamma)$ in an easy way as done in [7]. The reader is also referred to [8] for the study of problem (11).

We next can state a convergence result.

THEOREM 9 *Fix $i \in \{0, 1, 2\}$. For every $h > 0$ let $\bar{u}_h$ be a solution of (P_h^i). Then there exist subsequences $\{\bar{u}_h\}_{h>0}$ converging in the weak* topology of $L^\infty(\Gamma)$ that will be denoted in the same way. If $\bar{u}_h \rightharpoonup \bar{u}$ in the mentioned topology, then $\bar{u}$ is a solution of (P) and*

$$\lim_{h\to 0} J_h(\bar{u}_h) = J(\bar{u}) \ \ and \ \ \lim_{h \to 0} \|\bar{u} - \bar{u}_h\|_{L^2(\Gamma)} = 0.$$

Moreover, for $i \in \{0, 2\}$, $\lim_{h\to 0} \|\bar{u} - \bar{u}_h\|_{L^\infty(\Gamma)} = 0$.

The main result of the paper is the following.

THEOREM 10 *Let* $\bar{u}$ *be a solution of problem (P) such that* $J''(\bar{u})v^2 > 0$ *holds for all* $v \in C_{\bar{u}} \setminus \{0\}$ *and* $\bar{u}_h^i$ *a sequence of solutions of* (P_h^i) *converging in* $L^2(\Gamma)$ *to* $\bar{u}$. *Then*

1. *There exists a constant* $C > 0$ *and* $h_0 > 0$ *such that for* $0 < h < h_0$ $\|\bar{u} - \bar{u}_h^0\|_{L^2(\Gamma)} \le Ch$;
2. $\displaystyle\lim_{h\to 0} \frac{\|\bar{u} - \bar{u}_h^1\|_{L^2(\Gamma)}}{h} = 0$;
3. *For every* $0 < \varepsilon < 1/2$ *there exists a constant* $C > 0$ *and* $h_0 > 0$ *such that for* $0 < h < h_0$, $\|\bar{u} - \bar{u}_h^2\|_{L^2(\Gamma)} \le Ch^{3/2-\varepsilon}$.

In many practical cases when we make the full discretization using continuous piecewise linear controls ($i = 1$), the order of convergence observed for the controls in $L^2(\Gamma)$ is $h^{3/2}$. Let us show why. We will make two assumptions that are fulfilled in many situations:

(Q1) $l(x, y, u) = \ell(x, y) + e(x)u + \dfrac{\Lambda}{2}u^2$, where $\Lambda > 0$ and

- the function $\ell : \Gamma \times \mathbf{R} \longrightarrow \mathbf{R}$ is Lipschitz with respect to the first component, of class C^2 with respect to the second variable, $\ell(\cdot, 0) \in L^1(\Gamma)$, $\dfrac{\partial^2 \ell}{\partial y^2}(\cdot, 0) \in L^\infty(\Gamma)$ and for all $M > 0$ there exists a constant $C_{\ell,M} > 0$ such that

$$\left|\frac{\partial \ell}{\partial y}(x_2, y) - \frac{\partial \ell}{\partial y}(x_1, y)\right| \le C_{\ell,M}|x_2 - x_1|,$$

$$\left|\frac{\partial^2 \ell}{\partial y^2}(x, y_2) - \frac{\partial^2 \ell}{\partial y^2}(x, y_1)\right| \le C_{\ell,M}|y_2 - y_1|,$$

  for a.e. $x, x_i \in \Gamma$ and $|y|, |y_i| \le M$, $i = 1, 2$;

- the function $e : \Gamma \to \mathbf{R}$ is Lipschitz and satisfies the following approximation property: there exists $C_e > 0$ such that $\|e - \Pi_h e\|_{L^2(\Gamma)} \le C_e h^{3/2}$. This assumption is not very constraining. Although it is not true for Lipschitz functions in general, it is true for a very wide class of functions. For instance for Lipschitz functions that are piecewise in $H^{3/2}(\Gamma)$.

(Q2) If we name $\Gamma_s = \{x \in \Gamma : \bar{u}(x) = \alpha \text{ or } \bar{u}(x) = \beta\}$, then the number of points in $\partial\Gamma_s$ –the boundary of Γ_s in the topology of Γ– is finite.

THEOREM 11 *Suppose (Q1) and (Q2) are satisfied. Let* $\bar{u}$ *be a solution of problem (P) such that* $J''(\bar{u})v^2 > 0$ *holds for all* $v \in C_{\bar{u}} \setminus \{0\}$ *and* $\bar{u}_h$ *a sequence of solutions of* (P_h^1) *converging in* $L^2(\Gamma)$ *to* $\bar{u}$. *Then there exists* $C > 0$ *such that*

$$\|\bar{u}_h - \bar{u}\|_{L^2(\Gamma)} \leq Ch^{3/2}$$

REMARK 12 *Using the inverse inequality*

$$\|u_h\|_{L^\infty(\Gamma)} \leq Ch^{-1/2}\|u_h\|_{L^2(\Gamma)} \text{ for all } u_h \in U_h^1$$

We can get error estimates for continuous piecewise linear approximations for the control. This is important, since we have not been able to establish even uniform convergence up to now

About the proof of Theorem 10. For $i = 0$ see [7]. Using second order sufficient conditions, we prove that there exists $\nu > 0$ and $h_1 > 0$ such that for all $0 < h < h_1$

$$\nu\|\bar{u} - \bar{u}_h\|^2_{L^2(\Gamma)} \leq (J'(\bar{u}_h) - J'(\bar{u}))(\bar{u}_h - \bar{u}). \tag{12}$$

This is the most difficult part since we do not have uniform convergence of the states and we cannot apply the same techniques as in [3, 7].

Using (12) and first order optimality conditions (5) and (10) we have that

$$\nu\|\bar{u}_h - \bar{u}\|^2_{L^2(\Gamma)} \leq (J_h'(\bar{u}_h) - J'(\bar{u}))(u_h^* - \bar{u}) +$$

$$+J'(\bar{u})(u_h^* - \bar{u}) + (J_h'(\bar{u}_h) - J'(\bar{u}_h))(\bar{u} - \bar{u}_h).$$

For $i = 1$, $u_h^* = \Pi_h\bar{u}$ is the unique function in U_h^1 such that $\Pi_h\bar{u}(x_\Gamma^j) = \bar{u}(x_\Gamma^j)$ for $j = 1, \ldots, N(h)$. In this case first and second terms are of order $o(h^2)$. For $i = 2$ (semidiscretization), $u_h^* = \bar{u}$ and the first two terms are zero. Third term is more difficult. Since we do not know yet if $\{\bar{u}_h\}$ is bounded in $H^{1/2}(\Gamma)$ a direct proof as in the distributed case (see [3]) would lead to a bad estimate. With a small turnaround, we can prove that for every $\rho > 0$ and every $0 < \varepsilon \leq 1/2$ there exists $C_{\rho,\varepsilon} > 0$ independent of h such that the third term can be estimated by

$$\left(C_{\rho,\varepsilon}h^{3/2-\varepsilon} + \rho\|\bar{u}_h - \bar{u}\|_{L^2(\Gamma)}\right)\|\bar{u}_h - \bar{u}\|_{L^2(\Gamma)}.$$

We must take ρ small enough to conclude the proof.□

References

[1] N. Arada, E. Casas, and F. Tröltzsch. Error estimates for the numerical approximation of a semilinear elliptic control problem. *Comput. Optim. Appl.*, 23(2):201–229, 2002.

[2] S. C. Brenner and L. R. Scott. *The mathematical theory of finite element methods*, volume 15 of *Texts in Applied Mathematics*. Springer-Verlag, New York, 1994.

[3] E. Casas. Using piecewise linear functions in the numerical approximation of semilinear elliptic control problems. *AiCM*, 2005.

[4] E. Casas and M. Mateos. Second order optimality conditions for semilinear elliptic control problems with finitely many state constraints. *SIAM J. Control Optim.*, 40(5):1431–1454 (electronic), 2002.

[5] E. Casas and M. Mateos. Uniform convergence of the FEM. Applications to state constrained control problems. *Comput. Appl. Math.*, 21(1):67–100, 2002. Special issue in memory of Jacques-Louis Lions.

[6] E. Casas and M. Mateos. Error estimates for the numerical approximation of Neumann control problems. Submitted.

[7] E. Casas, M. Mateos, and F. Tröltzsch. Error estimates for the numerical approximation of boundary semilinear elliptic control problems. *Computational Optimization and Applications*, 31(2):193–219, 2005.

[8] E. Casas and J.-P. Raymond. The stability in $W^{s,p}(\Gamma)$ spaces of the L^2-projections on some convex sets of finite element function spaces. To appear.

[9] M. Hinze. A variational discretization concept in control constrained optimization: the linear-quadratic case. *Comput. Optim. Appl.*, 30(1):45–61, 2005.

[10] C. Meyer and A. Rösch. Superconvergence properties of optimal control problems. *SIAM Journal on Control and Optimization*, 43(3):970–985, 2005.

AUTONOMOUS UNDERWATER VEHICLES: SINGULAR EXTREMALS AND CHATTERING

M. Chyba [1] and T. Haberkorn [1]
[1] *Department of Mathematics, University of Hawaii, Honolulu, Hawaii 96822,*
*{mchyba,haberkor}@math.hawaii.edu**

Abstract In this paper, we consider the time minimal problem for an Autonomous Underwater Vehicle. We investigate, on a simplified model, the existence of singular extremals and discuss their optimality status. Moreover, we prove that singular extremals corresponding to the angular acceleration are of order 2. We produce in this case a semi-canonical form of our Hamiltonian system and we can conclude the existence of chattering extremals..

keywords: Underwater Vehicles, Time optimal, Singular Extremals, Chattering.

1. Introduction

In this paper, we consider the time minimal problem for autonomous underwater vehicle. We first describe the general model and give the equations of motion. However, due to the space limitation of this paper, and to highlight the properties of our system without tedious and lengthly calculations, we restrict ourselves to a simplified model. The general situation will be described in a forthcoming article. Based on the maximum principle we define bang-bang and singular trajectories. A first analysis of the singular extremals of the simplified model has been done in [2] and the existence of chattering extremals for a given pair of initial and final configurations at rest was discussed in [4] using numerical computations. In this paper, first we extend the previous results under the assumption that the vehicle is symmetric, in particular we show that if an extremal is bang-bang with respect to one of the linear velocities but singular in the two other controls then it cannot be time optimal. Moreover, the major addition to our previous results is that we provide a semi-canonical form in the situation of a singular extremal with respect to the angular velocity and bang with respect to the linear velocities. From [13] the existence of chat-

*Authors supported by NSF grant DMS-030641

Please use the following format when citing this chapter:

Chyba, M., and Haberkorn, T., 2006, in IFIP International Federation for Information Processing, Volume 202, Systems, Control, Modeling and Optimization, eds. Ceragioli, F., Dontchev, A., Furuta, H., Marti, K., Pandolfi, L., (Boston: Springer), pp. 103-113.

tering extremals in our problem follows. We supplement our conclusions with numerical computations.

2. The Model

We consider a 6 degrees of freedom Autonomous Underwater Vehicle (AUV). We denote by $\eta = (x, y, z, \phi, \theta, \psi)^t$ the position and orientation of the vehicle in the earth-fixed reference frame, where (ϕ, θ, ψ) are the classical Euler angles. The velocities $\nu = (u, v, w, p, q, r)^t$ are taken with respect to the body-fixed frame (see Figure 1). In the sequel χ represents the state and velocity variables: $\chi = (\eta, \nu)$.

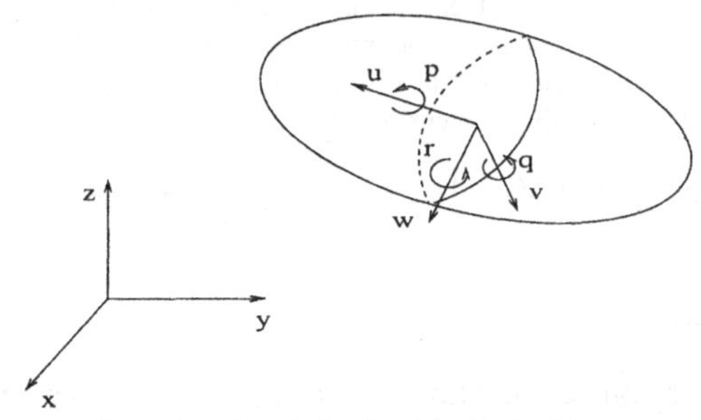

Figure 1. Earth-fixed and body-fixed frame

With those notations, the equations of motion can be written as, see [7]:

$$\begin{aligned} \dot{\eta} &= J(\eta)\nu \\ \dot{\nu} &= M^{-1}(\tau - [C(\nu) + D(\nu)]\nu - g(\eta)) \end{aligned} \tag{1}$$

where $J(\eta)$ is the transformation matrix between the earth-fixed coordinate system and the body-fixed one. M is the inertia matrix of the vehicle. The matrix C takes into account for the Coriolis and centrifugal forces and is assumed to be skew-symmetric. The matrix D stands for the damping forces. The column vector g represents the restoring forces and moment: gravity and buoyancy. Finally, τ is the control.

To reflect the fact that the thrusters have limited power, we assume:

$$\tau \in \mathcal{U} = \{\tau \in \mathbf{R}^6 | \alpha_i \le \tau_i \le \beta_i, i = 1, \cdots, 6\}, \tag{2}$$

where $\alpha_i < 0 < \beta_i$. An admissible control is a measurable bounded function τ defined on some time interval $[0, T]$ such that $\tau(t) \in \mathcal{U}$ for a.e. $t \in [0, T]$.
Due to the space limitation of this paper, we will restrict ourselves to a simplified model. The general situation will be studied in a forthcoming article. First, we assume the vehicle to be totally symmetric (hence M is diagonal), with coinciding centers of gravity and buoyancy (hence g is zero). The center of the

body-fixed frame is taken at the center of gravity and the damping forces are neglected. Finally, we restrict the vehicle to move in the xz-plane. Under these assumptions, the equations of motion become:

$$\begin{aligned} \dot{x} &= u\cos\theta + w\sin\theta && (3)\\ \dot{z} &= w\cos\theta - u\sin\theta && (4)\\ \dot{\theta} &= q && (5)\\ \dot{u} &= -qw + \tau_1/m && (6)\\ \dot{w} &= uq + \tau_3/m && (7)\\ \dot{q} &= \tau_5/I && (8) \end{aligned}$$

where m is the mass of the vehicle and I is the first inertia momentum around the y-axis. For our numerical computations, we will assume $m = 20$ and $I = 0.1$. We have the following bounds on the controls:

$$|\tau_i| \leq 1,\ i = 1, 3, 5$$

REMARK 1 *The equations of motion for the velocity variables (6) -(8) are equivalent to the equations of motion for the control problem of rotation of a rigid body around its gravicenter assuming that the body is axially symmetric. The existence of chattering trajectories for the time optimal problem in this case has been studied in [13]. We provide here a generalization of these results by allowing translational motions to the body.*

3. The Maximum Principle

Let χ_0, $\chi_T \in \mathbf{R}^6$ be prescribed initial and final configurations for our simplified model. The maximum principle for the minimum time problem, [9], states that if $\bar{\tau} : [0, T] \rightarrow \mathcal{U}$ is an admissible time optimal control such that the corresponding trajectory $\bar{\chi}(.)$ steers the vehicle from χ_0 to χ_T, then there exists an absolutely continuous vector function $\lambda : [0, T] \rightarrow \mathbf{R}^6$, $\lambda(t) \neq 0$ for all t, such that, almost everywhere in $[0, T]$ we have:

$$\begin{aligned} \dot{\bar{\chi}}_i(t) &= \frac{\partial H}{\partial \lambda_i}(\bar{\chi}(t), \lambda(t), \bar{\tau}(t)) && (9)\\ \dot{\lambda}_i(t) &= -\frac{\partial H}{\partial \chi_i}(\bar{\chi}(t), \lambda(t), \bar{\tau}(t)) && (10) \end{aligned}$$

for $i = 1, \cdots, 6$, where

$$H = \lambda^t \begin{bmatrix} J(\eta)\nu \\ M^{-1}(\tau - C(\nu)\nu) \end{bmatrix} \quad (11)$$

is the Hamiltonian function. Additionally, the maximum condition holds:

$$H(\bar{\chi}(t), \lambda(t), \bar{\tau}(t)) = max_{\omega \in \mathcal{U}} H(\bar{\chi}(t), \lambda(t), \omega) \tag{12}$$

Moreover, the maximum of the Hamiltonian is constant along the solutions of (9-10) and must satisfy $H(\lambda(t), \bar{\chi}(t), \bar{\tau}(t)) = \lambda_0$, $\lambda_0 \geq 0$. A solution (χ, λ, τ) of the maximum principle, in the sense just stated, is called an extremal and the vector function $\lambda(\cdot)$ an adjoint vector. For our model, the Hamiltonian system (9-10) gives the following dynamics for the adjoint vector:

$$\begin{aligned}
\dot{\lambda}_x &= 0 && (13)\\
\dot{\lambda}_z &= 0 && (14)\\
\dot{\lambda}_\theta &= -\lambda_x(w\cos\theta - u\sin\theta) + \lambda_z(u\cos\theta + w\sin\theta) && (15)\\
\dot{\lambda}_u &= -\lambda_x \cos\theta + \lambda_z \sin\theta - \lambda_w q && (16)\\
\dot{\lambda}_w &= -\lambda_x \sin\theta - \lambda_z \cos\theta + \lambda_u q && (17)\\
\dot{\lambda}_q &= -\lambda_\theta + \lambda_u w - \lambda_w u && (18)
\end{aligned}$$

The maximization condition (12) implies that the control $\bar{\tau}$ satisfies ($i = 1, 3, 5$):

$$\tau_i(t) = -1,\ if\ \varphi_i(t) < 0\ and\ \tau_i(t) = 1\ if\ \varphi_i(t) > 0 \tag{19}$$

with $\varphi_1 = \lambda_u/m$, $\varphi_3 = \lambda_w/m$, $\varphi_5 = \lambda_q/I$. Those functions are called the switching functions. If along the extremal, we have $\varphi_i(t) \neq 0$ almost everywhere on $[0, T]$, then the corresponding control τ_i takes its values in $\{-1, 1\}$ and the extremal is said to be bang-bang with respect to τ_i. However, if there exists a non trivial interval $[t_1, t_2]$ on which $\varphi_i(t)$ is identically zero, the control τ_i is said to be singular on $[t_1, t_2]$ and the corresponding extremal is called τ_i-singular. Let q be such that $\frac{d^{2q}}{dt^{2q}}\phi_i$ is the lowest order derivative in which τ_i appears explicitely with a nonzero coefficient. We defined q as the order of the singular control τ_i. The above definition lies on the fact that it is a well known result, see [10] for instance, that a singular control τ_i first appears explicitely in an even order derivative of ϕ_i. Assume that the component u_i of the control is bang-bang; then $t_s \in [0, T]$ is called a switching time for u_i if, for each interval of the form $]t_s - \varepsilon, t_s + \varepsilon[\bigcap[t_1, t_2]$, $\varepsilon > 0$, there is no constant c such that $u_i(t) = c$ for almost all $t \in [t_1, t_2]$.

Optimal trajectories are usually concatenations of bang-bang pieces with singular pieces. The Maximum principle does not give any direct information about these concatenations and it is known that in some cases such a concatenation may involve chattering, see [8] for the first example of such optimal trajectories.

4. Singular trajectories

In this section we focus on the singular trajectories for our model. It is a well known result, see for instance [1, 11], that for a fully actuated controlled mechanical systems all controls cannot be singular at the same time. In other words, we have:

THEOREM 2 *Along an extremal, all controls τ_1, τ_3 and τ_5 cannot be singular at the same time.*

The next step is to study extremals such that 2 controls are singular at the same time.

THEOREM 3 *Along a τ_1, τ_5-singular extremal, the control τ_3 is bang-bang with at most one switching. Along such extremal, we have that the singular controls are identically 0 as well as the angular velocity q while the linear velocity u is constant. Moreover, if τ_3 has one switching then we can deduce $u \equiv 0$ along the extremal. Furthermore, these extremals are not time optimal.*

Proof. Along a τ_1, τ_5-singular extremals we must have $\lambda_u \equiv 0$ and $\lambda_q \equiv 0$. Theorem 2 implies that τ_3 cannot be singular. In [2] it is proved that there is at most one switching for the control τ_3 and that if there is one we must have $u = q \equiv 0$ and both singular components of the controls are identically zero. Assume now τ_3 is constant along the extremal. It follows from (16) and from $\lambda_u \equiv 0$ that $q\lambda_w = -\lambda_x \cos\theta + \lambda_z \sin\theta$. Coupled with the fact that $\ddot{\lambda}_q = w(\lambda_x \cos\theta - \lambda_z \sin\theta) + qw\lambda_w - \lambda_w \tau_1/m \equiv 0$ we obtain $\lambda_w \tau_1 = 0$. In particular, since $\lambda_w \neq 0$ we have that τ_1 is constant and we can differentiate λ_q up to order 6:

$$\lambda_q^{(6)} = -4q^3(\lambda_x \sin\theta + \lambda_z \cos\theta)\tau_3 - 3q(-2\lambda_x \cos\theta + 2\lambda_z \sin\theta - q\lambda_w)\frac{\tau_3 \tau_5}{mI}$$

Using the second derivative of λ_q and of λ_u, we can reduce the previous expression to $\lambda_q^{(6)} = -5q^2\lambda_w \frac{\tau_3\tau_5}{mI}$. However, we must have $\lambda_q^{(6)} \equiv 0$. Using the fact that τ_3 is constant, we deduce that $q^2\tau_5 = 0$. Since moreover $\dot{q} = \tau_5/I$, the two last equalities implies that along the extremal we have $\tau_5 \equiv 0$. Now, if we combine the first and the third derivative of λ_u we get $2q^3\lambda_w = -(\lambda_x \sin\theta + \lambda_z \cos\theta)\frac{\tau_5}{I}$ which leads to $q \equiv 0$. The equations of motion imply then that u is constant along the extremal and we can deduce that $\ddot{\lambda}_w \equiv 0$. For the non optimality of such extremals, one only has to see that they correspond to a pure translational motion that, according to [3], are not optimal.

REMARK 4 *This theorem generalizes the results of [2] to the symmetric situation. Moreover, the theorem remains true if we exchange the role of τ_1 and τ_3.*

Since the non-optimality of the extremals studied in Theorem 3 as been established, let us provide numerically a fastest trajectory. Along a pure translational motion, if we start from χ_0 at the origin and follow the τ_1, τ_5-singular extremal for 2 seconds, we arrive at $\chi_f = (0, 0.1, 0, 0, 0.1, 0)$. Applying an optimization method to find a time optimal trajectory steering χ_0 to χ_f gives a minimum time $t_f^{\min} \approx 1.7604$ s which is clearly better than the one of the τ_1, τ_5-singular extremal. Figure 2 shows the computed time-optimal trajectory. Figure 2 only shows a singularity of τ_5 at the middle of the trajectory which corresponds to the time when the vehicle is properly oriented to use efficiently both its translational thrusters.

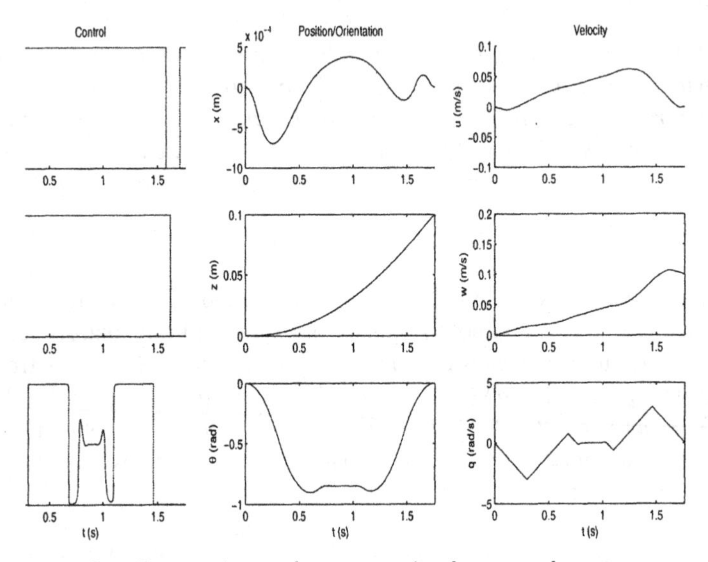

Figure 2. *Faster trajectory than a τ_1, τ_5-singular extremal*

REMARK 5 *The optimization method consists in the discretization of the optimal control problem using a Heun integration scheme for the dynamic and discretizing both state and control. Then we write the obtained nonlinear optimization problem in* AMPL *[6] and apply the large-scale nonlinear optimization software* IpOpt *[12].*

We now turn to the situation when τ_5 is bang-bang while the two other controls are singular.

THEOREM 6 *Along a τ_1, τ_3-singular extremals, the control τ_5 is bang-bang with at most one switching.*

Proof. This result is a direct consequence of the form of the differential equations for the adjoint vector.

It could be interesting to exhibit a time optimal trajectory that is τ_1, τ_3-singular since it is kind of counter intuitive not to use the full translational power. Actually, if we numerically integrate the Hamiltonian system starting, for instance, from $\chi_0 = (0,0,0,0,0,0)$, take $\tau_1 = -\tau_3 = 0.5$ and $\lambda_x(0) = \lambda_z(0) = \lambda_u(0) = \lambda_w(0) = 0$ we will have a trajectory that is extremal. Additionally, we set $\lambda_\theta(0) = 0.5$ and $\lambda_q(0) = 2$ and integrate from $t_0 = 0$ to $T = 8$ s to get one switching time for τ_5 at $t_s = 4$ s. The switching is from 1 to -1 (because $\lambda_q(t) = -\lambda_\theta(0)t + \lambda_q(0)$). Now, if we use the obtained final configuration χ_T and apply an optimization method to solve the minimum time transfer from χ_0 to χ_T we find that the minimum transfer time is $t_f^{\min} \approx 8.0005$ s which corresponds to T up to the unavoidable round-off errors. Figure 3 shows the two trajectories.

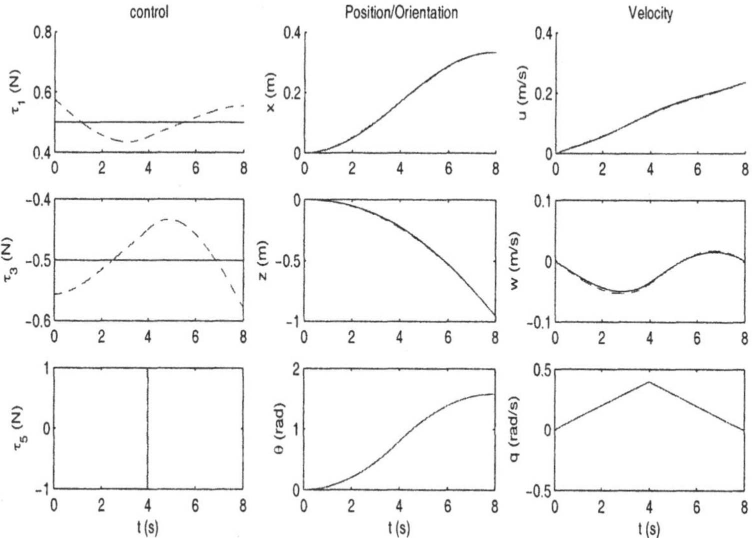

Figure 3. *Comparison between the τ_1, τ_3-singular trajectory computed by integration of the Hamiltonian system (the dotted one) and the one obtained by application of the nonlinear optimization solver (the plain one).*

On Figure 3 one can see that the two trajectories share some similarities but that they are not the same. Indeed, the control strategy is not the same since we do not have τ_1 and τ_3 constant along the time optimal trajectory computed numerically at the contrary of what was prescribed in our integration of the Hamiltonian system. However, we have a switching at the same time on both trajectories.

Let us now turn to extremals that are singular for one control only. For simplicity we will assume the two other controls to be constant.

THEOREM 7 *Assume τ_3, τ_5 to be bang along an extremal. Then, if the extremal is τ_1-singular it is of infinite order.*

Proof. To determine the order of a τ_1-singular extremal, we must find the order of the derivative at which τ_1 arises for the first time. A simple computation show that:

$$\dot{\lambda}_u = -\lambda_x \cos\theta + \lambda_z \sin\theta - \lambda_w q \tag{20}$$

$$\ddot{\lambda}_u = 2q(\lambda_x \sin\theta + \lambda_z \cos\theta) - \lambda_w \frac{\tau_5}{I} \tag{21}$$

We can easily generalize and see that the only control that can appear in any derivatives of the function λ_u is the control τ_5 and the result follows.

REMARK 8 *Theorem 7 still holds if we exchange the role of τ_1 and τ_3.*

Things work differently for the control τ_5.

THEOREM 9 *Assume τ_1, τ_3 to be bang along an extremal. Then, if the extremal is τ_5-singular, it is of intrinsic order 2.*

Proof. Let us define $H = H_0 + \varphi_5 \tau_5$. Then, the first two derivatives of the switching function $\varphi_5 = \lambda_q / I$ are given by:

$$\frac{d\varphi_5}{dt} = \{H_0, \varphi_5\}, \qquad \frac{d^2\varphi_5}{dt^2} = \{H_0, \{H_0, \varphi_5\}\} + \tau_5\{\varphi_5, \{H_0, \varphi_5\}\}. \tag{22}$$

Since, $\{\varphi_5, \{H_0, \varphi_5\}\} = 0$, we can keep differentiating and we have

$$\frac{d^3\varphi_5}{dt^3} = \{H_0, \{H_0, \{H_0, \varphi_5\}\}\}, \tag{23}$$

$$\frac{d^4\varphi_5}{dt^4} = \{H_0, \{H_0, \{H_0, \{H_0, \varphi_5\}\}\}\} + \tau_5\{\varphi_5, \{H_0, \{H_0, \{H_0, \varphi_5\}\}\}\}. \tag{24}$$

Explicitely, we have

$$\ddot{\lambda}_q = \frac{-\lambda_w \tau_1 + \lambda_u \tau_3}{m}, \qquad \lambda_q^{(4)} = A_4 + \tau_5 B_4 \tag{25}$$

where $A_4 = q(2\lambda_x \cos\theta - 2\lambda_z \sin\theta + q\lambda_w)\frac{\tau_1}{mI} + q(2\lambda_x \sin\theta + 2\lambda_z \cos\theta - q\lambda_u)\frac{\tau_3}{mI}$ and $B_4 = -\frac{\lambda_w \tau_3 + \lambda_u \tau_1}{mI^2}$. Since along the extremal the controls τ_1 and τ_3 are bang, and such that at least one is nonsingular, we have that $B_4 = -\frac{|\lambda_w| + |\lambda_u|}{mI^2}$ is strictly negative. It follows that the extremal is of order 2 and from (25), we have that the order is intrinsic (which means that the coefficient in front of τ_5 is identically 0 and not only along the singular extremal).

REMARK 10 *We saw that $B_4 < 0$ so the strict Kelley's necessary condition, [13], for a singular control to be optimal holds.*

Following the procedure described in [13] we put the Hamiltonian system (9-10) under a semi-canonical form in the case of a τ_5-singular extremal, the other controls being assumed to be bang. Since this extremal is of intrinsic order two, the four first coordinates of the new system (δ, ξ) are $\delta = (\delta_1, \delta_2, \delta_3, \delta_4)$ where $\delta_1 = \lambda_q/I, \delta_2 = \dot{\lambda}_q/I = (-\lambda_\theta + \lambda_u w - \lambda_w u)/I, \delta_3 = \ddot{\lambda}_q/I = (-\lambda_w \tau_1 + \lambda_u \tau_3)/(mI), \delta_4 = \lambda_q^{(3)}/I = ((\lambda_x \cos\theta - \lambda_z \sin\theta + q\lambda_w)\tau_3 - (\lambda_x \sin\theta + \lambda_z \cos\theta - \lambda_u q)\tau_1)/(mI)$.
To completely define a new coordinate system we need to find ξ such that the Jacobian $D(\delta, \xi)/D(\chi, \lambda)$ is of full rank. We suggest

$$\left\{\begin{array}{lllllll} \xi_1 & = & x & , & \xi_5 & = & \lambda_x \cos\theta - \lambda_z \sin\theta \\ \xi_2 & = & z & , & \xi_6 & = & \lambda_x \sin\theta + \lambda_z \cos\theta \\ \xi_3 & = & \theta & , & \xi_7 & = & \lambda_\theta \\ \xi_4 & = & u & , & \xi_8 & = & \lambda_u \end{array}\right. \tag{26}$$

The corresponding $D(\delta, \xi)/D(\chi, \lambda)$ is then of full rank and the canonical Hamiltonian system is

$$\left\{\begin{array}{l} \dot{\delta}_1 = \delta_2, \qquad \dot{\delta}_2 = \delta_3, \qquad \dot{\delta}_3 = \delta_4 \\ \dot{\delta}_4 = q(2\xi_5 + 2\xi_6 + q\lambda_w - q\xi_8)/(mI) - (\xi_8 \tau_1 + \lambda_w \tau_3)\tau_5/(mI^2) \\ \dot{\xi}_1 = \xi_4 \cos\xi_3 + w \sin\xi_3, \qquad \dot{\xi}_2 = w\cos\xi_3 - \xi_4 \sin\xi_8, \qquad \dot{\xi}_3 = q \\ \dot{\xi}_4 = -qw + \tau_1/m, \qquad \dot{\xi}_5 = -q\xi_6, \qquad \dot{\xi}_6 = q\xi_5 \\ \dot{\xi}_7 = \xi_4\xi_6 - w\xi_5, \qquad \dot{\xi}_8 = -\xi_5 - \lambda_w q \end{array}\right. \tag{27}$$

where

$$\left\{\begin{array}{lll} \lambda_w & = & (\xi_8 \tau_3 - mI\delta_3)/\tau_1 \\ w & = & (\xi_7 + \lambda_w \xi_4 - I\delta_2)/\xi_8 \\ q & = & (mI\delta_4 - \xi_5 \tau_3 + \xi_6 \tau_1)/(\lambda_w \tau_3 + \xi_8 \tau_1) \end{array}\right. \tag{28}$$

Since we were able to reduce our system to a semi-canonical form and since from Remark 10 the Kelley's condition holds, it is now possible to apply the results from [13]. In there, the authors describe the behavior of all extremals in the vicinity of the singular manifold S defined by $S = \{(\chi, \lambda) | \delta_i = 0, i = 1, \cdots, 4\}$. In particular, we can conclude that for each point (χ_0, λ_0) in S there exists a 2-dimensional integral manifold of the Hamiltonian system such that the behavior of the solutions inside this manifold is similar to the one of the chattering arcs in the Fuller problem (we have as well the existence of untwisted chattering arcs). To be more specific, there is a one-parameter family of solutions of system (27) reaching (χ_0, λ_0) in a finite time such that the switching times for the control τ_5 are infinite and follow a geometric progression. It is important to notice that this result does not imply the optimality of such trajectories neither it does imply that, assuming τ_1, τ_3 constants, every junction between a τ_5-singular and a τ_5 bang-bang trajectory involves chattering in the control. Indeed, in order for such junction to have chattering the control must be discontinuous, see [5]. this realized for instance of at the junction the angular velocity vanishes: $q = 0$. On Figure 4, we produce an example of a smooth junction between a τ_5-singular and a τ_5 bang-bang solution. In [4], the reader can see an example of a chattering junction computed in the non symmetrical case.

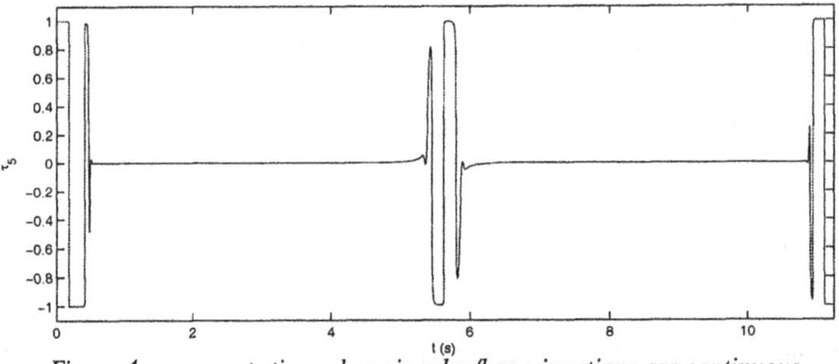

Figure 4. τ_5 *w.r.t. time when singular/bang junctions are continuous.*

Figure 4 corresponds to χ_0 at the origin and $\chi_f = (2, 1, 0.5, 0, 0, 0)$. Controls τ_1 and τ_3 are purely bang-bang.

5. Conclusion

In this paper, we produce an example of a system involving both, singular extremals of infinite order and singular extremals of intrinsic second order. The existence of chattering extremals has been proved based on a semi-canonical form for our Hamiltonian system along the singular extremals of order 2. In a forthcoming article we will study a more general model of underwater vehicle and will describe precisely the conditions under which a controlled mechanical system has singular extremals of order 2 and then is a potential candidate for chattering solutions. Another direction we intend to take is to use high order necessary condition to discuss the optimality of singular extremals of infinite order.

References

[1] M. Chyba, N.E. Leonard, E.D. Sontag. Singular trajectories in the multi-input time-optimal problem: Application to controlled mechanical systems. *Journal on Dynamical and Control Systems* 9(1):73-88, 2003.

[2] M. Chyba, N.E. Leonard, E.D. Sontag). Optimality for underwater vehicles. In *Proceedings of the 40th IEEE Conf. on Decision and Control*, Orlando, 2001.

[3] M. Chyba. Underwater vehicles: a surprising non time-optimal path.In *42th IEEE Conf. on Decision and Control*, Maui 2003.

[4] M.Chyba, H. Maurer, H.J. Sussmann, G. Vossen. *Underwater Vehicles: The Minimum Time Problem.* In *Proceedings of the 43th IEEE Conf. on Decision and Control*, Bahamas, 2004.

[5] J.P. Mcdanell and W.F. Powers. *Necessary conditions for joining optimal singular and nonsingular subarcs.* SIAM J. Control, 4(2): 161-173, 1971.

[6] R. Fourer, D.M. Gay, B.W. Kernighan. AMPL: A Modeling Language for Mathematical Programming. Duxbury Press, Brooks-Cole Publishing Company,1993.

[7] T.I.Fossen. Guidance and control of ocean vehicles. Wiley, New York, 1994

[8] A.T. Fuller. Study of an optimum nonlinear control system. *J. Electronics Control*, 15:63-71, 1963.

[9] L.S. Pontryagin, B. Boltyanski, R. Gamkrelidze, E. Michtchenko. The Mathematical Theory of Optimal Processes. Interscience, New-York, 1962.

[10] H.M. Robbins. A generalized Legendre-Clebsh condition for the singular cases of optimal control. *IBM J. Res. Develop* 11:361-372, 1967.

[11] E.D. Sontag, H.J. Sussmann. Time-Optimal Control of Manipulators. In *IEEE Int. Conf. on Robotics and Automation.*, San Francisco: 1962-1697, 1986.

[12] A. Waechter, L. T. Biegler. On the Implementation of an Interior-Point Filter-Line Search Algorithm for Large-Scale Nonlinear Programming. *Research Report RC 23149, IBM T.J. Watson Research Center*, Yorktown, New-York.

[13] M.I. Zelikin, V.F. Borisov. Theory of Chattering Control. *Birkhäuser*, Boston, 1994.

OPTIMIZATION OF A COUPLED FORCE INTENSITY BY HOMOGENIZATION METHODS

M. Codegone [1]

[1] *Dipartimento di Matematica, Politecnico di Torino, Corso Duca degli Abruzzi 24, 10129 Torino (Italy), marco.codegone@polito.it*

Abstract In this paper, in the framework of a problem related to an elastic non homogeneous medium, we deal with a periodic coupled force $(f(x)/\varepsilon^\alpha)\vec{F}(x/\varepsilon)$ with intensity of order $1/\varepsilon^\alpha$. The parameter ε is connected with the period of the non homogeneity of the medium and with the periodicity of the coupled force. The determination of the parameter α is the target of our study to obtain an effect in the microscopic equation. The homogenization technique is used in order to study the equation: $-(\partial/\partial x_j)(a_{ijkh}(x/\varepsilon)e_{kh}(\vec{u}^{\varepsilon,\alpha})) = (f(x)/\varepsilon^\alpha)F_i(x/\varepsilon) + G_i(x, x/\varepsilon)$, where $G_i(x, x/\varepsilon)$ is the volume applied force. The limit, when $\varepsilon \to 0$, of $\vec{u}^{\varepsilon,\alpha}(x)$, in the sense of two scale convergence, is $(\vec{u}^{0,\alpha}(x), \vec{u}^{1,\alpha}(x,y))$ and the microscopic equation becomes: $-(\partial/\partial y_j)(a_{ijkh}(y)e_{khx}(\vec{u}^{0,\alpha}(x))) - (\partial/\partial y_j)(e_{khy}(\vec{u}^{1,\alpha}(x,y))) = f(x)F_i(y)$ if $\alpha = 1$, $-(\partial/\partial y_j)(a_{ijkh}(y)e_{khx}(\vec{u}^{0,\alpha}(x)) + e_{khy}(\vec{u}^{1,\alpha}(x,y))) = 0$ if $0 < \alpha < 1$. When $\alpha > 1$ the solutions are not uniformly bounded respect to ε.

keywords: Coupled forces, Homogenization, Elasticity.

1. Introduction

In this paper, in the framework of a problem related to an elastic non homogeneous medium (see [2] and [7]), we deal with a periodic coupled force

$$\frac{f(x)}{\varepsilon^\alpha}\vec{F}\left(\frac{x}{\varepsilon}\right)$$

with intensity of order $1/\varepsilon^\alpha$ (see [10], [11] and [12]). The parameter ε is related to the period of the non homogeneity of the medium and to the periodicity of the coupled force. For other questions related to the homogenization theory see [3], [5] and [6]. The determination of the parameter α is the target of our study to obtain an effect in the microscopic equation. Let Ω be a regular domain of $\mathbf{R}^3$ and let $a_{ijkh}(y)$ be the symmetric, coercive and Y-periodic elasticity tensor:

$$a_{ijkh} = a_{jikh} = a_{khij}, \quad a_{ijkh}\,e_{ij}\,e_{kh} \geq c\,e_{ij}\,e_{ij}. \tag{1}$$

Please use the following format when citing this chapter:

Codegone, M., 2006, in IFIP International Federation for Information Processing, Volume 202, Systems, Control, Modeling and Optimization, eds. Ceragioli, F., Dontchev, A., Furuta, H., Marti, K., Pandolfi, L., (Boston: Springer), pp. 115-125.

We denote by $\vec{F}(y)$ the couple force Y-periodic and such that:

$$\frac{1}{|Y|}\int_Y \vec{F}(y)\mathrm{d}y = 0 \tag{2}$$

and by $f(x)$ a continuous scalar function representing the intensity of the coupled force. We indicate by $G(x,y)$ the external force Y-periodic in the y variable. The problem is expressed by the equation:

$$-\frac{\partial}{\partial x_j}\left(a_{ijkh}\left(\frac{x}{\varepsilon}\right)e_{kh}\left(\vec{u}^{\varepsilon,\alpha}\right)\right) = \frac{f(x)}{\varepsilon^\alpha}F_i\left(\frac{x}{\varepsilon}\right) + G_i\left(x,\frac{x}{\varepsilon}\right), \tag{3}$$

in Ω, with the boundary condition

$$\vec{u}^{\varepsilon,\alpha}|_{\partial\Omega} = 0. \tag{4}$$

We prove that, if $0 < \alpha \leq 1$ the solutions $u^{\varepsilon,\alpha}(x)$ are uniformly bounded:

$$\|\vec{u}^{\varepsilon,\alpha}(x)\|_{(H_0^1(\Omega))^3} \leq C.$$

Then taking as test function:

$$\vec{v}(x,y) = \vec{v}^0(x) + \varepsilon\vec{v}^1(x,\frac{x}{\varepsilon})$$

with:

$$\vec{v}^0(x) \in (H_0^1(\Omega))^3$$

and with

$$\vec{v}^1(x,y) \in (L^2(\Omega;H^1_{per}/\mathbf{R}))^3$$

the two-convergence limit gives us the following answers:
if $\alpha = 1$ then

$$\int_\Omega\int_Y a_{ijkh}(y)\left[e_{khx}(\vec{u}^{0,1}(x)) + e_{khy}(\vec{u}^{1,1}(x,y))\right]$$
$$\left[e_{khx}(\vec{v}^0(x)) + e_{khy}(\vec{v}^1(x,y))\right]\mathrm{d}x\mathrm{d}y +$$
$$-\int_\Omega\int_Y f(x)\vec{F}(y)\vec{v}^1(x,y)\mathrm{d}x\mathrm{d}y - \int_\Omega\int_Y \vec{G}(x,y)\vec{v}^0(x)\mathrm{d}x\mathrm{d}y = 0,$$

moreover if $0 < \alpha < 1$ then

$$\int_\Omega\int_Y a_{ijkh}(y)\left[e_{khx}(\vec{u}^{0,\alpha}(x)) + e_{khy}(\vec{u}^{1,\alpha}(x,y))\right]$$
$$\left[e_{khx}(\vec{v}^0(x)) + e_{khy}(\vec{v}^1(x,y))\right]\mathrm{d}x\mathrm{d}y +$$

$$-\int_{\Omega}\int_{Y}\vec{G}(x,y)\vec{v}^{0}(x)\mathrm{d}x\mathrm{d}y=0,$$

An integration by parts of the preceding formulae gives the following two limit problems: if $\alpha = 1$ then

$$\begin{cases} -\frac{\partial}{\partial y_j}\left[a_{ijkh}(y)\left[e_{khx}(\vec{u}^{0,1}(x))+e_{khy}(\vec{u}^{1,1}(x,y))\right]\right]-f(x)F_i(y)=0\\ -\frac{\partial}{\partial x_j}\left[\int_Y a_{ijkh}(y)\left[e_{khx}(\vec{u}^{0,1}(x))+e_{khy}(\vec{u}^{1,1}(x,y))\right]\mathrm{d}y\right]+\\ \qquad -\int_Y G_i(x,y)\mathrm{d}y=0,\end{cases}$$

and $0 < \alpha < 1$ then

$$\begin{cases} -\frac{\partial}{\partial y_j}\left[a_{ijkh}(y)\left[e_{khx}(\vec{u}^{0,\alpha}(x))+e_{khy}(\vec{u}^{1,\alpha}(x,y))\right]\right]=0\\ -\frac{\partial}{\partial x_j}\left[\int_Y a_{ijkh}(y)\left[e_{khx}(\vec{u}^{0,\alpha}(x))+e_{khy}(\vec{u}^{1,\alpha}(x,y))\right]\mathrm{d}y\right]+\\ \qquad -\int_Y G_i(x,y)\mathrm{d}y=0.\end{cases}$$

In order to discuss the influence of the coupled force at the microscopic level we remarque and discuss the difference of the two microscopic equations: if $\alpha = 1$

$$-\frac{\partial}{\partial y_j}\left(a_{ijkh}(y)\,e_{khx}\left(\vec{u}^{0,1}(x)\right)+e_{khy}\left(\vec{u}^{1,1}(x,y)\right)\right)=f(x)F_i(y) \tag{5}$$

and if $0 < \alpha < 1$

$$-\frac{\partial}{\partial y_j}\left(a_{ijkh}(y)\,e_{khx}\left(\vec{u}^{0,\alpha}(x)\right)+e_{khy}\left(\vec{u}^{1,\alpha}(x,y)\right)\right)=0. \tag{6}$$

We remark that the equation (6) is a standard microscopic equation, and there the external force $f(x)F_i(y)$ does not appear. On the contrary the equation (5) has the external force $f(x)F_i(y)$, then we can conclude that the optimum coupled force that works at microscopic level is of type $(f(x)/\varepsilon^{\alpha})\,\vec{F}(x/\varepsilon)$, with $\alpha = 1$. In this case the macroscopic equation becomes:

$$-\frac{\partial}{\partial x_j}\left(a_{ijkh}^{hom}e_{khx}\left(\vec{u}^{0,1}(x)\right)\right)=\Phi_{ij}\frac{\partial}{\partial x_j}f(x)+\tilde{G}_i(x) \tag{7}$$

where Φ_{ij} is obtained starting from the microscopic problem and where a_{ijkh}^{hom} are the classical homogenized coefficients and

$$\tilde{G}_i(x)=\frac{1}{|Y|}\int_Y G(x,y)\,\mathrm{d}\,x.$$

2. Statement of the problem with $\varepsilon > 0$ and $\alpha > 0$

In order to study the equation (3), we introduce the variational formulation in the Hilbert space $\left(H_0^1(\Omega)\right)^3$. In this setting the existence is proved of a unique solution for ε and α fixed and a uniform bound is obtained, with respect to ε if $0 < \alpha \le 1$. In view of these targets we multiply the equation (3) by $\vec{v}^0(x)$ with

$$\vec{v}^0(x) \in \left(H_0^1(\Omega)\right)^3$$

and we integrate by parts. We get the following weak formulation:

$$\text{find} \quad \vec{u}^{\varepsilon,\alpha}(x) \in \left(H_0^1(\Omega)\right)^3$$

such that

$$\int_\Omega a_{ijkh}(x/\varepsilon) e_{khx}(\vec{u}^{\varepsilon,\alpha}) e_{ijx}(\vec{v}^0) \mathrm{d}\, x = \tag{8}$$

$$= \int_\Omega \frac{f(x)}{\varepsilon^\alpha} \vec{F}\left(\frac{x}{\varepsilon}\right) \cdot \vec{v}^0 \, \mathrm{d}\, x + \int_\Omega \vec{G}\left(x, \frac{x}{\varepsilon}\right) \cdot \vec{v}^0 \, \mathrm{d}\, x$$

We suppose that a_{ijkh} satisfy the hypothesis of ellipticity and symmetry:

$$a_{ijkh} = a_{jikh} = a_{khij} \qquad m\xi_{ij}\xi_{ij} \le a_{ijkh}\xi_{ij}\xi_{kh} \le M\xi_{ij}\xi_{ij}, \tag{9}$$

with $m > 0$. Moreover $a_{ijkh}(y)$ is Y-periodic of period Y and

$$f(x)\vec{F}(y), \; \vec{G}(x,y) \in \left(L^2\left(\Omega; C_{per}^0(Y)\right)\right)^3 \tag{10}$$

where $C_{per}^0(Y)$ is the space of continuous periodic functions of period Y. It is known (see [1]) that, if (10) holds, then it follows that:

$$f(x)\vec{F}(x/\varepsilon), \; \vec{G}(x, x/\varepsilon) \in \left(L^2(\Omega)\right)^3.$$

By hypothesis (9), we get:

$$m e_{ij}(\vec{u}^{\varepsilon,\alpha}) e_{ij}(\vec{u}^{\varepsilon,\alpha}) \le \int_\Omega a_{ijkh}(x/\varepsilon) e_{kh}(\vec{u}^{\varepsilon,\alpha}) e_{ij}(\vec{u}^{\varepsilon,\alpha}) \mathrm{d}\, x,$$

then, by Korn inequalities (see, for instance, [4] and [8]) we get:

$$\|\vec{u}^{\varepsilon,\alpha}(x)\|^2_{(H_0^1(\Omega))^3} \le C \int_\Omega a_{ijkh}(x/\varepsilon) e_{kh}(\vec{u}^{\varepsilon,\alpha}) e_{ij}(\vec{u}^{\varepsilon,\alpha}) \mathrm{d}\, x. \tag{11}$$

On the other hand, we need to have an estimate of the right-hand side of the equation (8):

$$\left| \int_\Omega \frac{f(x)}{\varepsilon^\alpha} \vec{F}\left(\frac{x}{\varepsilon}\right) \cdot \vec{u}^{\varepsilon,\alpha} \, \mathrm{d}\, x + \int_\Omega \vec{G}\left(x, \frac{x}{\varepsilon}\right) \cdot \vec{u}^{\varepsilon,\alpha} \, \mathrm{d}\, x \right|$$
$$\le K \|\vec{u}^{\varepsilon,\alpha}\|_{(H_0^1(\Omega))^3}. \tag{12}$$

Taking $\vec{G}(x, x/\varepsilon) \in \left(L^2\left(\Omega; C^0_{per}\right)\right)^3$ we get

$$\int_\Omega \vec{G}\left(x, \frac{x}{\varepsilon}\right) \cdot \vec{u}^{\,\varepsilon,\alpha} \, \mathrm{d}x \leq K_1 \left\|\vec{u}^{\,\varepsilon,\alpha}\right\|_{\left(H_0^1(\Omega)\right)^3}.$$

The complete inequality (12) will be proved in the next section.

The estimates (11) and (12), by Lax-Milgram lemma, give the existence of a unique solution, for fixed ε and $0 < \alpha \leq 1$, of the weak formulation (8) and moreover give a uniform, with respect to ε, estimate of following norm of $\vec{u}^{\,\varepsilon,\alpha}$:

$$\left\|\vec{u}^{\,\varepsilon,\alpha}(x)\right\|_{\left(H_0^1(\Omega)\right)^3} \leq K. \tag{13}$$

The estimate (13) permits us to apply the result of the two scale convergence method (see, for instance, [1], [4] and [9]) to obtain that there exist two functions

$$\vec{u}^{\,0,\alpha}(x) \in \left(H_0^1(\Omega)\right)^3 \quad \text{and} \quad \vec{u}^{\,1,\alpha}(x,y) \in \left(L^2(\Omega, H^1_{\mathrm{per}}/\mathbf{R})\right)^3$$

such that:

$$\vec{u}^{\,\varepsilon,\alpha}(x) \rightharpoonup \vec{u}^{\,0,\alpha}(x) \quad \text{in} \quad \left(H_0^1(\Omega)\right)^3 \quad \text{weakly}$$

and

$$e_{ij}(\vec{u}^{\,\varepsilon,\alpha}) \rightharpoonup e_{ijx}(\vec{u}^{\,0,\alpha}(x)) + e_{ijy}(\vec{u}^{\,1,\alpha}(x,y))$$

where we wrote:

$$e_{ijx}(\vec{v}) = \frac{1}{2}\left(\frac{\partial v_i}{\partial x_j} + \frac{\partial v_j}{\partial x_i}\right) \quad \text{and} \quad e_{ijy}(\vec{v}) = \frac{1}{2}\left(\frac{\partial v_i}{\partial y_j} + \frac{\partial v_j}{\partial y_i}\right).$$

We introduce the Hilbert space

$$\mathbf{V} = \left(H_0^1(\Omega)\right)^3 \times \left(L^2(\Omega, H^1_{\mathrm{per}}/\mathbf{R})\right)^3, \tag{14}$$

with the scalar product

$$\left((\vec{u}^{\,0,\alpha}, \vec{u}^{\,1,\alpha}), (\vec{v}^{\,0}, \vec{v}^{\,1})\right) =$$

$$= \int_\Omega \int_Y \left[e_{ijx}(\vec{u}^{\,0,\alpha}) + e_{ijy}(\vec{u}^{\,1,\alpha})\right] \left[e_{ijx}(\vec{v}^{\,0}) + e_{ijy}(\vec{v}^{\,1})\right] \mathrm{d}x\mathrm{d}y. \tag{15}$$

Multiplying the equation (3) by

$$\vec{v}(x, x/\varepsilon) = \vec{v}^{\,0}(x) + \varepsilon \vec{v}^{\,1}(x, x/\varepsilon) \quad \text{with} \quad \vec{v}(x, x/\varepsilon) \in \mathbf{V}$$

the problem (8), in the context of the two scale convergence theory, can be written as follows

$$\int_\Omega a_{ijkh}(x/\varepsilon)e_{khx}(\vec{u}^{\varepsilon,\alpha}(x))\left[e_{ijx}(\vec{v}^0(x))+e_{ijy}(\vec{v}^1(x,x/\varepsilon))+\right.$$

$$\left.+\ \varepsilon e_{ijx}(\vec{v}^1(x,x/\varepsilon))\right]\mathrm{d}\,x=$$

$$=\int_\Omega \frac{f(x)}{\varepsilon^\alpha}\vec{F}\left(\frac{x}{\varepsilon}\right)\cdot\left[\vec{v}^0+\varepsilon\vec{v}^1(x,x/\varepsilon)\right]\mathrm{d}\,x+ \tag{16}$$

$$+\int_\Omega \vec{G}\left(x,\frac{x}{\varepsilon}\right)\cdot\left[\vec{v}^0+\varepsilon\vec{v}^1(x,x/\varepsilon)\right]\mathrm{d}\,x$$

3. Estimate of the term with the couple force

The coupled force $(f(x)/\varepsilon^\alpha)\vec{F}(x/\varepsilon)$ at the right-hand side of the equation (3) has an intensity that increases as ε goes to zero. Then is not evident that the term with the coupled force may be bounded:

$$\left|\frac{1}{\varepsilon^\alpha}\int_\Omega f(x)\vec{F}\left(\frac{x}{\varepsilon}\right)\cdot\vec{u}^{\varepsilon,\alpha}\,\mathrm{d}\,x\right|\le C_1\|\vec{u}^{\varepsilon,\alpha}\|_{\left(H_0^1(\Omega)\right)^3} \tag{17}$$

with C_1 independent of ε. The hypothesis (2) that $\vec{F}(x/\varepsilon)$ has zero mean value is essential. We prove the estimate (17), if

$$0<\alpha\le 1,$$

in the following lemma:

LEMMA 1 *Let* $f(x)\in C^1(\bar{\Omega})$, $0<\alpha\le 1$, $\vec{F}(y)\in\left(H^1_{per}(Y)\right)^3$ *and* $\vec{u}^\varepsilon\in\left(H_0^1(\Omega)\right)^3$ *with* Ω *bounded region of* $\mathbf{R}^n$ *and* $Y=[0,1]^n$. *Moreover let* $\vec{F}(y)$ *be periodic with period* Y *and suppose that the condition (2) of zero mean value holds. Then there exists a constant* C_1 *such that the estimate (17) is fulfilled.*

Proof. We consider the vector $\vec{y}_k\in\mathbf{Z}^n$ and any translation $\vec{y}_k+Y$ of the fundamental period $Y=[0,1]^n$. We scale down the translated period with the parameter ε and we say $Y_k^\varepsilon=\varepsilon(\vec{y}_k+Y)$. There are a finite number N of the Y_k^ε periods such that $\bigcup_{k=1}^N Y_k^\varepsilon\supseteq\Omega$. There are many period Y_k^ε that are strictly included in Ω, for instance $\bigcup_{k=1}^{N-T}Y_k^\varepsilon\subset\Omega$, moreover there are a number T of periods Y_k^ε such that $\{\bigcup_{k=N-T+1}^N Y_k^\varepsilon\}\bigcap\partial\Omega\neq\emptyset$. We now extend to zero the function $\vec{u}^{\varepsilon,\alpha}$ in the periods $Y_k^\varepsilon\bigcap\{\mathbf{R}^n\backslash\Omega\}$, with $N-T+1\le k\le N$. Now we decompose the integral over Ω in a sum of integral over the cells Y_k^ε:

$$\left|\frac{1}{\varepsilon^\alpha}\int_\Omega f(x)\vec{F}\left(\frac{x}{\varepsilon}\right)\cdot\vec{u}^{\varepsilon,\alpha}(x)\,\mathrm{d}\,x\right| \leq \frac{1}{\varepsilon^\alpha}\sum_{k=1}^{N}\left|\int_{Y_k^\varepsilon} f(x)\vec{F}\left(\frac{x}{\varepsilon}\right)\cdot\vec{u}^{\varepsilon,\alpha}(x)\,\mathrm{d}\,x\right| \tag{18}$$

In any period Y_k^ε we choose a point x_k where $f(x)$ attains the maximum and adding and subtracting $f(x_k)$ we get:

$$\begin{aligned} &\frac{1}{\varepsilon^\alpha}\sum_{k=1}^{N}\left|\int_{Y_k^\varepsilon} f(x)\vec{F}\left(\frac{x}{\varepsilon}\right)\cdot\vec{u}^{\varepsilon,\alpha}(x)\,\mathrm{d}\,x\right| \leq \\ &\leq \frac{1}{\varepsilon^\alpha}\sum_{k=1}^{N}\left|\int_{Y_k^\varepsilon} f(x_k)\vec{F}\left(\frac{x}{\varepsilon}\right)\cdot\vec{u}^{\varepsilon,\alpha}(x)\,\mathrm{d}\,x\right| + \\ &+ \frac{1}{\varepsilon^\alpha}\sum_{k=1}^{N}\int_{Y_k^\varepsilon} |f(x)-f(x_k)|\left|\vec{F}\left(\frac{x}{\varepsilon}\right)\right| |\cdot\vec{u}^{\varepsilon,\alpha}(x)|\,\mathrm{d}\,x \end{aligned} \tag{19}$$

In the k-cell Y_k^ε we take the mean value of $\vec{u}^{\varepsilon,\alpha}(x)$:

$$\mathcal{M}\left(\vec{u}^{\varepsilon,\alpha}(x)\right) = \frac{1}{|Y_k^\varepsilon|}\int_{Y_k^\varepsilon}\vec{u}^{\varepsilon,\alpha}(x)\,\mathrm{d}\,x. \tag{20}$$

By the condition (2) of zero mean value and by the preceding formula (20), the first term at the right-hand side of inequality (19) becomes

$$\begin{aligned} &\frac{1}{\varepsilon^\alpha}\sum_{k=1}^{N}\left|\int_{Y_k^\varepsilon} f(x_k)\vec{F}\left(\frac{x}{\varepsilon}\right)\cdot\vec{u}^{\varepsilon,\alpha}(x)\,\mathrm{d}\,x\right| = \\ &= \frac{1}{\varepsilon^\alpha}\sum_{k=1}^{N}\left|\int_{Y_k^\varepsilon} f(x_k)\vec{F}\left(\frac{x}{\varepsilon}\right)\cdot\left(\vec{u}^{\varepsilon,\alpha}(x)-\mathcal{M}\left(\vec{u}^{\varepsilon,\alpha}(x)\right)\right)\mathrm{d}\,x\right| \end{aligned} \tag{21}$$

Now we apply the Cauchy-Schwartz inequality in the right-hand side of (21) to obtain

$$\begin{aligned} &\frac{1}{\varepsilon^\alpha}\sum_{k=1}^{N}\left|\int_{Y_k^\varepsilon} f(x_k)\vec{F}\left(\frac{x}{\varepsilon}\right)\cdot\left(\vec{u}^{\varepsilon,\alpha}(x)-\mathcal{M}\left(\vec{u}^{\varepsilon,\alpha}(x)\right)\right)\mathrm{d}\,x\right| \leq \\ &\leq \frac{1}{\varepsilon^\alpha}\sum_{k=1}^{N}|f(x_k)|\left\|\vec{F}\left(\frac{x}{\varepsilon}\right)\right\|_{L^2(Y_k^\varepsilon)}\left\|\vec{u}^{\varepsilon,\alpha}(x)-\mathcal{M}\left(\vec{u}^{\varepsilon,\alpha}(x)\right)\right\|_{L^2(Y_k^\varepsilon)} \end{aligned} \tag{22}$$

In the last factor at the right-hand side of relation (22) we apply a Korn inequality for functions with zero mean value; it is known that in the Korn inequalities (as in the Poincaré-Wirtinger inequality for the scalar case, see for instance [4] and [8]) the constant $c(\varepsilon) = c \cdot \varepsilon$ is related to the measure ε of the region Y_k^ε

$$\|\vec{u}^{\varepsilon,\alpha}(x) - \mathcal{M}\left(\vec{u}^{\varepsilon,\alpha}(x)\right)\|_{L^2(Y_k^\varepsilon)} \leq$$
$$\leq \; c \cdot \varepsilon \left\| \sum_{i,j=1}^{n} e_{ij}\left(\vec{u}^{\varepsilon,\alpha}(x)\right) \right\|_{L^2(Y_k^\varepsilon)} \leq c \cdot \varepsilon \, \|\vec{u}^{\varepsilon,\alpha}(x)\|_{H^1(Y_k^\varepsilon)} \tag{23}$$

The inequalities (21), (22) and (23) give us

$$\frac{1}{\varepsilon^\alpha} \sum_{k=1}^{N} \left| \int_{Y_k^\varepsilon} f(x_k) \vec{F}\left(\frac{x}{\varepsilon}\right) \cdot \vec{u}^{\varepsilon,\alpha}(x) \, \mathrm{d}x \right| \leq$$
$$\leq \; \frac{1}{\varepsilon^\alpha} \sum_{k=1}^{N} |f(x_k)| \left\| \vec{F}\left(\frac{x}{\varepsilon}\right) \right\|_{L^2(Y_k^\varepsilon)} \cdot c \cdot \varepsilon \, \|\vec{u}^{\varepsilon,\alpha}\|_{H^1(Y_k^\varepsilon)} \leq$$
$$\leq \; \varepsilon^{1-\alpha} C_3 \, \{\max_{x \in \Omega} |f(x)|\} \sum_{k=1}^{N} \left\| \vec{F}\left(\frac{x}{\varepsilon}\right) \right\|_{L^2(Y_k^\varepsilon)} \sum_{k=1}^{N} \|\vec{u}^{\varepsilon,\alpha}\|_{H^1(Y_k^\varepsilon)} \leq$$
$$\leq \; \varepsilon^{1-\alpha} C_4 \, \{\max_{x \in \Omega} |f(x)|\} \left\| \vec{F}\left(\frac{x}{\varepsilon}\right) \right\|_{L^2(\Omega)} \|\vec{u}^{\varepsilon,\alpha}\|_{H^1(\Omega)} \leq$$
$$\leq \; \varepsilon^{1-\alpha} C_5 \, \|\vec{u}^{\varepsilon,\alpha}(x)\|_{H^1(\Omega)} \tag{24}$$

In view to obtain the estimate of the couple force, in the second term at the right-hand side of formula (19), we apply the Lipschtz inequality to the function $f(x) \in C^1$ in the cell Y_k^ε

$$|f(x) - f(x_k)| \leq C_6 \cdot \varepsilon \qquad x \in Y_k^\varepsilon$$

and then we have

$$\frac{1}{\varepsilon^\alpha} \sum_{k=1}^{N} \int_{Y_k^\varepsilon} |f(x) - f(x_k)| \left| \vec{F}\left(\frac{x}{\varepsilon}\right) \right| |\vec{u}^{\varepsilon,\alpha}(x)| \, \mathrm{d}x \leq$$
$$\leq \; \frac{1}{\varepsilon^\alpha} \cdot C_6 \cdot \varepsilon \sum_{k=1}^{N} \int_{Y_k^\varepsilon} \left| \vec{F}\left(\frac{x}{\varepsilon}\right) \right| |\vec{u}^{\varepsilon,\alpha}(x)| \, \mathrm{d}x. \tag{25}$$

Adding on the cell from 1 to N and applying the Cauchy-Schwartz inequality, we have:

$$\frac{1}{\varepsilon^\alpha} \cdot C_6 \cdot \varepsilon \sum_{k=1}^{N} \int_{Y_k^\varepsilon} \left| \vec{F}\left(\frac{x}{\varepsilon}\right) \right| |\vec{u}^{\varepsilon,\alpha}(x)| \, \mathrm{d}x \leq$$

$$\leq \varepsilon^{1-\alpha} C_6 \int_\Omega \left| \vec{F}\left(\frac{x}{\varepsilon}\right) \right| |\vec{u}^{\varepsilon,\alpha}(x)| \, \mathrm{d}x \leq$$

$$\leq \varepsilon^{1-\alpha} C_6 \left\| \vec{F}\left(\frac{x}{\varepsilon}\right) \right\|_{L^2(\Omega)} \|\vec{u}^{\varepsilon,\alpha}(x)\|_{L^2(\Omega)} \leq$$

$$\leq \varepsilon^{1-\alpha} C_6 \left\| \vec{F}\left(\frac{x}{\varepsilon}\right) \right\|_{L^2(\Omega)} \|\vec{u}^{\varepsilon,\alpha}(x)\|_{H^1(\Omega)} . \tag{26}$$

The relations (24) and (26) give a uniform estimate for any $\varepsilon \in (0,1]$ if $0 < \alpha \leq 1$. If $\alpha > 1$ the estimate is uniform only for $0 < k \leq \varepsilon \leq 1$, and we can not obtain a convergent subsequence of $\vec{u}^{\varepsilon,\alpha}(x)$ as ε goes to zero. The proof of the estimate of the couple force is then complete.

4. Two-scale convergence limit

A vector valued functions $\vec{u}^{\varepsilon,\alpha}(x)$ in $\left(L^2(\Omega)\right)^3$ is said to two-scale converge to a limit $\vec{u}^{0,\alpha}(x,y)$ belonging to $\left(L^2(\Omega \times Y)\right)^3$ if, for any function $\varphi(x,y)$ in $\left(\mathcal{D}\left(\Omega; C^\infty_{per}(Y)\right)\right)^3$, we have

$$\lim_{\varepsilon \to 0} \int_\Omega \vec{u}^{\varepsilon,\alpha}(x) \varphi\left(x, \frac{x}{\varepsilon}\right) \mathrm{d}x \longrightarrow \int_\Omega \int_Y \vec{u}^{0,\alpha}(x,y) \varphi(x,y) \mathrm{d}x \mathrm{d}y.$$

By estimate (13), the bounded sequence $\vec{u}^{\varepsilon,\alpha}(x)$ weakly converges in $\left(H^1_0(\Omega)\right)^3$ to a limit $\vec{u}^{0,\alpha}(x)$. Then (see, for instance [1], [4] or [7]), if $\vec{u}^{\varepsilon,\alpha}(x)$ two-scale converges to $\vec{u}^{0,\alpha}(x)$, there exists a function $\vec{u}^{1,\alpha}(x,y)$ in $\left(L^2(\Omega, H^1_{\mathrm{per}}(Y)/\mathbf{R})\right)^3$ such that, up to a subsequence, $e_{ijx}(\vec{u}^{\varepsilon,\alpha}(x))$ two-scale converges:

$$e_{ijx}(\vec{u}^{\varepsilon,\alpha}(x)) \rightharpoonup e_{ijx}(\vec{u}^{0,\alpha}(x)) + e_{ijy}(\vec{u}^{1,\alpha}(x,y)).$$

The weak formulation (16) of the problem (3), in the context of the two-scale convergence, gives the limit problem in the following variational form:

$$\text{find} \qquad (u^{0,\alpha}, u^{1,\alpha}) \in \left(H^1_0(\Omega)\right)^3 \times \left(L^2(\Omega, H^1_{\mathrm{per}}(Y)/\mathbf{R})\right)^3 = \mathbf{V}$$

such that

$$\int_\Omega \int_Y a_{ijkh}(y) \left[e_{khx}(\vec{u}^{0,\alpha}(x)) + e_{khy}(\vec{u}^{1,\alpha}(x,y))\right] \cdot$$

$$\cdot \left[e_{khx}(\vec{v}^0(x)) + e_{khy}(\vec{v}^1(x,y))\right] \mathrm{d}x\mathrm{d}y =$$

$$= \delta_{\alpha 1} \int_\Omega \int_Y f(x) \vec{F}(y) \cdot \vec{v}^1(x,y) \mathrm{d}x\mathrm{d}y + \int_\Omega \int_Y \vec{G}(x,y) \cdot \vec{v}^0(x) \mathrm{d}x\mathrm{d}y, \tag{27}$$

$\forall(v_0, v_1) \in \mathbf{V}$ and where

$$\delta_{\alpha 1} = \begin{cases} 1 & \text{if} \quad \alpha = 1 \\ 0 & \text{if} \quad \alpha \neq 1. \end{cases}$$

By the coercivity (9) of the bilinear form defined by the left-hand side of (27), we can apply the Lax-Milgram lemma and we have a well posed formulation of the limit two-scale problem. Integrating by parts the equation (27) in x and y and choosing $\vec{v}^0(x) = 0$ we get the microscopic equation

$$-\frac{\partial}{\partial y_j} \left\{ a_{ijkh}(y) \left[e_{khx} \left(\vec{u}^{0,\alpha}(x) \right) + e_{khy} \left(\vec{u}^{1,\alpha}(x,y) \right) \right] \right\} = f(x) \vec{F}_i(y) \quad (28)$$

By the linearity of the equation (28), we solve the following problem in the period Y

$$\begin{cases} -\dfrac{\partial}{\partial y_j} \left[a_{ijkh}(y)\, e_{khy} \left(\vec{w}^{lm,\alpha}(y) \right) \right] = \dfrac{\partial a_{ijlm}(y)}{\partial y_j} & \text{in } Y \\ \vec{w}^{lm,\alpha}(y) & Y - \text{periodic} \end{cases} \quad (29)$$

and, if $\alpha = 1$, we solve the further equation:

$$\begin{cases} -\dfrac{\partial}{\partial y_j} \left[a_{ijkh}(y)\, e_{khy} \left(\vec{v}^{\alpha}(y) \right) \right] = \vec{F}_i(y) & \text{in } Y \\ \vec{v}^{\alpha}(y) & Y - \text{periodic} \end{cases} \quad (30)$$

The solution of problems (29) and (30) gives the solution of the microscopic equation (28)

$$\vec{u}^{1,\alpha}(x,y) = e_{khx} \left(\vec{u}^{0,\alpha}(x) \right) \vec{w}^{kh,\alpha}(y) + \delta_{\alpha 1} f(x) \vec{v}^{\alpha}(y) \quad (31)$$

We remark that, at the microscopic level, the couple force has, if $\alpha = 1$, an important role.

As in classical homogenization we define

$$a_{ijkh}^{\text{hom},\alpha} = \int_Y a_{ijlm}(y) \left[\delta_{lk} \delta_{mh} + e_{lmy} \left(\vec{w}^{kh,\alpha}(y) \right) \right] \mathrm{d}y \quad (32)$$

Moreover we define

$$\Phi_{ij}^{\alpha} = \int_Y a_{ijkh}(y)\, e_{khy} \left(\vec{v}^{\alpha} \right) \mathrm{d}y \quad (33)$$

At this point we write the limit homogenized equation

$$\begin{cases} -\dfrac{\partial}{\partial x_j}\left[a_{ijkh}^{\mathrm{hom},\alpha}(y)e_{khx}\left(\vec{u}^{0,\alpha}(x)\right)\right] = \\ \qquad = \delta_{\alpha 1}\,\Phi_{ij}^{\alpha}\,\dfrac{\partial}{\partial x_j}\left(f(x)\right) + \displaystyle\int_Y \vec{G}_i(x,y) \quad \text{in} \quad \Omega \\ \vec{u}^{0,\alpha}(x) = 0 \qquad\qquad\qquad\qquad\qquad\qquad \text{on} \quad \partial\Omega \end{cases} \tag{34}$$

REMARK 2 *We remark that if $a_{ijkh}(y) = \bar{a}_{ijkh}$ is constant in the period Y, then Φ_{ij}^{α} in formula (33) is zero. In same way if the intensity $f(x)$ of the couple force is constant, then Φ_{ij}^{α} in formula (33) is zero. Moreover if $0 < \alpha < 1$, then Φ_{ij}^{α} in formula (33) is zero. Then the effect of the force $(f(x)/\varepsilon^{\alpha})F(y)$ may work at the microscopic level, if the medium is non homogeneous and $f(x)$ is non constant and if $\alpha = 1$.*

References

[1] G. Allaire. Homogenization and two scale convergence. *SIAM J. Math. Anal.* 23:1482-1518, 1992.

[2] A.Bensoussan, J.L. Lions, G. Papanicolaou. *Asymptotic Analysis for Periodic Structures.* North-Holland, 1978.

[3] V. Chiadò Piat, M. Codegone. Scattering problem in a perforated domain. *Rev. R. Acad. Cien. Serie A Mat.* 97:447-454, 2003.

[4] D. Cioranescu, P. Donato. *An Introduction to Homogenization.* Oxford University Press, Oxford, 1999.

[5] M. Codegone. Problème d'homogénéisation en théorie de la diffraction. *C.R. Acad.Sc. Paris.* 288:387-389, 1979.

[6] M. Codegone. G-Convergence and Scattering Problems. *Boll. U.M.I.* 6.1-A:367-375, 1982.

[7] V.V. Jikov, S.M. Kozlov, O.A. Oleinik. *Homogenization of Differential Operators and Integral Functionals.* Springer-Verlag, Berlin Heidelberg, 1994.

[8] O.A. Oleinik, A.S. Shamaev, G.A. Yosifian. *Mathematical problems in elasticity and homogenization.* Studies in mathematics and its applications, Elsevier Science Publishers, Amsterdam, 1992.

[9] G. Nguetseng. A general convergence result for a functional related to the theory of homogenization. *SIAM J. Math. Anal.* 20:608-623, 1989.

[10] J. Sanchez-Hubert, E. Sanchez-Palencia. *Vibration and Coupling of Continuous Systems Asymptotic Methods.* Springer-Verlag, Berlin, 1989.

[11] E. Sanchez-Palencia. Comportement local et macroscopique d'un type de milieux physiques hétérogènes. *Internat. J. Engrg. Sci.* 12:331-351, 1974.

[12] E. Sanchez-Palencia. *Non Homogeneous Media and Vibration Theory.* Lecture Notes in Physics, Springer-Verlag, Berlin, 1980.

A NUMERICAL STUDY FOR GROWING SANDPILES ON FLAT TABLES WITH WALLS

M. Falcone [1] and S. Finzi Vita[1]
[1] *Università di Roma La Sapienza, Dipartimento di Matematica, Roma, Italy,*
{*falcone,finzi*} *@mat.uniroma1.it**

Abstract We continue our study on the approximation of a system of partial differential equations recently proposed by Hadeler and Kuttler to model the dynamics of growing sandpiles on a flat bounded table. The novelty here is the introduction of (infinite) walls on the boundary of the domain and the corresponding modification of boundary conditions for the standing and for the rolling layers. An explicit finite difference scheme is introduced and new boundary conditions are analyzed. We show experiments in 1D and 2D which characterize the steady-state solutions.

keywords: granular matter, hyperbolic systems, finite differences schemes

1. Introduction

In this paper we continue our study on numerical methods for the simulation of growing sandpiles, started in [5] and [7] for the so-called open table problem (we refer to those papers for a list of recent references on the modelling of granular matter). The prototype situation is the following. A flat bounded table (a bounded open domain Ω of $\mathbb{R}^2$) is initially empty and the sand is poured on a subset of Ω according to the value of a nonnegative function $f(x,t)$ (the source). The pile grows in height with a slope which is always lower than a characteristic value, which naturally depends on the the physical properties of the granular matter.

Among the pde-models recently studied for the description of this phenomenon (see for example [1, 3, 8]), we will focus our attention on the one proposed by Hadeler and Kuttler [8], based on two interacting variables: the standing layer u and the rolling layer v. The relations between u and v are

*Paper written with financial support of the MIUR Project COFIN 2003 "Modellistica Numerica per il Calcolo Scientifico ed Applicazioni Avanzate".

Please use the following format when citing this chapter:

Falcone, M., and Finzi Vita, S., 2006, in IFIP International Federation for Information Processing, Volume 202, Systems, Control, Modeling and Optimization, eds. Ceragioli, F., Dontchev, A., Furuta, H., Marti, K., Pandolfi, L., (Boston: Springer), pp. 127-137.

described by the system of nonlinear partial differential equations

$$v_t = \nabla \cdot (v\nabla u) - (1 - |\nabla u|)v + f, \qquad \text{in } \Omega \times (0,T) \tag{1}$$
$$u_t = (1 - |\nabla u|)v, \qquad \text{in } \Omega \times (0,T) \tag{2}$$
$$u(\cdot,t) = 0 \quad \text{on } \partial\Omega\,, \qquad u(\cdot,0) = 0 \qquad \text{in } \Omega\,. \tag{3}$$

It is important to note that existence and uniqueness results are difficult to prove even when the table is "open", *i.e.* when the sand can fall down from every point of the boundary $\partial\Omega$ and we have the homogeneous boundary condition (3). However, in that situation, a partial characterization of the stationary solutions of that system has been recently given by Cannarsa and Cardialiaguet [2]. In [5] we proposed an approximation scheme for the above system analyzing its properties in the one dimensional case and showing that the scheme mimics several characteristics of the continuous model. More recently Finzi Vita [7] has compared this model with other approaches proposed by Aronsson, Evans and Wu [1] and by Prigozhin [10] and has developed several tests with the three approaches. It is interesting to note that although the three models have theoretically the same set of admissible equilibria the evolution in time of the piles is rather different.

In this paper we want to study a model for a growing pile on a table Ω whose boundary $\partial\Omega$ can be split into two parts: Γ_0 which is the subset of the boundary where the sand can fall down the table and Γ_1 which is the subset of the boundary where the sand is blocked by a wall. This corresponds to the assumption that the walls are sufficiently high to guarantee that the pile is not trespassing it (for simplicity let us assume that they have infinite height). In this situation we have to impose mixed boundary conditions on $\partial\Omega = \Gamma_0 \cup \Gamma_1$. The boundary condition on Γ_0 is trivial and corresponds to the usual homogeneous Dirichlet boundary condition on u in (3). The boundary condition on Γ_1 deserves some analysis and this will be done later in this paper. Also for the model with walls it is interesting to study the steady-state solutions of the two-layer model and several questions arise: what is the equilibrium configuration (if any) for u and v? is there any representation formula for them? what is the singular set for u? In this paper we will try to answer some of these questions, making use of the scheme introduced in [5] (modified with a suitable boundary conditions at the walls), also showing that such modified scheme produces reasonable results which seem to match the physical behavior of the growing piles. A partial analysis of the equilibrium configurations for the same problem (called the tray problem) as well as for the obstacle problem when the table is a surface $z = g(x, y)$ can be found in [9]. However, the main emphasis there is on the equivalence of different types of solutions with the maximal volume solution.

2. The two-layers model for the open table

Following the fundamental theories of de Gennes [3] on granular matter, Hadeler and Kuttler [8] introduced a system of two partial differential equations to govern the growth of a sandpile, seen as the superposition of two distinct layers, the standing and the rolling ones. In their model, an eikonal type equation for the standing layer u forming heaps and slopes is coupled to an advection type equation for the small rolling layer v running down the slope. This leads to system (1)-(2), which has to be complemented with the initial and the Dirichlet boundary conditions (3).

Note that for this model positive rolling layers are allowed during the evolution even before the corresponding standing layer becomes critical. The equilibria $(\overline{u}, \overline{v})$ of (1)-(3)for a constant in time source f solve the system

$$-\nabla \cdot (v\nabla u) = f\,, \quad (1-|\nabla u|)v = 0 \quad \text{in } \Omega\,, \quad u|_{\partial\Omega} = 0\,. \tag{4}$$

Solutions of (4) have been recently characterized by Cannarsa and Cardaliaguet [2] for the 2-dimensional problem by means of viscosity solution techniques. In their result the boundary $\partial\Omega$ is assumed to be sufficiently regular. In order to state their main result let us introduce some notations and definitions. Let $d(x) = \inf\{|x-z| : z \in \partial\Omega\}$ be the distance function from the boundary of Ω, let $\mathcal{R}$ be the set of discontinuity of ∇d (usually called the *ridge*) and for every $x \in \Omega \setminus \mathcal{R}$ let $k(x)$ denote the curvature of $\partial\Omega$ at the projection of x onto it (note that the projection is unique for every x not belonging to $\mathcal{R}$). Finally, let $\tau(x) = \min\{t \geq 0 : x + t\nabla d(x) \in \overline{\mathcal{R}}\}$ indicate the so-called normal distance to $\overline{\mathcal{R}}$.

THEOREM 1 *Let $\Omega \subset R^2$ be a bounded domain with C^2 boundary and $f \geq 0$ a continuous function in Ω. Then, there exists a solution of system (4) given by the pair (u, v), where $u = d$ in Ω, $v = 0$ in $\mathcal{R}$ and*

$$v(x) = \int_0^{\tau(x)} f(x + t\nabla d(x))\frac{1-(d(x)+t)k(x)}{1-d(x)k(x)}\,dt, \quad \forall x \in \Omega\backslash\overline{\mathcal{R}}\,. \tag{5}$$

Such a solution is unique in the sense that if (u', v') is another solution, then $v' = v$ in Ω and $u' = d$ in $\{x \in \Omega : v > 0\}$.

By the above result there exists a unique equilibrium solution v for any given source f, and an integral representation formula is given for it outside $\mathcal{R}$ (where $v = 0$). On the contrary, for u there is no uniqueness, although every solution must coincide with d in the set where $v > 0$. Then, if we denote by D_f the set of positivity of f in Ω, two distinct cases can occur:
(*a*) $\mathcal{R} \subset D_f$: the unique possible asymptotic equilibrium for u is essentially

given by the distance function d.
(b) $\mathcal{R}$ is not contained in D_f: the model has an equilibrium configuration $\overline{u}$ which also depends on the source intensity (as clearly shown by the numerical experiments discussed in [5]) and for which a rigorous mathematical characterization is not known. In this situation, that equilibrium configuration is larger then the one predicted by the model by Prygozhin [10], whereas in the first situation they coincide.

In [5] we have proposed a finite difference approximation for the system (1)-(3) in the one dimensional case, which for $\Omega = (0,1)$ corresponds to

$$v_t = (vu_x)_x - (1-|u_x|)v + f = vu_{xx} + v_x u_x - (1-|u_x|)v + f \quad (6)$$

$$u_t = (1-|u_x|)v \quad (7)$$

$$u(0,t) = u(1,t) = 0\,, \qquad u(x,0) = 0 \quad x \in \Omega\,. \quad (8)$$

Now let us consider the space grid $x_i = (i-1)h$, for $i = 1,..,N$ and $h = \Delta x = 1/(N-1)$. We denote by D^+u_i and D^-u_i respectively the right and the left discrete derivative for u at x_i.

Then we discretize the derivative of u at a node x_i by the difference with maximal absolute value, namely $Du_i \equiv \text{maxmod}(D^-u_i, D^+u_i)$, whereas the corresponding derivative of v is approximated by the upwind (with respect to Du_i) finite difference, *i.e.*

$$\overline{D}v_i \equiv \begin{cases} D^+v_i & \text{if } Du_i > 0\,,\ D^+u_i > 0 \\ D^-v_i & \text{if } Du_i < 0\,,\ D^-u_i < 0 \\ 0 & \text{otherwise}\,. \end{cases}$$

We discretize time by a fixed step Δt and use the standard notation, where z_i^n denotes the approximate value of a function $z(x,t)$ for $t = t_n \equiv n\Delta t$ and $x = x_i$. Using backward Euler for the time derivatives, we obtain our fully explicit scheme for the solution of (6)-(8)

$$v_i^{n+1} = v_i^n + \Delta t\left[v_i^n D^2 u_i^n + \overline{D}v_i^n Du_i^n - (1-|Du_i^n|)v_i^n + f_i\right] \quad (9)$$

$$u_i^{n+1} = u_i^n + \Delta t(1-|Du_i^n|)v_i^n \quad (10)$$

$$u_i^0 = v_i^0 = 0 \quad \forall i\,, \quad u_1^n = u_N^n = 0 \quad \forall n\,, \quad (11)$$

where $D^2u_i^n$ denotes the classical 3-point discretization of the second derivative u_{xx} at x_i. In [5], where several experiments have been reported, it has been proved that the discrete model mimics some properties of the continuous model if the Courant number $\lambda = \Delta t/\Delta x$ is sufficiently small. Namely, the approximation u^n is always nonnegative and monotonically increasing in time, the approximation v^n stays nonnegative and the bound $|Du^n| \leq 1$ is always satisfied. Moreover, using standard arguments one can show that the above scheme is consistent with the continuous system and that the local error is $O(\Delta x, \Delta t)$.

3. The two-layers model with walls

As we said in the introduction, the main difference with respect to the original model of the open table in [8] is the introduction of the wall on a subset of the boundary. Let us consider the one dimensional case and take $\Omega = (0,1)$, open at the left-hand side and with an infinite wall at $x = 1$, *i.e.* $\Gamma_0 = \{0\}$ and $\Gamma_1 = \{1\}$. In order to derive the correct boundary condition on Γ_1, we remark that at the equilibrium configuration we must have

$$0 = \frac{d}{dt}\int_0^1 (\overline{u} + \overline{v})\mathrm{d}x = \int_0^1 (\overline{u}_t + \overline{v}_t)\mathrm{d}x = \tag{12}$$

$$= \int_0^1 (\overline{v}\,\overline{u}_x)_x \mathrm{d}x + \int_0^1 f\mathrm{d}x = \overline{v}(1)\overline{u}_x(1) - \overline{v}(0)\overline{u}_x(0) + \int_0^1 f\mathrm{d}x\,, \tag{13}$$

where we have used (6) and (7) to obtain (13). Note that the last two terms in (13) have a precise physical meaning: the second is the sand falling down from the table at $x = 0$, the third the sand falling on the table from the source in the unit time. If the system is in equilibrium, the incoming sand has to balance the sand leaving the table. Then, the necessary boundary condition to be satisfied at $x = 1$ will be the complementarity equation

$$\overline{v}(1)\overline{u}_x(1) = 0\,. \tag{14}$$

Then, reasoning as in [8], it is easy to determine explicit formulas for the equilibria of such modified problem. If we denote by $D_f = [x_1, x_2]$ the support of the source term, we can distinguish between two cases:
(*a*) $x_2 = 1$ (D_f touches the wall): the unique possible asymptotic equilibrium for u is essentially given by the distance function from Γ_0, and we have

$$u(x) = x\,, \quad v(x) = \int_x^1 f(z)dz\,, \quad \forall x \in \Omega\,.$$

(*b*) $x_2 < 1$ (no sand flows along the wall): then the equilibria are uniquely determined in $[0, x_2]$, and we have

$$u(x) = x\,, \quad v(x) = \int_x^{x_2} f(z)dz\,, \quad \forall x \in [0, x_2]\,,$$

whereas in $(x_2, 1]$ we can only say that $v = 0$. The system is not able to characterize u where $v = 0$.

We see that condition (14) is satisfied in both cases simply because $v(1) = 0$, but such reduced condition would not be the right one for the dynamic model, since it would block the growth of u at the wall (see (7)). We have then considered (14) as the boundary condition on Γ_1 for the evolutive two-layers system, even if this is not the only possible choice.

From the numerical point of view, we had to modify the scheme (9)-(11) at x_N in order to implement the wall boundary condition (14). We tried various ways to implement that condition, some of them unsatisfactory:

- *If* $Du_N^n > 0$ *then* $v_N^n = 0$. When $x_2 = 1$ the scheme finds the correct equilibria, but the growth process is not uniform near the wall. When $x_2 < 1$, the stationary u is linear in $(x_2, 1)$.
- *If* $Du_N^n > 0$ *then* $v_N^n = 0$, *else* $u_N^n = u_{N-1}^n$. The dynamics is now correct, except for the values at the wall in the case $x_2 = 1$.
- *If* $Du_N^n > 0$ *then* $v_N^n = 0$, *else* $u_N^n = (4u_{N-1}^n - u_{N-2}^n)/3$. The second order difference now produces correct values on the wall in both cases. The stationary u is nonlinear in $(x_2, 1)$.

We have then used in the tests of Section 4.1 the scheme (9)-(11) replacing the homogeneous Dirichlet boundary condition $u_N^n = 0$ by

$$\text{if } Du_N^n > 0 \text{ then } v_N^n = 0, \text{ else } u_N^n = (4u_{N-1}^n - u_{N-2}^n)/3\,, \quad \forall n\,, \tag{15}$$

for different choices of the source support $D_f = [x_1, x_2]$.

Let us now examine the 2-dimensional model. With similar arguments the right boundary condition at Γ_1 for the equilibria can be derived using Green's theorem, obtaining

$$\int_{\Gamma_1} v\,\frac{\partial u}{\partial n}\,d\sigma = 0\,. \tag{16}$$

Again we have chosen for the dynamic model the point-wise condition $v\frac{\partial u}{\partial n} = 0$ on Γ_1. However, in two dimensions this problem is much more complex than the open table problem. The extremal points of Γ_1 are usually singular (discontinuity) points for the rolling layer, since at those points infinite transport rays meet. Moreover, the ridge set $\mathcal{R}$, *i.e.* the singular set of the distance function from Γ_0, can change very much from case to case. In general, it is the union of several arcs which can reach the boundary at the open vertices or at some (regular) points of the wall-boundary Γ_1.

The two-dimensional extension of the scheme (9)-(11) for the open table is straightforward and its main features are preserved. We introduce in the domain Ω (for simplicity the square $(0,1)\times(0,1)$) an $N \times N$ uniform grid of nodes $x_{i,j}$, and, as usual, we denote by $(u_{i,j}^n, v_{i,j}^n)$ the components of the discrete solutions on node $x_{i,j}$ at time $t = t_n$. Then, the scheme can be written as

$$\begin{aligned} v_{i,j}^{n+1} = v_{i,j}^n + \Delta t \Big[& v_{i,j}^n D^2 u_{i,j}^n + \overline{D} v_{i,j}^n \cdot D u_{i,j}^n \\ & -(1 - |Du_{i,j}^n|) v_{i,j}^n + f_{i,j} \Big] \end{aligned} \tag{17}$$

$$u_{i,j}^{n+1} = u_{i,j}^n + \Delta t(1 - |Du_{i,j}^n|)v_{i,j}^n \tag{18}$$
$$u_{i,j}^0 = v_{i,j}^0 = 0 \quad \forall i,j \tag{19}$$
$$u_{i,j}^n = 0 \text{ on } \partial\Omega \quad \forall n\,, \tag{20}$$

where now $D^2u_{i,j}$ is the standard five-points discretization of the Laplace operator, and the discrete gradient vectors $Du_{i,j}$ and $\overline{D}v_{i,j}$ are defined, component by component, as in the one-dimensional case.

The application of such scheme to the wall problem requires a modification of boundary conditions. In analogy with the one-dimensional case discussed in the previous section, the boundary condition (16) on Γ_1 is implemented in the scheme by replacing condition (20) with the following ones

$$\forall (i,j) \text{ such that } x_{i,j} \in \Gamma_1 : \text{if } Du_{i,j}^n \cdot \nu_{i,j} > 0 \text{ then } v_{i,j}^n = 0, \tag{21}$$
$$\text{else } u_{i,j}^n = u_{((i,j)-\nu_{i,j})}^n \text{ and } v_{i,j}^n = v_{((i,j)-\nu_{i,j})}^n, \tag{22}$$
$$\forall (i,j) \text{ such that } x_{i,j} \in \overline{\Gamma}_0 : u_{i,j}^n = 0, \tag{23}$$
$$\forall \text{ vertex } x_{i,j} \in \Gamma_0 : v_{i,j}^n = 0\,, \tag{24}$$

where $\nu_{i,j}$ denotes the outward normal unit vector at $x_{i,j}$. We have found that Neumann boundary conditions both on u and v (see (22)) yield better results in the tests. The technical condition (24) on the open vertices is necessary to avoid locking at those points. When singular points are present on Γ_1, numerical tests show that instability rapidly occurs in time at nodes close to them. In order to prevent this phenomenon we found that it is enough to impose at every time iteration the explicit gradient constraint which is known to hold at the equilibrium:

$$|Du_{i,j}^n| = min(|Du_{i,j}^n|, 1)\,, \quad \forall (i,j)\,, \forall n\,. \tag{25}$$

4. Numerical tests

4.1 1-dimensional tests

For a constant source term f over $D_f = (x_1, x_2) \subset \overline{\Omega}$, we describe the results of some experiments.

For $x_2 = 1$ (case (*a*) of Section 3), we know that there is only a possible equilibrium, $u(x) \equiv x$ in Ω. In Figure 1 (top) we see two examples of that situation, where the scheme is able to detect the correct dynamics for u and v (circles indicate the profile of the rolling layer). Different values of x_1 can only influence the final shape of the rolling layer, according to the explicit formula (5). We remark that the first order difference in u becomes strictly positive near the wall, so that the boundary condition (15) of the scheme reduces definitively to $v_N^n = 0$.

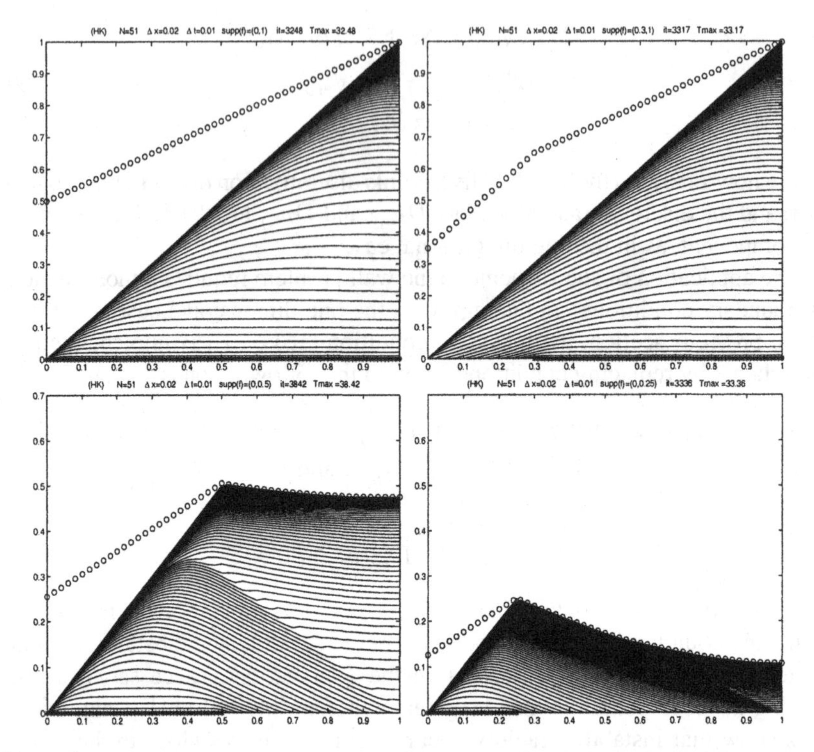

Figure 1. (1D) Growing standing layers on a table closed at the right-hand side, for different source supports D_f: a) [0,1]; b) [0.3,1]; c) [0,0.5]; d) [0,0.25].

For $x_2 < 1$ (case (*b*) of Section 3), we know that the equilibrium solution u in $[x_2, 1]$ is not uniquely determined, so that it can be strongly influenced by the boundary condition at the wall. Our condition (15) in that case becomes equivalent to a zero normal derivative for u, as shown in the two examples of Figure 1 (bottom). We can see that the rolling layer is correctly equal to zero in that region. The final value of u_N is directly proportional to the intensity of the source and to x_2.

4.2 2-dimensional tests

We have implemented the two-dimensional scheme (17)-(19),(21)-(24),(25) in the case of a constant source term f with $D_f = \Omega$, being Ω the unit square. In the first example Γ_1 coincides with one side of the square Ω (see Figure 2). In this case there are no singular points on the boundary, and the stationary rolling layer can be proved to be continuous. The scheme produces the correct dynamics and equilibria, even without the gradient condition (25).

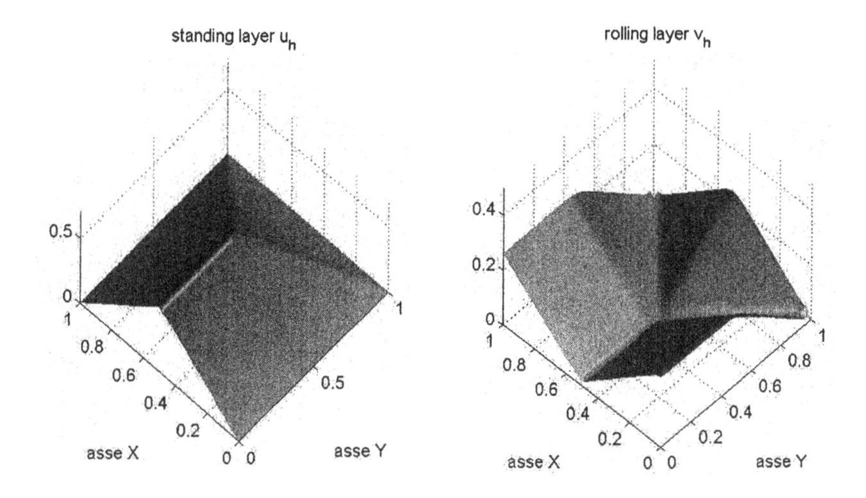

Figure 2. (2D) $f \equiv 0.5$, $D_f = \Omega$, $N = 41 : \Gamma_1 = \{0 < x < 1,\ y = 0\}$.

In the other three examples (Figures 3, 4 and 5) it is possible to see the effect of the boundary singular points. The time iterates u^n converges towards the correct distance function from Γ_0 (left), whereas the rolling layer v^n very quickly grows around the singular points in the normal direction to the boundary (the black areas in the pictures, where v is shown from above). An equilibrium configuration is always reached, and the zero level set of the final rolling layer is able to detect efficiently the ridge set (in white).

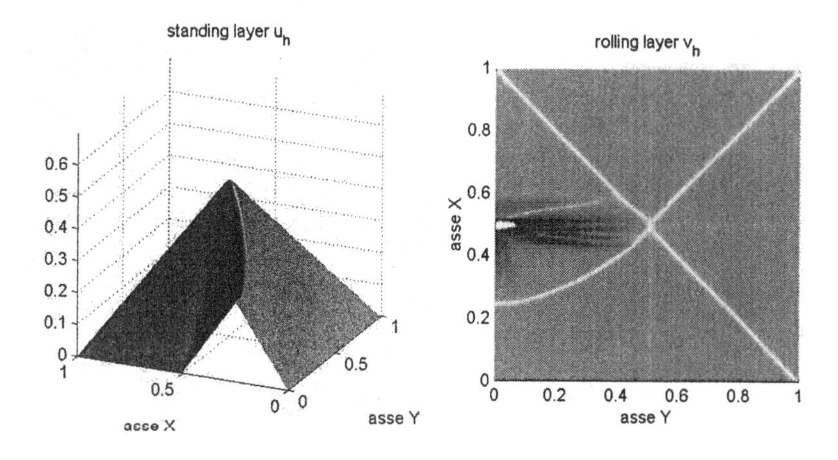

Figure 3. (2D) $f \equiv 0.5$, $D_f = \Omega$, $N = 41 : \Gamma_1 = \{0 < x < 0.5,\ y = 0\}$.

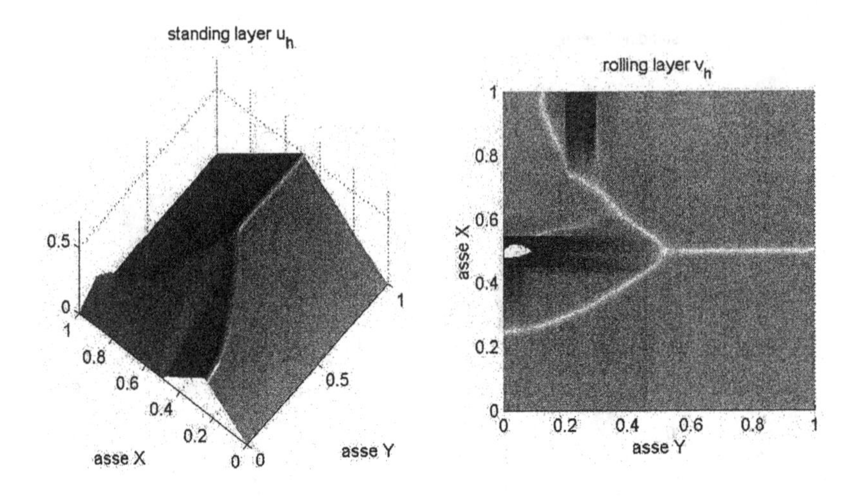

Figure 4. (2D) $f \equiv 0.5$, $D_f = \Omega$, $N = 41$: $\Gamma_1 = \{0 < x < 0.5,\ y = 0\} \cup \{x = 0,\ 0 < y < 0.25\} \cup \{0 < x < 1,\ y = 1\}$.

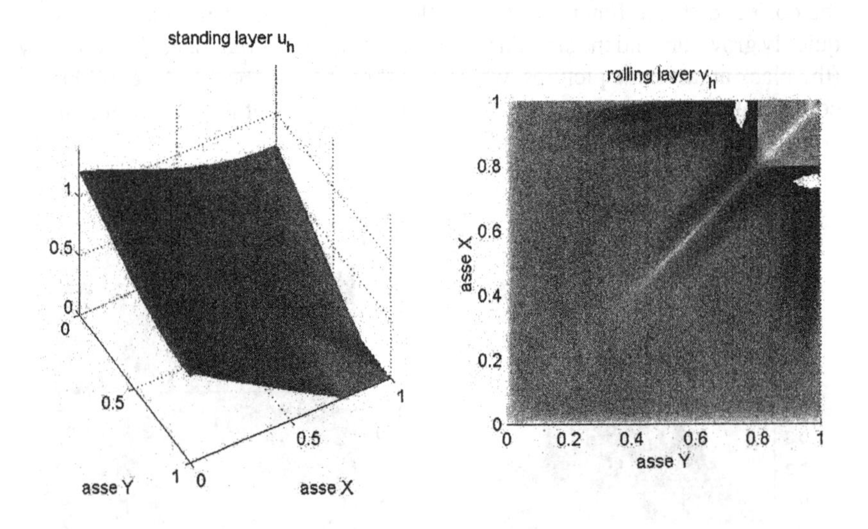

Figure 5. (2D) $f \equiv 0.5$, $D_f = \Omega$, $N = 41$: $\Gamma_0 = \{0.75 < x < 1,\ y = 1\} \cup \{x = 1,\ 0.75 < y < 1\}$.

5. Conclusions and perspectives

We end this paper drawing some partial conclusions on the approximation of the two-layers model that is proposed here. The numerical results show that the scheme is reasonably accurate for the approximation of the stationary standing layer as well as for the approximation of the evolution of u and v. The simulation seems to be close to the physical behavior of u and v for all times $t \in [0, T]$. Another interesting feature of the scheme is the fact that it is rather flexible to handle easily the open and the wall problem. The scheme gives also an accurate reconstruction of the ridge without spurious oscillations and this is due to the up-wind discretization of the nonlinear terms of the system.
In the future, we intend to pursue our study completing the analysis of stability in order to obtain a general convergence theorem and attack other problems related to the modelling of granular matter, *e.g.* the silos and the obstacle problem. Finally, an interesting open problem is to prove an analogue of the representation formula (5) for the wall problem.

References

[1] G. Aronsson, L.C. Evans and Y. Wu. Fast/slow diffusion and growing sandpiles. *J. Diff. Equations* 131:304-335, 1996.

[2] P. Cannarsa and P. Cardaliaguet. Representation of equilibrium solutions to the table problem for growing sandpiles. *JEMS* 6:435-464, 2004.

[3] P.G. de Gennes. Granular matter. In *Summer School on Complex Systems, Varenna, Lecture Notes Società Italiana di Fisica* 1996.

[4] L.C. Evans, M. Feldman and R.F. Gariepy. Fast/slow diffusion and collapsing sandpiles. *J. Diff. Equations* 137:166-209, 1997.

[5] M. Falcone and S. Finzi Vita. A finite-difference approximation of a two-layer system for growing sandpiles, Dipartimento di Matematica, Preprint, May 2005 also available at `http://cpde.iac.rm.cnr.it/preprint.php`.

[6] M. Falcone and S. Finzi Vita. Convergence of a finite-difference approximation of a two-layer system for growing sandpiles, in preparation.

[7] S. Finzi Vita. Numerical simulation of growing sandpiles. In *Control Systems: Theory, Numerics and Applications, CSTNA2005*. e-published by SISSA, PoS (`http://pos.sissa.it`), 2005.

[8] K.P. Hadeler and C. Kuttler. Dynamical models for granular matter. *Granular Matter* 2:9-18, 1999.

[9] K.P. Hadeler, C. Kuttler and I. Gergert. Dirichlet and obstacle problems for granular matter. Preprint, University of Tübingen, 2002.

[10] L. Prigozhin. Variational model of sandpile growth. *Euro. J. Appl. Math.* 7:225-235, 1996.

DATA FUSION AND FILTERING VIA CALCULUS OF VARIATIONS

L. Fatone,[1] P. Maponi,[2] and F. Zirilli[3]

[1] *Dipartimento di Matematica Pura ed Applicata, Università di Modena e Reggio Emilia, Via Campi 213/b, 41100 Modena (MO), Italy, fatone.lorella@unimo.it,**, [2] *Dipartimento di Matematica e Informatica, Università di Camerino, Via Madonna delle Carceri, 62032 Camerino (MC), Italy, pierluigi.maponi@unicam.it,* [3] *Dipartimento di Matematica "G. Castelnuovo", Università di Roma "La Sapienza", Piazzale Aldo Moro 2, 00185 Roma, Italy, f.zirilli@caspur.it*

Abstract We study the problem of urban areas detection from satellite images. In particular, we consider two types of satellite images: SAR (Synthetic Aperture Radar) images and optical images. We describe a simple algorithm for the detection of urban areas. We show that the performance of the detection algorithm can be improved using a fusion procedure of the SAR and optical images considered. The fusion algorithm presented in this paper is based on a simple use of ideas taken from calculus of variations and it makes possible to do together the filtering and the data fusion steps. Some numerical examples obtained processing real data are reported at the end of the paper. In the website `http://web.unicam.it/matinf/fatone/w1` several animations relative to these numerical examples can be seen.

keywords: Data fusion, Optimization algorithms, Urban areas detection

1. Introduction

The exploration of the Earth surface is an important use of remotely sensed data obtained from instruments on board of artificial satellites. These data can be used for many different purposes. In this paper we restrict our attention to the study of urban areas using SAR images and satellite optical images. Taking advantage of the physical properties of the electromagnetic waves with frequency in the range of SAR sensors and of optical sensors we adopt a well known urban areas detection algorithm based on the fact that urban areas can be recognized as the parts of the images containing the more brilliant pixels.

*The research on SAR/optical data fusion has been carried out with the support of ESA-ESRIN through the Esrin contract No. 13796/99/I-DC "Study of optical/SAR complementarity and data fusion techniques" granted to the Università di Camerino, Italy. The authors thank A.R. Conn, N.I.M. Gould and Ph.L. Toint for making available, free of charge, the optimization package LANCELOT.

Please use the following format when citing this chapter:

Fatone, L., Maponi, P., and Zirilli, F., 2006, in IFIP International Federation for Information Processing, Volume 202, Systems, Control, Modeling and Optimization, eds. Ceragioli, F., Dontchev, A., Furuta, H., Marti, K., Pandolfi, L., (Boston: Springer), pp. 139-149.

We can use the urban areas detection algorithm on SAR images and on optical images separately or, when the two images refer to the same scene, we can use jointly these two kinds of images in a fusion procedure to obtain new synthetic images, that is the fused images. The detection algorithm applied to the fused images should have a better performance than the one obtained on the original images considered separately. We show that this is really the case on some examples using real data. Moreover in this paper we propose a fusion procedure to combine the SAR images and the satellite optical images. Note that in order to perform meaningfully the fusion procedure the SAR and optical images to be fused must be not only relative to the same scene but also co-registered.
Many different authors have considered the fusion problem of SAR and optical images and the problem of urban areas detection, see, for example, [11], [6], [4], [7]. Image fusion is a special case of data fusion. For a general survey of this field see [1] and the references quoted there. The use of data fusion techniques is common practice in many fields different from remote sensing, such as inverse scattering, [5], and medical imaging, [8].
The mathematical formulation of the fusion procedure proposed here is based on a constrained optimization problem. This problem improves the mathematical formulation of the SAR/optical fusion procedure previously considered by the authors, see [4] and the web site `http://web.unicam.it/matinf/fatone/esrin.asp`, where two different steps were performed: 1) segmentation and denoising of the images; 2) fusion of the images obtained from step 1). The fusion procedure considered in this paper performs these two steps together, through the formulation of a new optimization problem. This is obtained using a new objective function and adding some suitable constraints. The change made in the objective function is based on some simple ideas taken from calculus of variations. Mathematical models that simulate the functioning of the SAR and optical processors and describe the measuring processes are used in the constraints added to the optimization problem. These two facts correspond to a significant improvement of the work presented in [6], and [4]. The simple mathematical models of the SAR and optical sensors used to process the images considered are based on some classical results on remote sensing theory, see for example [9], [3], [10], and are given by integral relations between the measured data and the unknown measured quantities. We note that more sophisticated models of the SAR and the optical sensors can be integrated in the fusion procedure proposed. Moreover many different ad hoc detection algorithms can be used to characterize urban areas in the fused images.
In section 2 the mathematical models used to interpret the SAR and optical images are described. In section 3 we recall the urban areas detection algorithm presented in [4] and we present the improved version of the fusion procedure proposed here. In section 4 some implementation details of the algorithms presented in sections 2, 3 and some numerical experiments are reported. In

the website `http://web.unicam.it/matinf/fatone/w1` some animations relative to these numerical experiments can be seen.

2. The mathematical models used to interpret SAR and optical images

In this section we introduce two simple mathematical models to interpret satellite images with the purpose of showing how to integrate these models in the fusion procedure presented in Section 3.
Let us begin with the model for SAR images used here. We refer to the papers [9], [3] for a more detailed discussion. We denote with $\mathbf{R}$ the real line, and with $\mathbf{R}^2$ the two dimensional real Euclidean space. Let $A_1(\underline{\xi})$, $\underline{\xi} = (\xi_1, \xi_2)^t \in \mathbf{R}^2$ be the amplitude of the electromagnetic signal at the SAR frequency emitted from $\underline{\xi} \in \mathbf{R}^2$; the superscript t means transposed. Note that in this model $A_1(\underline{\xi})$, $\underline{\xi} = (\xi_1, \xi_2)^t \in \mathbf{R}^2$ is a random variable. We have that in the position $\underline{x} \in \mathbf{R}^2$ the intensity SAR image $U_1(\underline{x})$ is given by:

$$U_1(\underline{x}) = (\mathbf{H}_S\langle A_1^2\rangle)(\underline{x}) = \int_{\mathbf{R}^2} \left|h_S(\underline{\xi} - \underline{x})\right|^2 \langle A_1^2(\underline{\xi})\rangle \, d\underline{\xi}, \tag{1}$$

where $\langle\cdot\rangle$ denotes the expected value of $\cdot$ and:

$$\left|h_S(\underline{\xi})\right|^2 = \frac{1}{\nu}\left[\frac{R_1}{\xi_1}\sin\left(\pi\frac{\xi_1}{R_1}\right)\frac{R_2}{\xi_2}\sin\left(\pi\frac{\xi_2}{R_2}\right)\right]^2, \ \underline{\xi} \in \mathbf{R}^2, \tag{2}$$

where $\nu > 0$ is a normalization constant that makes the integral of $|h_S|^2$ over $\mathbf{R}^2$ equal to one, and $R_1, R_2 > 0$ are the resolutions of the SAR image along the two cartesian coordinates.
Let us consider now the model for optical images, see [10] for a more detailed discussion. We denote with $A_2(\underline{\xi})$ the amplitude of the electromagnetic signal at the optical frequency emitted from $\underline{\xi} \in \mathbf{R}^2$. This quantity is modeled as a random variable. Arguing as above we have that in the position $\underline{x} \in \mathbf{R}^2$ the optical image $U_2(\underline{x})$ is given by:

$$U_2(\underline{x}) = (\mathbf{H}_O\langle A_2^2\rangle)(\underline{x}) = \int_{\mathbf{R}^2} \left|h_O(\underline{\xi} - \underline{x})\right|^2 \langle A_2^2(\underline{\xi})\rangle d\underline{\xi}. \tag{3}$$

Let F be the Fourier transform operator, $\underline{\omega} \in \mathbf{R}^2$ be the conjugate variable in the Fourier transform of $\underline{\xi} \in \mathbf{R}^2$. Note that $F(|h_O|^2)$ can be written as:

$$F(|h_O|^2)(\underline{\omega}) = T(\underline{\omega})B(\underline{\omega}), \ \underline{\omega} \in \mathbf{R}^2, \tag{4}$$

where T represents the functioning of the instrument and B represents the perturbation due to the presence of the atmosphere; this perturbation is not negligible when modeling optical measurements. In [10] the following expressions of $T(\underline{\omega})$ and $B(\underline{\omega})$ are proposed:

$$T(\underline{\omega}) = \frac{1}{A}\int_{\mathbf{R}^2} Q_0(\lambda(\underline{\xi}+\underline{\omega}))Q_0(\lambda\underline{\xi})\,d\underline{\xi},\ B(\underline{\omega}) = e^{-3.44\left(\frac{\lambda\|\underline{\omega}\|}{r_0}\right)^{5/3}} \tag{5}$$

($\underline{\omega}$ is in $\mathbf{R}^2$). λ is the wavelength of the electromagnetic signal considered, $r_0 > 0$ is a parameter that controls the perturbation induced by the atmosphere on the image, Q_0 is the characteristic function of the disk with center the origin and with radius equal to the radius of the instrument lens, and A is the area of the instrument lens. We note that the usual representation in pixels of the measured images that we denote with u_1, u_2 can be regarded as a piecewise constant approximation of the functions U_1, U_2 given by (1), (3). The value assumed by u_1, u_2 on a pixel can be understood as an "average" of U_1, U_2 respectively in the area corresponding to the pixel considered. These approximations are denoted with u_1, u_2 respectively. Moreover the support of the densities $\langle A_1^2\rangle$, $\langle A_2^2\rangle$ of the convolution equations (1), (3) is restricted to the support R of the region covered by the measured images. In the sequel these restricted densities are denoted with χ_1, χ_2 respectively. Note that this restriction operation is justified since the convolution kernels appearing in equations (1), (3) have significant decay properties.

3. The fusion procedure

Let us describe the urban areas detection algorithm used here, see [4] for more details. Let u be an image representing either a SAR image or an optical image or a synthetic image obtained from the fusion algorithm described in the sequel. The image u by virtue of the usual pixel structure can be regarded as a piecewise constant real function defined on a rectangular region R of the two-dimensional Euclidean space $\mathbf{R}^2$. We use the same notation u to denote the image and the associated function, that is $u : R \to \mathbf{R}$. We note that the values of this function are the usual gray levels of the image. Let μ and σ^2 be respectively the mean and the variance of u. The urban areas detection algorithm used here consists of the following steps. Given a suitable positive integer n, the image u is partitioned in non intersecting subimages having dimensions $n \times n$ pixels. Eventually we adjust the dimensions of the subimages lying on the right end side and on the upper end side of u to fit with the dimensions of u. Let L be the number of the subimages considered. For $l = 1, 2, \ldots, L$ let μ_l be the mean value associated to the gray levels of the l-th subimage. Since urban areas produce very brilliant texture (above all in SAR images) we assume that subimages containing urban areas are characterized by the fact of having a high value of the parameter:

$$D_l = (\mu_l - \mu)/\sigma. \tag{6}$$

Given a threshold $\Upsilon > 0$, for $l = 1, 2, \ldots, L$ the l-th subimage is recognized as containing an urban area when:

$$D_l > \Upsilon. \tag{7}$$

Let us describe the fusion procedures of SAR and optical images. In [4] (see also `http://web.unicam.it/matinf/fatone/esrin.asp`) we have proposed the following fusion procedure made of two steps: 1) *denoising of SAR and optical images*; 2) *fusion of the denoised images produced in* 1).

Let $\hat{u}_1$, $\hat{u}_2$ be two images defined in the rectangular region R representing the SAR and the optical image to be fused respectively. We assume that the images $\hat{u}_1, \hat{u}_2$ refer to the same scene, are coregistered, and that they have the same pixel structure. Step 1) consists in the solution of an initial boundary value problem, that is, for $i = 1, 2$, given the positive real numbers a_i, T_i, we compute the solution $v_i(\underline{x}, t)$, $\underline{x} \in R$, $t \in [0, T_i]$, of the following problem:

$$\begin{cases} \frac{\partial v_i}{\partial t} = \operatorname{div}\left(g_{a_i}(||\nabla v_i||)\, \nabla v_i\right), & (\underline{x}, t) \in R \times (0, T_i), \\ v_i(t = 0) = \hat{u}_i, & \underline{x} \in R, \\ \frac{\partial v_i}{\partial \underline{n}} = 0, & (\underline{x}, t) \in \partial R \times (0, T_i), \end{cases} \tag{8}$$

where $\operatorname{div}(\cdot)$ and $\nabla(\cdot)$ are the divergence and the gradient of $\cdot$ with respect to $\underline{x}$ respectively, $||\cdot||$ is the Euclidean norm of $\cdot$, R is an open set, that we have assumed to be a rectangle, ∂R is the boundary of the domain R, $\underline{n}(\underline{x})$ is the exterior unit normal vector to ∂R in $\underline{x} \in \partial R$, $\frac{\partial \cdot}{\partial \underline{n}}$ is the usual directional derivative of $\cdot$ with respect to the direction $\underline{n}$, and finally $g_{a_i}(\eta) = \frac{1}{1+\eta^2/a_i^2}$, $\eta \in \mathbf{R}$, is the diffusion coefficient of problem (8). For $i = 1, 2$ and $0 \leq \tilde{t} \leq T_i$ we denote with $v_i(\tilde{t})$ the function $v_i(\underline{x}, \tilde{t})$, $\underline{x} \in R$. For later convenience we define $V_1 = v_1(T_1)$, $V_2 = v_2(T_2)$. For $i = 1, 2$, V_i is the denoised version of $\hat{u}_i$. Step 2) consists in the solution of an optimization problem. Let us consider a "structure" operator $S_{\tau_1,\tau_2}(\eta)$, $\eta \in \mathbf{R}$, that is an approximation of the Heaviside function $H(\eta - \tau)$, $\eta \in \mathbf{R}$, where $0 < \tau_1 < \tau < \tau_2$. For example $S_{\tau_1,\tau_2}(\eta) = H(\eta - \tau)$ for $\eta \in \mathbf{R} \setminus (\tau_1, \tau_2)$ and $S_{\tau_1,\tau_2}(\eta)$ is a twice continuously differentiable function for $\eta \in \mathbf{R}$ strictly increasing for $\eta \in [\tau_1, \tau_2]$. Given the denoised images V_1, V_2 and a suitable norm $|||\cdot|||$, the "fused" images V_1^*, V_2^* are obtained as the minimizer of the following problem:

$$\min_{w_1, w_2} \{|||S_{\tau_1,\tau_2}(||\nabla w_1||) - S_{\tau_1,\tau_2}(||\nabla w_2||)|||^2 + \lambda_1 |||w_1 - V_1|||^2 + \lambda_2 |||w_2 - V_2|||^2\}, \tag{9}$$

where λ_1, λ_2 are suitable positive penalization parameters and w_1, w_2 are functions defined on R. We note that problem (9) is a fusion procedure, in fact problem (9) tries to change the variables w_1, w_2 in order to obtain structures $S_{\tau_1,\tau_2}(||\nabla w_1||)$, $S_{\tau_1,\tau_2}(||\nabla w_2||)$ that are closer than $S_{\tau_1,\tau_2}(||\nabla V_1||)$ and

$S_{\tau_1,\tau_2}(||\nabla V_2||)$, i.e. the initial structures, while the penalization terms force w_1 and w_2 to remain close to the denoised data V_1 and V_2 respectively.
Now we examine a new fusion procedure that unifies these two steps. Note that the denoising procedure (8) is well-suited for this purpose. More precisely we consider the following optimization problem: given $\epsilon > 0$, $\hat{\lambda}_1 > 0$, $\hat{\lambda}_2 > 0$, the measured images $\hat{u}_1$, $\hat{u}_2$ and a suitable norm $|||\cdot|||$, the fused densities χ_1^*, χ_2^* are obtained as the minimizer of the following problem:

$$\begin{aligned} &\min_{\chi_1,\chi_2\geq 0}\{|||S_{\tau_1,\tau_2}(||\nabla\chi_1||) - S_{\tau_1,\tau_2}(||\nabla\chi_2||)|||^2 + J(\chi_1,\chi_2)\}, \\ &\text{subject to: } \hat{\lambda}_1||\mathbf{H}_S\chi_1 - \hat{u}_1|| + \hat{\lambda}_2||\mathbf{H}_O\chi_2 - \hat{u}_2|| \leq \epsilon, \end{aligned} \tag{10}$$

where $\mathbf{H}_S$, $\mathbf{H}_O$ are the integral operators defined in formulae (1), (3) respectively, χ_1, χ_2 are functions defined on R, the nonegativity constraints for χ_1 and χ_2 must be understood as pointwise inequalities and are dictated by elementary physics, and

$$J(\chi_1,\chi_2) = l_1\int_R \ln\left(1+\frac{||\nabla\chi_1(\underline{x})||^2}{a_1^2}\right)d\underline{x} + l_2\int_R \ln\left(1+\frac{||\nabla\chi_2(\underline{x})||^2}{a_2^2}\right)d\underline{x}, \tag{11}$$

where l_1, l_2 are suitable positive parameters and a_1, a_2 are the parameters appearing in problem (8). From the solution χ_1^*, χ_2^* of problem (10) the "fused" images U_1^*, U_2^* are computed as follows:

$$U_1^*(\underline{x}) = (\mathbf{H}_S\chi_1^*)(\underline{x}),\ \underline{x}\in R, \quad U_2^*(\underline{x}) = (\mathbf{H}_O\chi_2^*)(\underline{x}),\ \underline{x}\in R. \tag{12}$$

Minimizing the objective function of problem (10) means to make small the quantity $|||S_{\tau_1,\tau_2}(||\nabla\chi_1||) - S_{\tau_1,\tau_2}(||\nabla\chi_2||)|||$, this corresponds to fuse the densities, χ_1, χ_2, and to make small $J(\chi_1,\chi_2)$, this corresponds to denoise the images following a procedure similar to the one defined in (8). In fact the trajectory $(v_1(\underline{x},t), v_2(\underline{x},t))$, $\underline{x}\in R$ defined for $t > 0$ by (8) when we choose $\hat{u}_1, \hat{u}_2$, respectively as initial data is the steepest descent trajectory passing through $(\hat{u}_1,\hat{u}_2)$ associated to the minimization of the functional $J(\chi_1,\chi_2)$. More in detail let $v : R \to \mathbf{R}$ be a sufficiently regular function and let $f : \mathbf{R} \to \mathbf{R}$ be a smooth function on $\mathbf{R}$. Let $\tilde{J}(v) = \int_R f(||\nabla v(\underline{x})||)\, d\underline{x}$ be a functional depending on the function v. Let $\Phi = \{\phi : R \to \mathbf{R}$: ϕ is a sufficiently regular function$\}$. From calculus of variations we know that the minimizers v^* of $\tilde{J}$ must satisfy the first order necessary condition $\frac{\delta\tilde{J}}{\delta v}(v^*) = 0$, where $\frac{\delta\cdot}{\delta v}$ means functional derivative of $\cdot$, i.e. $\int_R f'(||\nabla v^*(\underline{x})||)\frac{\nabla v^*(\underline{x})\nabla\phi(\underline{x})}{||\nabla v^*(\underline{x})||}\, d\underline{x} = 0$, $\forall\phi\in\Phi$, where f' denotes the derivative of f. Using the divergence theorem and assuming that $f'(\eta)/\eta \neq 0$, $\eta\in\mathbf{R}$ we have:

$$-\mathrm{div}\left(f'(||\nabla v^*(\underline{x})||)\frac{\nabla v^*(\underline{x})}{||\nabla v^*(\underline{x})||}\right) = 0,\ \underline{x}\in R, \quad \frac{\partial v^*(\underline{x})}{\partial\underline{n}(\underline{x})} = 0,\ \underline{x}\in\partial R. \tag{13}$$

Note that $v^* = constant$ satisfies (13) and comparing the boundary value problem (13) with problem (8) we can see that when $f'(\eta)/\eta = g_{a_i}(\eta)$, $\eta \in \mathbf{R}$, $i = 1, 2$, problem (8) defines the steepest descent trajectory associated to the functional $\tilde{J}$ going through the point $\hat{u}_i$, $i = 1, 2$. We note that in this case the minimizer v^* of $\tilde{J}$, obtained as steady state solution of problem (8), is $v_i^* = \int_R \hat{u}_i(\underline{x})d\underline{x}$, $i = 1, 2$. From $f'(\eta)/\eta = g_{a_i}(\eta)$, $\eta \in \mathbf{R}$, we have: $f(\eta) = f_i(\eta) = a_i^2/2 \ln(1 + \eta^2/a_i^2) + c_i$, $\eta \in \mathbf{R}$, $i = 1, 2$, where $c_i, i = 1, 2$ are arbitrary constants that we choose equal zero.
We note that problem (10), (11), (1), (3) differs from problem (8), (9) mainly for two reasons: the use of the constraint corresponding to the mathematical models of the SAR and optical processors that replaces the penalization terms appearing in (9) and the new term J in the objective function. Due to these differences problem (10), (11), (1), (3) can be seen as a refinement of problem (8), (9). In fact the use of the mathematical models (1), (3) that represent the measurement processes of the SAR and optical images makes problem (10), (11), (1), (3) more realistic than the problem considered previously. The parameters l_1, l_2 of problem (10) are tied to the parameters T_1, T_2 of problem (8) in the following sense: large values of l_1, l_2 correspond to large values of T_1, T_2, small values of l_1, l_2 correspond to small values of T_1, T_2. When the parameters l_1, l_2 take appropriate values minimizing the objective function of problem (10) corresponds both to solve problem (8) (due to the presence of the term J) and to make the fusion of the structures of χ_1 and χ_2 (due to the presence of the term $|||S_{T_1,T_2}(||\nabla\chi_1||) - S_{T_1,T_2}(||\nabla\chi_2||)|||^2$). In fact this last term appears also in problem (9). That is solving problem (10) corresponds to performing step 1) (denoising) and 2) (fusion) of the fusion procedure used in [6] and in [4] together.

4. The numerical experience

We show some numerical results obtained using the fusion procedure and the urban areas detection algorithm proposed previously. These results are obtained processing real satellites data. In particular we consider two pairs of SAR/optical images, made of ERS SAR-average amplitude (range looks=azimuth looks=1) images and optical (one of the four SPOT-4 channels) images. These images have been provided to us by ESA-ESRIN, Frascati-Italy, with the authorization of SPOT Image. Each image is mono-channel, orthorectified using a Digital Elevation Model (DEM) of the observed scene. Moreover the SAR and optical images are coregistered. The first pair of images corresponds to a peri-urban area in the south of Paris, see Fig. 1 (a), (c). These images have 170×180 pixels and each pixel is $20m \times 20m$. The second pair corresponds to an area in the north of Paris that contains a part of the Roissy Charles de Gaulle airport, see Fig. 2 (a), (c). These images have 180×180 pixels and each pixel is

$20m \times 20m$. The $20m \times 20m$ resolution of the images has been obtained pre-processing the original ERS and SPOT images. This pre-processing consists in resampling the images to have the same pixel of $20m \times 20m$ in both and coregistering the SAR and the optical images taking care of the DEM of the scene represented in the images. This pre-processing has been carried out by SPOT Image. We note that in these figures the white color represents high values of the pixel variable, i.e. gray level = 255, and the black color represents low values of the pixel variable, i.e. gray level = 0.

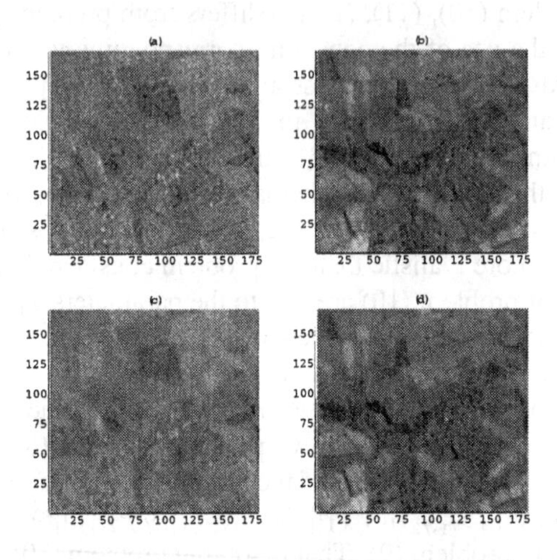

Figure 1. A peri-urban area south of Paris: (a) SAR image, (b) density corresponding to the SAR image obtained from the fusion procedure, (c) optical image, (d) density corresponding to the optical image obtained from the fusion procedure.

Problem (10) has been discretized using the natural pixels structure of the images and using the finite differences approximation of the derivatives and the rectangular quadrature formulae for the integrals appearing in (10). Moreover the finite dimensional optimization problem obtained in this way from problem (10) has been solved using the optimization software package LANCELOT (see [2]). Finally in the numerical solution of this optimization problem, due to the difficulties arising from the highly nonlinear function S_{τ_1,τ_2}, the fusion procedure is performed via an iterative process, where in each iteration is solved a problem of type (10) with a different choice of τ_1, τ_2 running from 1, 200 to τ_1^*, τ_2^* where the values of τ_1^*, τ_2^* are specified later. In the first optimization problem we use the images $\hat{u}_1$, $\hat{u}_2$, as initial guess for χ_1, χ_2, in the following optimization problems we use as initial guess the solution obtained in the previous optimization problem. Moreover in the constraints of problem (10), due to the large dimensions of the matrices obtained from the discretization of

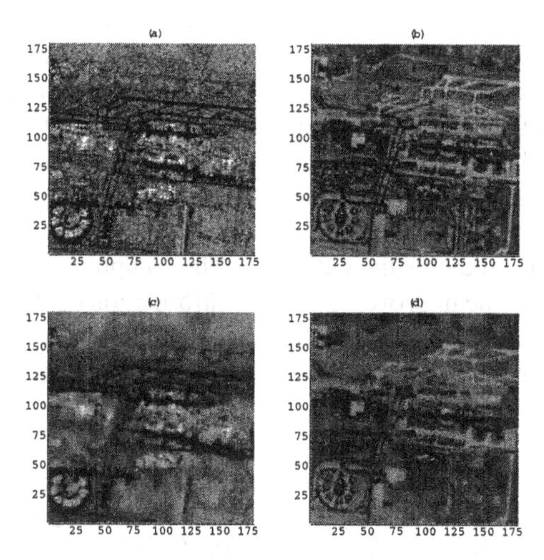

Figure 2. An area north of Paris containing a part of the Roissy Charles de Gaulle airport: (a) SAR image, (b) density corresponding to the SAR image obtained from the fusion procedure, (c) optical image, (d) density corresponding to the optical image obtained from the fusion procedure.

the integral operator $\mathbf{H}_S$ and $\mathbf{H}_O$, given a positive integer γ, we group together the pixels of the $\gamma \times \gamma$ subimages of the images considered. More precisely, each image is divided in non intersecting subimages having $\gamma \times \gamma$ pixels, and each subimage is considered as a new pixel in the constraints of problem (10). Moreover we define the gray level of this new pixel to be the average of the gray levels of the real pixels belonging to the $\gamma \times \gamma$ subimage. We note that in problem (10) the action of $\mathbf{H}_S$ on χ_1 can be easily computed from formulae (1), (2), and that the action of $\mathbf{H}_O$ on χ_2 can be easily computed from formulae (3), (4), (5). We note that the computation of the inverse Fourier transform of $T(\underline{\omega})$, $\underline{\omega} \in \mathbf{R}^2$ defined in (5) can be done explicitly in terms of the Bessel function of order one.
The numerical results of the fusion procedure shown in this section are obtained using the following values of the parameters mentioned previously: $R_1 = R_2 = 20m$, $A = 0.28m$, $r_0 = \lambda^{6/5}$, $\lambda = 5 \cdot 10^{-7}m$, $\tau_1^* = 1$, $\tau_2^* = 10$, $\hat{\lambda}_1 = \hat{\lambda}_2 = 1$, $l_1 = 10^{-6}$, $l_2 = 10^{-7}$, $a_1 = 0.5$, $a_2 = 6$, $\gamma = 10$. Finally the parameter ϵ is the upperbound for the weighted sum of the Euclidean norm of the remainders of the linear systems coming from the constraints appearing in (10) when we choose $\chi_i = \hat{u}_i$, $i = 1, 2$. Note that the (b), (d) parts of these figures do not show the same physical quantities than the (a), (c) parts of the figures. In fact in (a), (c) are reported the measured SAR and the measured optical images respectively, in (b), (d) are reported the densities χ_1^*, χ_2^* solution

of problem (10) that are related to the images via the mathematical models (1), (3). These densities are used in the urban areas detection algorithms. We note that the computational cost of the proposed fusion procedure is given by the computational cost of the optimization problem (10). More specifically the highly nonlinear character of the objective function and the huge number of independent variables coming from the discretized version of the unknowns of the problem make the solution of the nonlinear optimization problem a challenging task both from the mathematical and the computational point of view.
We want to point out the improvement of the information content of the images consequence of the fusion procedure from the point of view of urban areas detection. Let u be a generic image. We consider the parameters D_l, $l = 1, 2, \ldots,$ L defined in (6) associated to u. Moreover we denote $\overline{U}(u)$ the average of the parameters D_l corresponding to the subimages of u containing urban areas on the basis of the test (7) and we denote $\overline{V}(u)$ the average of the remaining parameters D_l, corresponding to subimages that, according to the test (7), do not correspond to urban areas. We introduce the following performance indices for $i = 1, 2$ see [4]:

$$\alpha_i = (\overline{U}(\chi_i^*) - \overline{U}(\hat{u}_i))/\overline{U}(\hat{u}_i), \quad \beta_i = (\overline{V}(\chi_i^*) - \overline{V}(\hat{u}_i))/\overline{V}(\hat{u}_i). \tag{14}$$

In section 3 we proposed to detect urban areas as the more brilliant areas in the image. So that it is easy to see that indices α_1, α_2 and β_1, β_2 give a measure of the improvement obtained by the fusion procedure in the urban areas detection algorithm and that positive values of the indices correspond to positive improvement. In the numerical experience that we report has been used $n = 10$, $\Upsilon = 0.25$ and we have obtained: $\alpha_1 = 0.35$, $\beta_1 = 0.51$, $\alpha_2 = 0.02$, $\beta_2 = 0.05$ for the first pair of images (see Fig. 1); $\alpha_1 = 0.23$, $\beta_1 = 0.38$, $\alpha_2 = 0.09$, $\beta_2 = 0.18$ for the second pair of images (see Fig. 2).
Finally we report another numerical experiment that shows the improvement obtained with the fusion procedure in the understanding of the scene contained in Fig. 2 (a), (c). In this experiment we compare the content of Fig. 2 (a), (c), the measured images, and the content of Fig. 2 (b), (d), the densities obtained from the fusion procedure, with a high resolution optical image of the same scene that we call ground truth (i.e.: an IRS-1C image). The IRS-1C image of the scene considered has pixels of size is $5m \times 5m$. We have co-registered manually the IRS-1C image with the images shown in Fig. 2 and we have computed the center of mass of several "objects" of size approximately 10 pixels (1 pixel=$20m \times 20m$) in the images of Fig. 2 and in the IRS-1C image. The centers of mass associated to the densities obtained from the fusion procedure appear to be more accurate than those computed from the measured images when compared with the centers of mass computed from the ground truth. The accuracy gained is of the order of a few meters (i.e. $5 - 10m$). This experiment suggests that the fusion procedure introduced here can be an useful

ingredient in a more ambitious procedure that includes for example automatic target recognition and location. Some animations relative to these numerical experiments can be seen in `http://web.unicam.it/matinf/fatone/w1`.

References

[1] Special issue on data fusion. *Proceedings of the IEEE.* 85:1-208, 1997.

[2] A.R. Conn, N.I.M. Gould, Ph.L. Toint. *LANCELOT: A Fortran Package for Large Scale Nonlinear Optimization (Release A).* Berlin, Springer-Verlag, 1992.

[3] M. Costantini, A. Farina, F. Zirilli. The fusion of different resolution SAR images. *Proceedings of the IEEE.* 85:139-146, 1997

[4] L. Fatone, P. Maponi, F. Zirilli. Fusion of SAR/Optical images to detect urban areas. *Proceedings of the IEEE/ISPRS Joint Workshop on Remote Sensing and Data Fusion over Urban Areas, Roma Italy.* 217-221, 2001.

[5] L. Fatone, P. Maponi, F. Zirilli. An image fusion approach to the numerical inversion of multifrequency electromagnetic scattering data. *Inverse Problems.* 17:1689-1702, 2001.

[6] L. Fatone, P. Maponi, F. Zirilli. Data fusion and nonlinear optimization. *SIAM News.* 35(1):4;10, 2002.

[7] C. Gouinaud, F. Tupin, H. Maître. Potential and use of radar images for characterization and detection of urban areas. *IGARSS'96. Lincoln, Nebraska, USA.* 1:474-476, 1996.

[8] K. Kneöaurek, M. Ivanovic, J. Machac, D.A. Weber. Medical image registration. *Europhysics News.* 31:5-8, 2000.

[9] C.J. Oliver. Information from SAR Images. *Journal of Physics D-Applied Physics.* 24:1493-1514, 1991.

[10] F. Roddier. The effects of atmospheric turbulence in optical astronomy. In *Progress in Optics, North-Holland.* Wolf, E. (Ed.) 19:281-376, 1981.

[11] A.H. Soldberg. Multisource classification of remotely sensed data: fusion of Landsat TM and SAR images. *IEEE Transactions on Geoscience and Remote Sensing.* 32:768-776, 1994.

MODELLING AND FAST NUMERICAL METHODS FOR GRANULAR FLOWS

E. Ferrari,[1] G. Naldi,[2] and G. Toscani[3]
[1] *University of Ferrara - Department of Mathematics, Ferrara, Italy, elisa.ferrari@unife.it* *,
[2] *University of Milan - Department of Mathematics, Milan, Italy, giovanni.naldi@mat.unimi.it*
[3] *University of Pavia - Department of Mathematics, Pavia, Italy, giuseppe.toscani@unipv.it*

Abstract In this work we discuss the development of fast algorithms for the inelastic Boltzmann equation describing the collisional motion of a granular gas. In such systems the collisions between particles occur in an inelastic way and are characterized by a coefficient of restitution which in the general case depends on the relative velocity of the collision. In the quasi-elastic approximation the granular operator is replaced by the sum of an elastic Boltzmann operator and a nonlinear friction term. Fast numerical methods based on a suitable spectral representation of the approximated model are then presented.

keywords: Inelastic Boltzmann equation, Spectral methods, Granular gases, Fast algorithms.

1. Introduction

The inelastic Boltzmann equation describes the evolution of materials composed of many small discrete grains, in which the mean free path of the grains is much larger than the typical particle size. Similar as molecular gases, granular gases can in fact be described at a mesoscopic level within the concepts of classical statistical mechanics. Many recent papers (see for example [1, 2, 5, 12] and the references therein), consider Boltzmann-like equations for partially inelastic rigid spheres. Once initialized with a certain velocity distribution, granular gases cool down due to inelastic collisions of their particles. The dissipation of kinetic energy causes a series of non-trivial effects, as formation of clusters and other spatial structures [11], non-Maxwellian velocity distributions, anomalous diffusion, and others.

*The financial support of the European network HYKE, funded by the EC as contract HPRN-CT-2002-00282 and of the project NUMSTAT 2005, Comitato dei Sostenitori, funded by the University of Ferrara is acknowledged.

Please use the following format when citing this chapter:

Ferrari, E., Naldi, G., and Toscani, G., 2006, in IFIP International Federation for Information Processing, Volume 202, Systems, Control, Modeling and Optimization, eds. Ceragioli, F., Dontchev, A., Furuta, H., Marti, K., Pandolfi, L., (Boston: Springer), pp. 151-161.

In a granular gas, the microscopic dynamics of grains is governed by the restitution coefficient h which relates the normal components of the particle velocities before and after a collision. If one assumes that the grains are identical perfect spheres of diameter $\sigma > 0$, (x, v) and $(x - \sigma n, w)$ are their states before a collision, where $n \in \mathbf{S}^2$ is the unit vector along the center of both spheres, the post collisional velocities (v^*, w^*) are such that

$$(v^* - w^*) \cdot n = -h((v - w) \cdot n). \tag{1}$$

Thanks to (1), and assuming the conservation of momentum, one finds the change of velocity for the colliding particles as

$$v^* = v - \frac{1}{2}(1+h)((v-w)\cdot n)n, \quad w^* = w + \frac{1}{2}(1+h)((v-w)\cdot n)n. \tag{2}$$

For elastic collisions one has $h = 1$, while for inelastic collisions h decreases with increasing degree of inelasticity. In the first part of this note we briefly review the basic ideas behind the kinetic modelling of dissipative collisions.

From a numerical viewpoint, similarly to the classical rarefied gas dynamics case, the solution of the inelastic Boltzmann equation represents a real challenge. This is mostly due to the high dimensionality of the equation but also to the inelastic collision dynamics which preclude the use of the fast Boltzmann solvers recently presented in [2] for elastic collisions. We will see in the last part of this note, how in the quasi-elastic approximation we can recover fast algorithms also in the inelastic case.

2. Modelling dissipative collisions

The main difference between the classical Boltzmann equation for elastic rigid spheres and its dissipative version is contained in the binary collision among particles. In (2) the only parameter which can contain the description of the inelastic collision is the coefficient of restitution.

In the literature, essentially for simplicity, the restitution coefficient is frequently assumed to be a physical constant. A constant restitution coefficient however does not describe realistic situations. In fact, the restitution coefficient may depend on the relative velocity in such a way that collisions with small relative velocity are close to be elastic. The simplest physically correct description of dissipative collisions is based on the assumption that the spheres are composed by viscoelastic material, which is in good agreement with experimental data. The velocity-dependent restitution coefficient for viscoelastic spheres of diameter $\sigma > 0$ and mass m reads

$$h = 1 - C_1 A\alpha^{2/5}|(v-w)\cdot n|^{1/5} + C_2 A^2 \alpha^{4/5}|(v-w)\cdot n|^{2/5} \pm \dots \tag{3}$$

with

$$\alpha = \frac{3\sqrt{3}}{2} \frac{\sqrt{\sigma}Y}{m(1-\nu^2)}, \tag{4}$$

where Y is the Young modulus, ν is the Poisson ratio, and A depends on dissipative parameters of the material. The constant C_1 and C_2 can be explicitly computed. The impact velocity dependence (3) of the restitution coefficient $h = h((v-w)\cdot n)$ has been recently obtained by generalizing Hertz's contact problem to viscoelastic spheres. We skip here details that can be found in the literature (see [5] and the references therein). What is important in what follows, is that real situations of microscopic collisions between grains can be described in general assuming that the coefficient of restitution satisfies

$$1 - h = 2\bar{\beta}\gamma\left(|(v-w)\cdot n|\right), \tag{5}$$

where $\gamma(\cdot)$ is a given function and $\bar{\beta}$ is a parameter which is small in presence of small inelasticity. For example, for small values of α, the velocity dependence of the restitution coefficient in a collision of viscoelastic spheres can be expressed at the leading order as in (5), choosing $\gamma(r) = r^{1/5}$.

3. The Boltzmann equation

Following the standard procedures of kinetic theory [8], the evolution of the distribution function can be described by the Boltzmann-Enskog equation for inelastic hard spheres, which for the force-free case reads [4]

$$\frac{\partial f}{\partial t} + v \cdot \nabla_x f = G(\rho)\bar{Q}(f,f)(x,v,t), \tag{6}$$

where $\bar{Q}$ is the so-called granular collision operator, which describes the change in the density function due to creation and annihilation of particles in binary collisions

$$\bar{Q}(f,f)(v) = 4\sigma^2 \int_{\mathbf{R}^3}\int_{\mathbf{S}^2_+} (q\cdot n)\left\{\chi f(v^{**})f(w^{**}) - f(v)f(w)\right\} dw\, dn. \tag{7}$$

In (6)

$$\rho(x,t) = \int_{\mathbf{R}^3} f(x,v,t)\, dv$$

is the density, and the function $G(\rho)$ is the statistical correlation function between particles, which accounts for the increasing collision frequency due to the excluded volume effects. We refer to [7] for a detailed discussion of the meaning of the function G.

In (7), $q = (v-w)$, and $\mathbf{S}^2_+$ is the hemisphere corresponding to $q\cdot n > 0$. The velocities (v^{**}, w^{**}) are the pre-collisional velocities of the so-called inverse collision, which results with (v, w) as post-collisional velocities. The factor χ

in the gain term appears respectively from the Jacobian of the transformation $dv^{**}dw^{**}$ into $dvdw$ and from the lengths of the collisional cylinders $h|q^{**}\cdot n| = |q\cdot n|$. For a constant restitution coefficient, $\chi = h^{-2}$. This enlightens a second remarkable difference between the elastic and the inelastic collision operators. While the Jacobian of the elastic collision is equal to unity, allowing for the exchangeability of the rule of the pre- and post-collisional velocities, in the inelastic case the Jacobian is different from unity, and this implies a different role of pre- and post-collisional quantities.

To avoid the presence of the function χ, and to study approximations to the granular operator (7) it is extremely convenient to write the operator (7) in weak form. More precisely, let us define with $< \cdot , \cdot >$ the inner product in $L_1(\mathbf{R}^3)$. For all smooth functions $\varphi(v)$, it holds

$$\begin{aligned} &< \varphi , \bar{Q}(f,f) > = 4\sigma^2 \int_{\mathbf{R}^3} \varphi(v)\bar{Q}(f,f)(v)\, dv = \\ &2\sigma^2 \int_{\mathbf{R}^3}\int_{\mathbf{R}^3}\int_{\mathbf{S}^2} |q\cdot n| \left(\varphi(v^*) - \varphi(v)\right) f(v)f(w) dv\, dw\, dn. \end{aligned} \tag{8}$$

The last equality follows since the integral over the hemisphere $\mathbf{S}^2_+$ can be extended to the entire sphere $\mathbf{S}^2$, provided the factor $1/2$ is inserted in front of the integral itself. In fact changing n into $-n$ does not change the integrand.

Let (v', w') be the post collisional velocities in a elastic collision with (v, w) as incoming velocities,

$$v' = v - (q\cdot n)n , \quad w' = w + (q\cdot n)n. \tag{9}$$

Following [18], we rewrite the inelastic collision (2) in terms of the elastic collision (9) obtaining

$$v^* = v' + \frac{1}{2}(1-h)(q\cdot n)n , \quad w^* = w' - \frac{1}{2}(1-h)(q\cdot n)n. \tag{10}$$

If we assume that the coefficient of restitution can be described at the leading order by (5),

$$v^* - v' = \bar{\beta}\,\gamma\left(|q\cdot n|\right)(q\cdot n)n. \tag{11}$$

Let us consider a Taylor expansion of $\varphi(v^*)$ around $\varphi(v')$. Thanks to (11) we get

$$\varphi(v^*) = \varphi(v') + \bar{\beta}\,\nabla\varphi(v')\cdot\gamma\left(|q\cdot n|\right)(q\cdot n)n + O(\bar{\beta}^2) \tag{12}$$

If the collisions are nearly elastic, $\bar{\beta} << 1$, and we can cut the expansion (12) after the first-order term. Inserting (12) into (20) gives

$$< \varphi , \bar{Q}(f,f) > =< \varphi , Q(f,f) > + \bar{\beta} < \varphi , I(f,f) > . \tag{13}$$

It is a simple matter to recognize that in (13) $Q(f, f)$ is the classical Boltzmann collision operator for elastic hard spheres molecules [8],

$$Q(f,f)(v) = 2\sigma^2 \int_{\mathbf{R}^3} \int_{\mathbf{S}^2} |q \cdot n| \left\{ f(v')f(w') - f(v)f(w) \right\} dw\, dn. \quad (14)$$

In fact, the velocity v' into (13) is obtained from (v, w) through the elastic collision (9).

Let us now study in more detail the second contribution to the inner product (13). Using the properties of the transformation (9), we obtain

$$< \varphi, I(f,f) > =$$

$$2\sigma^2 \int_{\mathbf{R}^3} dv\, \varphi(v) \mathrm{div}_v \int_{\mathbf{R}^3} \int_{\mathbf{S}^2} n(q \cdot n)|q \cdot n| \gamma\left(|q \cdot n|\right) f(v')f(w') dw\, dn. \quad (15)$$

In fact, the transformation $dv\, dw$ into $dv'\, dw'$ given by (9) is such that $q' \cdot n = -q \cdot n$, while its Jacobian is equal to unity. The last equality follows from the divergence theorem. This shows that the granular correction is the nonlinear friction operator $\bar{\beta}\, I(f, f)(v)$, where

$$I(f,f)(v) = 2\sigma^2 \mathrm{div}_v \int_{\mathbf{R}^3} \int_{\mathbf{S}^2} n(q{\cdot}n)|q{\cdot}n| \gamma\left(|q \cdot n|\right) f(v')f(w') dw\, dn. \quad (16)$$

Finally, for nearly elastic granular collisions, with a restitution coefficient satisfying (5), the Enskog-Boltzmann equation can be modelled at the leading order as

$$\frac{\partial f}{\partial t} + v \cdot \nabla_x f = G(\rho) Q(f,f)(x,v,t) + G(\rho)\, \bar{\beta}\, I(f,f)(x,v,t), \quad (17)$$

where Q is the classical elastic Boltzmann collision operator, and I is a dissipative nonlinear friction operator which is based on elastic collisions between particles.

4. Fast methods

In this section we restrict ourselves to the study of the space homogeneous case

$$\frac{\partial f}{\partial t} = G(\rho) Q(f,f)(v,t) + G(\rho)\, \bar{\beta}\, I(f,f)(v,t). \quad (18)$$

This is motivated by the use of a splitting argument in the numerical solution of the kinetic equation. It is clear that all the main numerical difficulties are contained in the right hand side of (18). Here we will use a Carleman-like representation of the operators $Q(f, f)$ and $I(f, f)$, together with a suitable angular approximation, in order to derive spectral methods that can be evaluated through fast algorithms. We refer to [2, 10, 15] and references therein for further details on fast spectral methods.

For the sake of simplicity we will first derive the method for the classical operator $Q(f, f)$ and then we briefly describe how to extend it to the nonlinear friction term $I(f, f)$.

4.1 A Carlemann-like representation

Let us use the identity

$$\int_{\mathbf{S}^2} (u \cdot n)_+ \, \varphi\,(n(u \cdot n))\, dn = \frac{|u|}{4} \int_{\mathbf{S}^2} \varphi\left(\frac{u - |u|\, n}{2}\right) dn, \tag{19}$$

in order to write collision operator $Q(f, f)$ in the form

$$Q(f, f)(v) = \frac{\sigma^2}{2} \int_{\mathbf{R}^3} \int_{\mathbf{S}^2} |q| \left\{ f(v') f(w') - f(v) f(w) \right\} dw\, dn, \tag{20}$$

where now

$$v' = \frac{1}{2}(v + w) + \frac{1}{2}|q| n\,, \quad w' = \frac{1}{2}(v + w) - \frac{1}{2}|q| n. \tag{21}$$

Then we use a Carlemann-like representation which conserves more symmetries of the collision operator when one truncates it in a bounded domain.

As explained in [2] the basic identity we shall need is

$$\frac{1}{2} \int_{\mathbf{S}^2} F(|u| n - u)\, dn = \frac{1}{|u|} \int_{\mathbf{R}^3} \delta(2\, x \cdot u + |x|^2)\, F(x)\, dx. \tag{22}$$

Using (22) with $u = q = v - w$ and performing the change of variables $x \to x/2$ and $w \to y = w - v - x$ we can write

$$Q(f, f)(v) = 2\sigma^2 \int_{x \in \mathbf{R}^3} \int_{y \in \mathbf{R}^3} \delta(x \cdot y)$$
$$\left[f(v + y)\, f(v + x) - f(v + x + y)\, f(v) \right] dx\, dy.$$

Now let us consider the bounded domain $\mathcal{D}_T = [-T, T]^3$ $(0 < T < +\infty)$. Next we have to truncate the integration in x and y without affecting the action of the operator for compactly supported functions. Thus we set them to vary in $\mathcal{B}_S$, the ball of center 0 and radius S. For a compactly supported function f with support $\mathcal{B}_R$, we take $S = 2R$ in order to obtain all possible collisions. In fact we have

$$|x|^2 \leq |x|^2 + |y|^2 = |x + y|^2 = |q|^2 \leq (2R)^2,$$

thus $|x| \leq 2R$ and similarly we get $|y| \leq 2R$.

The operator now reads

$$Q^R(f,f)(v) = 2\sigma^2 \int_{x\in\mathcal{B}_{2R}} \int_{y\in\mathcal{B}_{2R}} \delta(x\cdot y) \\ [f(v+y)f(v+x) - f(v+x+y)f(v)]\, dx\, dy, \tag{23}$$

with $v \in \mathcal{B}_{\sqrt{2}R}$. The interest of this representation is to preserve the real collision kernel and its invariance properties. The next step consist in a suitable periodization of the operator on $\mathcal{D}_T$ which prevents intersections of the regions where f is different from zero. Note that in (23) the arguments of the integrands are contained into $\mathcal{B}_{3\sqrt{2}R}$. In fact we have that $|x| \le 2R$ and $|y| \le 2R$ imply $|x+y|^2 = |x|^2 + |y|^2 \le 8R^2$ (thanks to the orthogonality condition $x \cdot y = 0$ consequence of the δ function) and then $|x+y| \le 2\sqrt{2}R$. From this we get $|v+x+y| \le |v| + |x+y| \le \sqrt{2}R + 2\sqrt{2}R = 3\sqrt{2}R$. Thus we need to take $T \ge (3+\sqrt{2})R/\sqrt{2}$ as a bound for the periodization.

4.2 Spectral methods and fast algorithms

Now we use the representation Q^R to derive the spectral methods. In the rest of the paragraph, for simplicity, we take $G(\rho) = 1$ and we neglect the friction correction setting $I(f,f) = 0$ into (18). Following the same computation as in the classical spectral method [15] but using representation (23) we obtain the following set of ordinary differential equations on the Fourier coefficients

$$\frac{d\hat{f}_k(t)}{dt} = \sum_{\substack{l,m=-N \\ l+m=k}}^{N} \hat{\beta}(l,m)\, \hat{f}_l\, \hat{f}_m, \quad k = -N, ..., N \tag{24}$$

where now $\hat{\beta}(l,m) = \beta(l,m) - \beta(m,m)$ with

$$\beta(l,m) = 2\sigma^2 \int_{x\in\mathcal{B}_{2R}} \int_{y\in\mathcal{B}_{2R}} \delta(x\cdot y)\, e^{il\cdot x}\, e^{im\cdot y}\, dx\, dy. \tag{25}$$

In the sequel we shall focus on β, and one easily checks that $\beta(l,m)$ depends only on $|l|$, $|m|$ and $|l\cdot m|$.

The search for fast deterministic algorithms for a collision operator in $\mathbf{R}^d$, i.e. algorithms with a cost lower than $O(N^{2d+\kappa})$ (with typically $\kappa = 1$), consists mainly in identifying some convolution structure in the operator. If this is trivial for the loss part of the operator, for the gain part this is rather contradictory with the search for a conservative scheme in a bounded domain, since the boundary condition needed to prevent for the outgoing or ingoing collisions breaks the invariance.

The aim is to approximate each $\hat{\beta}(l,m)$ by a sum

$$\beta(l,m) \simeq \sum_{p=1}^{A} \alpha_p(l)\alpha'_p(m). \tag{26}$$

This gives a sum of A discrete convolutions and so the algorithm can be computed in $O(A\,N^d \log_2 N)$ operations by means of standard FFT techniques. To this purpose we shall use a further approximated collision operator where the number of possible directions of collision is reduced to a finite set.

We start from representation (23) and write x and y in spherical coordinates

$$Q^R(f,f)(v) = \frac{\sigma^2}{2} \int_{e\in\mathbf{S}^2} \int_{e'\in\mathbf{S}^2} \delta(e\cdot e')\,de\,de'$$
$$\left\{ \int_{-R}^{R} \int_{-R}^{R} \rho\,\rho' \, [f(v+\rho' e')f(v+\rho e) - f(v+\rho e+\rho' e')f(v)]\,d\rho\,d\rho' \right\}.$$

Let us denote with $\mathcal{A}$ a discrete set of orthogonal couples of unit vectors (e,e'), which is even, i.e. $(e,e') \in \mathcal{A}$ implies that $(-e,e')$, $(e,-e')$ and $(-e,-e')$ belong to $\mathcal{A}$ (this property on the set $\mathcal{A}$ is required to preserve the conservation properties of the operator). Now we define $Q^{R,\mathcal{A}}$ to be

$$\begin{aligned} Q^{R,\mathcal{A}}(f,f)(v) = \frac{\sigma^2}{2} \int_{(e,e')\in\mathcal{A}} \Bigg\{ \int_{-R}^{R} \int_{-R}^{R} \rho\,\rho' \\ [f(v+\rho' e')f(v+\rho e) - f(v+\rho e+\rho' e')f(v)]\,d\rho\,d\rho' \Bigg\}\,d\mathcal{A} \end{aligned} \tag{27}$$

where $d\mathcal{A}$ denotes a discrete measure on $\mathcal{A}$ which is also even in the sense that $d\mathcal{A}(e,e') = d\mathcal{A}(-e,e') = d\mathcal{A}(e,-e') = d\mathcal{A}(-e,-e')$. It is easy to check that $Q^{R,\mathcal{A}}$ has the same conservation properties as Q^R.

By taking a spherical parametrization (θ,φ) of $e \in \mathbf{S}^2_+$ and uniform grids of respective size M_1 and M_2 for θ and φ, we get

$$\beta(l,m) \simeq \frac{2\sigma^2\pi^2}{M_1 M_2} \sum_{p,q=0}^{M_1,M_2} \alpha_{p,q}(l)\,\alpha'_{p,q}(m),$$

where

$$\alpha_{p,q}(l) = \phi_R^3\left(l\cdot e_{(\theta_p,\varphi_q)}\right), \qquad \alpha'_{p,q}(m) = \psi_R^3\left(\Pi_{e^\perp_{(\theta_p,\varphi_q)}}(m)\right),$$

$$\phi_R^3(s) = \int_{-R}^{R} \rho\, e^{i\rho s}\,d\rho, \qquad \psi_R^3(s) = \int_0^\pi \sin\theta\,\phi_R^3(s\cos\theta)\,d\theta$$

and

$$(\theta_p, \varphi_q) = \left(\frac{p\pi}{M_1}, \frac{q\pi}{M_2} \right).$$

Typically we shall consider this expansion with $M = M_1 = M_2$ to avoid anisotropy in the computational grid. The computational cost of the algorithm is then $O(M^2 N^3 \log_2 N)$, compared to $O(N^6)$ of the usual spectral method. Thus one requires $M^2 \log N \ll N^3$ in order to speed up the schemes.

4.3 Extension to the nonlinear friction

Finally, by applying an analogous procedure, it's possible to approximate the friction operator $I(f, f)$ in (17) by its truncated version $I^R(f, f)$ and to extend the above spectral method to the full problem (18). The computations are similar and hereafter are shortly summarized. The idea is that one just applies the same representation and truncation as for Q, inside the divergence of I. The result has exactly the same form with another kernel. More precisely using identity (19) we can write

$$I(f, f)(v) = \frac{\sigma^2}{2} \mathrm{div}_v \int_{\mathbf{R}^3} \int_{\mathbf{S}^2} \left(\frac{|q| - |q|n}{2} \right) |q| \gamma\left(|q|\right) f(v') f(w') dw\, dn. \tag{28}$$

Next the Carlemann-like representation is obtained through (22) and yields

$$I(f, f)(v) = 2\sigma^2 \mathrm{div}_v \int_{\mathbf{R}^3} \int_{\mathbf{R}^3} x\gamma\left(|x|\right) \delta(x \cdot y) f(v + x) f(v + y) dx\, dy. \tag{29}$$

Periodization on $\mathcal{D}_T$ then gives

$$I^R(f, f)(v) = 2\sigma^2 \int_{\mathcal{B}_{2R}} \int_{\mathcal{B}_{2R}} \delta(x \cdot y) \gamma\left(|x|\right)\, x \cdot \nabla_v (f(v + x) f(v + y)) dx\, dy, \tag{30}$$

where now the arguments of the integrand are supported into $\mathcal{B}_{(2+\sqrt{2})R}$.

The major difference is that the resulting kernel is characterized by the vector

$$x\gamma(|x|). \tag{31}$$

This kernel clearly decouples since it does not depend on y and so the resulting spectral scheme, similarly to the previous section, can be computed with fast algorithms.

Thus, for the full model (18) we obtain the Fourier coefficients

$$\frac{d\hat{f}_k(t)}{dt} = G(\rho) \sum_{\substack{l,m=-N \\ l+m=k}}^{N} (\hat{\beta}(l, m) + \bar{\beta}\beta_I(l, m))\, \hat{f}_l\, \hat{f}_m, \quad k = -N, ..., N \tag{32}$$

where $\hat{\beta}(l,m)$ are given by (25) and the nonlinear friction coefficients are

$$\beta_I(l,m) = 2\sigma^2 ik \cdot \int_{x\in\mathcal{B}_{2R}} \int_{y\in\mathcal{B}_{2R}} x\gamma(|x|)\delta(x\cdot y)\, e^{il\cdot x}\, e^{im\cdot y}\, dx\, dy. \quad (33)$$

We omit the details of the fast solver which follows the lines of the one described for the elastic Boltzmann equation [9, 10, 2].

5. Conclusions

In this note we have summarized some recent results related to the modelling of granular gases and the development of fast algorithms. In particular we have seen how the method recently developed in [2] can be extended to nonlinear friction equations and to the quasi-elastic approximation even for non constant coefficient of restitution.

References

[1] N. Bellomo, M. Esteban, M. Lachowicz. Nonlinear kinetic equations with dissipative collisions. *Appl. Math. Letters* 8:46-52, 1995.

[2] D. Benedetto, E. Caglioti, M. Pulvirenti. A kinetic equation for granular media. *M2AN Math. Model. Numer. Anal.* 31:615-641, 1997.

[3] D. Benedetto, E. Caglioti, J.A. Carrillo, M. Pulvirenti. A non-maxwellian steady distribution for one-dimensional granular media. *J. Statist. Phys.* 91:979-990, 1998.

[4] A.V. Bobylev, J.A. Carrillo, I.M. Gamba. On some properties of kinetic and hydrodynamic equations for inelastic interactions *J. Statist. Phys.* 98:743-773, 2000.

[5] N.V. Brilliantov, T. Pöschel. Granular gases with impact-velocity dependent restitution coefficient. In *Granular Gases*: 100-124, Lecture Notes in Physics, Vol. 564, Springer-Verlag, Berlin, 2000.

[6] J.A. Carrillo, C. Cercignani, I.M. Gamba. Steady states of a Boltzmann equation for driven granular media. *Phys. Rev. E (3)* 62:7700-7707, 2000.

[7] C. Cercignani. Recent developments in the mechanism of granular materials. *Fisica Matematica e ingegneria delle strutture*, Pitagora Editrice, Bologna, 1995.

[8] C. Cercignani, R. Illner, M. Pulvirenti. The mathematical theory of dilute gases. *Applied Mathematical Sciences*, Vol. 106, Springer-Verlag, New-York, 1994.

[9] F.Filbet, C. Mouhot, L. Pareschi. Solving the Boltzmann equation in $N \log_2 N$. *SISC* (to appear), 2005.

[10] F.Filbet, C. Mouhot, L. Pareschi. Work in progress.

[11] I. Goldhirsch. Scales and kinetics of granular flows. *Chaos* 9:659-672, 1999.

[12] S. McNamara, W.R. Young. Kinetics of a one-dimensional granular medium in the quasielastic limit. *Phys. Fluids A* 5:34-45, 1993.

[13] C. Mouhot, L. Pareschi. Fast algorithms for computing the Boltzmann collision operator. *Math. Comp.* (to appear), 2005.

[14] G. Naldi, L. Pareschi, G. Toscani. Spectral methods for one-dimensional kinetic models of granular flows and numerical quasi elastic limit. *M2AN Math. Model. Numer. Anal.* 37:73-90, 2003.

[15] L. Pareschi, G. Russo. Numerical solution of the Boltzmann equation. Spectrally accurate approximation of the collision operator. *SIAM J. Numer. Anal.* 37:1217-1245, 2000.

[16] L. Pareschi, G. Toscani. Modelling and numerics of granular gases. In *Modeling and computational methods for kinetic equations*: 259-285, Series Model. Simul. Sci. Eng. Technol., Birkhauser, Boston, 2004.

[17] G. Toscani. One-dimensional kinetic models of granular flows, *M2AN Math. Model. Numer. Anal.* 34:1277-1292, 2000.

[18] G. Toscani. Kinetic and hydrodinamic models of nearly elastic granular flows, *Monatsch. Math.* 142:179-192, 2004.

LINEAR DEGENERATE PARABOLIC EQUATIONS IN BOUNDED DOMAINS: CONTROLLABILITY AND OBSERVABILITY

P. Cannarsa,[1] G. Fragnelli,[1] and J. Vancostenoble[2]

[1] *Dipartimento di Matematica, Università di Roma "Tor Vergata", via della Ricerca Scientifica, 1, 00133 ROMA, Italy, {cannarsa,fragnell}@mat.uniroam2.it**,

[2] *Laboratoire M.I.P., U.M.R. C.N.R.S. 5640, Université Paul Sabatier Toulouse III, 118 route de Narbonne, 31 062 Toulouse Cedex 4, France, vancoste@mip.ups-tlse.fr*

Abstract In this paper we study controllability properties of linear degenerate parabolic equations. Due to degeneracy, classical null controllability results do not hold in general. Thus we investigate results of 'regional null controllability', showing that we can drive the solution to rest at time T on a subset of the space domain, contained in the set where the equation is nondegenerate.

keywords: linear degenerate equations, regional null controllability, persistent regional null controllability.

1. Introduction

This paper is concerned with null controllability for the degenerate heat equation:

$$\begin{cases} u_t - (a(x)u_x)_x + b(t,x)u_x + c(t,x)u = h(t,x)\chi_{(\alpha,\beta)}(x), \\ u(t,0) = u(t,1) = 0, \\ u(0,x) = u_0(x), \end{cases} \tag{1}$$

where $(t,x) \in (0,T') \times (0,1)$, $u_0 \in L^2(0,1)$, $h \in L^2((0,T') \times (0,1))$, $0 \le \alpha < \beta \le 1$ and $T' > T > 0$ fixed. Moreover, assume that $b,\ c \in L^\infty((0,T') \times (0,1))$ and

$$\begin{gathered} a : [0,1] \to [0,+\infty) \text{ is } C[0,1] \cap C^1(0,1],\ \frac{1}{a} \in L^1(0,1), \\ a(0) = 0 \text{ and } a > 0 \text{ on } (0,1]. \end{gathered} \tag{2}$$

*Paper written with financial support of "Istituto Nazionale di Alta Matematica".

Please use the following format when citing this chapter:

Cannarsa, P., Fragnelli, G., and Vancostenoble, J., 2006, in IFIP International Federation for Information Processing, Volume 202, Systems, Control, Modeling and Optimization, eds. Ceragioli, F., Dontchev, A., Furuta, H., Marti, K., Pandolfi, L., (Boston: Springer), pp. 163-173.

Note that, under suitable assumptions on b, the problem is well-posed in the sense of semigroup theory, working in appropriate weighted spaces.

Interest in degenerate parabolic equations as the one above is motivated by applications to probability (see, e.g., [7]) as well as to physical problems (see, e.g., [11]). Moreover, while null controllability for nondegenerate parabolic operators of second order in bounded domains has been studied in several papers (see, e.g. [10, 8]), the same problem seems widely open in the case of degenerate equations.

We recall the standard notion of null controllability.

DEFINITION 1 **(i):** *A given initial condition* $u_0 \in L^2(0,1)$ *is null controllable in time* $T > 0$ *if there exists* $h \in L^2((0,T) \times (0,1))$ *such that the solution* u *of (1) satisfies* $u(T) \equiv 0$ *in* $(0,1)$.

(ii): *Equation* (1) *is null controllable in time* $T > 0$ *if for all* $u_0 \in L^2(0,1)$ *there exists* $h \in L^2((0,T) \times (0,1))$ *such that the solution* u *of* (1) *satisfies* $u(T) \equiv 0$ *in* $(0,1)$.

It is well-known that null controllability in any time $T > 0$ holds for equation (1) in the *nondegenerate* case, i.e., if a is assumed to be *positive on* $[0,1]$ (see for instance [10, 8]). On the contrary simple examples (see, e.g., [6]) show that null controllability fails due to the degeneracy of a.

In [6] and in [4], problem (1) is considered, under different assumptions on a, in the special case $b \equiv 0$ and $b \neq 0$, $c(t,x)u = f(t,x,u)$, respectively. In both cases the following notion of regional null controllability has been developed.

DEFINITION 2 (REGIONAL NULL CONTROLLABILITY, [6]) *Set* $b \equiv 0$. *Equation* (1) *is regional null controllable in time* T *if for all* $u_0 \in L^2(0,1)$, *and* $\delta \in (0, \beta - \alpha)$, *there exists* $h \in L^2((0,T) \times (0,1))$ *such that the solution* u *of* (1) *satisfies*

$$u(T,x) = 0 \text{ for } x \in (\alpha + \delta, 1). \tag{3}$$

The proof given in [6] to show that the solution of (1) satisfies (3) is based on an observability inequality for a suitable adjoint problem. Such an inequality is obtained by an appropriate use of cut-off functions and Carleman estimates (see, e.g., [1], [9], or [12]) for nondegenerate parabolic operators. In [4] and in the present paper the main feature of our approach is that we use a new method of proof. Indeed, instead of deducing null controllability from observability, we derive the result directly, using cut-off functions and the fact that equation of (1) is null controllable when x varies in any subinterval $I \subset\subset (0,1]$, where a in nondegenerate. Although, in the present paper, we have focussed our attention on linear equations, we believe that our approach can be extended to more general problems such as semilinear equations, higher space dimensions, and so on.

We note that global null controllability is a property stronger than (3) in the sense that it is automatically preserved with time. More precisely, if $u(T) \equiv 0$ in $(0,1)$ and if we stop controlling the system at time T, then for all $t \geq T$, $u(t) \equiv 0$ in $(0,1)$. On the contrary, regional null controllability is a weaker property: due to the uncontrolled part on $(0, \alpha + \delta)$, (3) is no more preserved with time if we stop controlling at time T. Thus, it is important to improve the previous result, as shown in [6] or [4], proving that the solution can be forced to vanish identically on $(\alpha + \delta, 1)$ during a given time interval (T, T'), i.e. that the solution is persistent regional null controllable.

DEFINITION 3 (PERSISTENT REGIONAL NULL CONTROLLABILITY, [6]) *Set $b \equiv 0$. Equation (1) is persistent regional null controllable in time $T' > T > 0$ if for all $u_0 \in L^2(0,1)$, and $\delta \in (0, \beta - \alpha)$, there exists $h \in L^2((0,T') \times (0,1))$ such that the solution u of (1) satisfies*

$$u(t,x) = 0 \text{ for } (t,x) \in (T,T') \times (\alpha + \delta, 1). \tag{4}$$

In the present paper, we extend the above definitions and results to the case of $b \neq 0$, that is

$$u_t - (a(x)u_x)_x + b(t,x)u_x + c(t,x)u = h(t,x)\chi_{(\alpha,\beta)}(x), \tag{5}$$

where the coefficients b and c satisfy suitable conditions so that the problem is well-posed. In particular, the coefficient b will be assumed to satisfy a bound of the form $|b(t,x)| \leq K\sqrt{a(x)}$, a condition which is well-known in the literature (see also Remark 5).

As an application of our null controllability results, we derive observability inequalities for a class of linear degenerate parabolic equations which includes the adjoint systems of certain optimal control problems considered in [6] (see Corollaries 9 and 10).

The paper is organized as follows: in sections 1 and 2 we discuss the well-posedness of equation (1), introducing function spaces and operators, and state our controllability results. The proofs of these results are given in section 3.

2. Well-posedness

In this section we make the following assumptions:

ASSUMPTION 4 *Let $0 \leq \alpha < \beta \leq 1$ and $T' > T > 0$ be fixed. Assume that*

$$a : [0,1] \to [0,+\infty) \text{ is } \mathcal{C}[0,1] \cap \mathcal{C}^1(0,1],\ \frac{1}{a} \in L^1(0,1), \\ a(0) = 0 \text{ and } a > 0 \text{ on } (0,1]; \tag{6}$$

$$b,\ c \in L^\infty((0,T') \times (0,1)); \tag{7}$$

$$\exists K > 0 \text{ such that } |b(t,x)| \leq K\sqrt{a(x)} \text{ for } (t,x) \in (0,T') \times (0,1). \tag{8}$$

Observe that (6) is, for example, satisfied by $a(x) := x^p$, $p < 1$.

REMARK 5 The assumption (8), with the other assumptions, ensures that the Markov process described by the operator $Cu := -(au_x)_x + bu_x$ in $[0,1]$ doesn't reach the point $x = 0$, while the point $x = 1$ is an absorbing barrier since $u(t,1) = 0$. This implies that, if we set the problem in $C([0,1])$ instead of $L^2(0,1)$, then we don't need a boundary condition at $x = 0$ (see, e.g., [7]).

Let us consider the linear degenerate parabolic equation on $(0,1)$:

$$\begin{cases} u_t - (a(x)u_x)_x + b(t,x)u_x + c(t,x)u = h(t,x)\chi_{(\alpha,\beta)}(x), \\ u(t,0) = u(t,1) = 0, \\ u(0,x) = u_0(x), \end{cases} \tag{9}$$

where $(t,x) \in (0,T') \times (0,1)$, $u_0 \in L^2(0,1)$ and $h \in L^2((0,T') \times (0,1))$.

For well-posedness, we introduce the following weighted spaces

$$H_a^1 := \{u \in L^2(0,1) \mid \ u \text{ locally absolutely continuous in } (0,1], \\ \sqrt{a}u_x \in L^2(0,1) \text{ and } u(1) = u(0) = 0\}, \tag{10}$$

and

$$H_a^2 := \{u \in H_a^1(0,1) |\ au_x \in H^1(0,1)\}, \tag{11}$$

with the norms

$$\|u\|_{H_a^1}^2 := \|u\|_{L^2(0,1)}^2 + \|\sqrt{a}u_x\|_{L^2(0,1)}^2,$$

and

$$\|u\|_{H_a^2}^2 := \|u\|_{H_a^1}^2 + \|(au_x)_x\|_{L^2(0,1)}^2.$$

We define the operator $(A, D(A))$ by

$$D(A) = H_a^2 \text{ and } \forall u \in D(A),\ Au := (au_x)_x. \tag{12}$$

We recall the following properties of $(A, D(A))$ (see [2] for a proof in the case $a(0) = a(1) = 0$, and [6] for the proof in our case):

PROPOSITION 6 *The operator* $A : D(A) \to L^2(0,1)$ *is a closed self-adjoint negative operator with dense domain.*

Hence, A is the infinitesimal generator of a strongly continuous semigroup e^{tA} on $L^2(0,1)$. Moreover, one can show that e^{tA} is analytic, even if we make no use of such a property. Since A is a generator, working in the spaces considered above, we have that (9) is well-posed in the sense of semigroup theory:

THEOREM 7 *Under Hypothesis 1, for every* $h \in L^2((0,T') \times (0,1))$ *and for every* $u_0 \in L^2(0,1)$*, there exists a unique weak solution* u *of* (9) *such that* $u \in C^0([0,T']; L^2(0,1)) \cap L^2(0,T'; H^1_a)$*. Moreover, if* $u_0 \in H^1_a(0,1)$*, then*

$$u \in \mathcal{U} := H^1(0,T'; L^2(0,1)) \cap L^2(0,T'; H^2_a) \cap C^0([0,T']; H^1_a),$$

and

$$\begin{aligned} \sup_{t\in[0,T']} (\|u(t)\|^2_{H^1_a}) + \int_0^{T'} \left(\|u_t\|^2_{L^2(0,1)} + \|(au_x)_x\|^2_{L^2(0,1)}\right) dt \\ \leq C(\|u_0\|^2_{H^1_a} + \|h\|^2_{L^2((0,T')\times(0,1))}), \end{aligned} \tag{13}$$

where C *is a positive constant.*

3. Controllability results

Assume that Assumption 4 1 is satisfied. Using the fact that a is nondegenerate on $(\alpha, 1)$ and a classical result known for linear nondegenerate parabolic equations in bounded domains (see for example [10, 8]), we will now give a direct proof of regional null controllability for the linear degenerate problem (9).

THEOREM 8 *Assume Assumption 4. Then the following holds.*
(i) Regional null controllability. *Given* $T > 0$, $u_0 \in L^2(0,1)$*, and* $\delta \in (0, \beta - \alpha)$*, there exists* $h \in L^2((0,T) \times (0,1))$ *such that the solution* u *of* (9) *satisfies*

$$u(T,x) = 0 \text{ for } x \in (\alpha + \delta, 1).$$

Moreover, there exists a constant $C_T > 0$ *independent of* u_0 *such that*

$$\int_0^T \int_0^1 h^2(t,x)dxdt \leq C_T \int_0^1 u_0^2(x)dx. \tag{14}$$

(ii) Persistent regional null controllability. *Given* $T' > T > 0$, $u_0 \in$

$L^2(0,1)$, and $\delta \in (0, \beta - \alpha)$, there exists $h \in L^2((0,T') \times (0,1))$ such that the solution u of (9) satisfies

$$u(t,x) = 0 \text{ for } (t,x) \in (T,T') \times (\alpha + \delta, 1).$$

Moreover, there exists a constant $C_{T,T'} > 0$ such that

$$\int_0^{T'} \int_0^1 h^2(t,x)dxdt \leq C_{T,T'} \int_0^1 u_0^2(x)dx.$$

This result was proved in [6] and in [4] in the case $b \equiv 0$ and $b \neq 0$, respectively, and $a \in C^1[0,1]$. In particular, in [6], the proof was based on suitable *regional* observability inequalities which constituted the major technical part of the paper. Here, following [4], we give a different proof: we can deduce directly (i) from the classical null controllability results known for *nondegenerate* parabolic equations. Then, (ii) follows from (i) (as in [6]). Recently in [3] the null controllability result stated in the previous theorem is improved in the sense that global null controllability is proved for the following equation

$$u_t - (a(x)u_x)_x + f(t,x,u) = h(t,x)\chi_{(\alpha,\beta)}(x), \qquad (t,x) \in (0,T) \times (0,1).$$

As an application of Theorem 12.(i), we will deduce directly the *regional* observability inequality found in [6]. Consider the adjoint problem associated to (9)

$$\begin{cases} \varphi_t + (a\varphi_x)_x + (b\varphi)_x - c\varphi = 0, & (x,t) \in (0,T) \times (0,1), \\ \varphi(t,0) = \varphi(t,1) = 0, & t \in (0,T). \end{cases} \tag{15}$$

Then the following corollary holds.

COROLLARY 9 *For all $\delta \in (0, \beta - \alpha)$ there exists a positive constant K_T such that, for all φ solution of (15) in $\mathcal{U}$,*

$$\int_0^1 \varphi^2(0,x)dx \leq K_T \left(\int_0^T \int_\alpha^\beta \varphi^2(t,x)dxdt + \int_0^{\alpha+\delta} \varphi^2(T,x)dx \right). \tag{16}$$

Similarly, as a consequence of the persistent regional null controllability result above one can deduce the second observability inequality given in [6] for the *non homogeneous* adjoint problem. Indeed, given the adjoint system

$$\begin{cases} \varphi_t + (a\varphi_x)_x + (b\varphi)_x - c\varphi = G(t,x)\chi_{(T,T')}(t), & (x,t) \in (0,T') \times (0,1), \\ \varphi(t,0) = \varphi(t,1) = 0, & t \in (0,T'), \end{cases} \tag{17}$$

one can prove the next result.

COROLLARY 10 *For all $\delta \in (0, \beta - \alpha)$ there exists a positive constant $K_{T'}$ such that, for all φ solution of* (17) *in* $\mathcal{U}$,

$$\int_0^1 \varphi^2(0,x)dx \le K_{T'} \left(\int_0^{T'} \int_\alpha^\beta \varphi^2(t,x)dxdt + \int_0^{\alpha+\delta} \varphi^2(T',x)dx + \int_T^{T'} \int_0^{\alpha+\delta} G^2(t,x)dxdt \right). \tag{18}$$

4. Proofs

First of all, we have to observe that the well-posedness of (9) follows from the fact that A generates a strongly continuous semigroup and the operator $B(t)$ defined as

$$B(t)u := -b(t,\cdot)u_x - c(t,\cdot)u$$

can be seen as a particular perturbation of A in $D((-A)^{\frac{1}{2}})$.

4.1 Regional null controllability

In this section, we prove point (i) of Theorem 12. Note that (ii) follows from (i) as in [6]. We now construct cut-off functions that will be used in the following. Let $\phi \in C^\infty([0,+\infty))$ be such that $0 \le \phi \le 1$, and

$$\begin{cases} \phi(x) = 0, & 0 \le x \le \alpha, \\ \phi(x) = 1, & \alpha + \delta \le x \le 1. \end{cases} \tag{19}$$

Set $\xi := 1 - \phi \in C^\infty([0,+\infty))$. Then $0 \le \xi \le 1$ and

$$\begin{cases} \xi(x) = 1, & 0 \le x \le \alpha, \\ \xi(x) = 0, & \alpha + \delta \le x \le 1. \end{cases} \tag{20}$$

1) Since there is no degeneracy on $(\alpha, 1)$, by classical results for linear nondegenerate parabolic equation in bounded domain (see for example [8]), we have that there exists $h_1 \in L^2((0,T) \times (\alpha,1))$ such that the solution v of

$$\begin{cases} v_t - (a(x)v_x)_x + b(t,x)v_x + c(t,x)v = h_1(t,x)\chi_{(\alpha,\beta)}(x), \\ v(t,\alpha) = v(t,1) = 0, \\ v(0,x) = u_0(x), \end{cases} \tag{21}$$

where $(t,x) \in (0,T) \times (\alpha,1)$, satisfies

$$v(T,\cdot) \equiv 0 \text{ on } (\alpha,1).$$

Moreover, there exists a constant $C > 0$ such that $\|h_1\|^2_{L^2((0,T)\times(\alpha,1))} \leq C\|u_0\|^2_{L^2(\alpha,1)}$. And so, we have:

$$\int_0^T \|v(t)\|^2_{L^2(\alpha,1)}dt + \int_0^T \|v_x(t)\|^2{}_{L^2(\alpha,1)}dt \leq C_T\|u_0\|^2_{L^2(\alpha,1)}. \tag{22}$$

Then $\tilde{v}(t,x) := \phi(x)v(t,x)$ is the solution of

$$\begin{cases} \tilde{v}_t - (a(x)\tilde{v}_x)_x + b(t,x)\tilde{v}_x + c(t,x)\tilde{v} = \tilde{h}_1(t,x)\chi_{(\alpha,\beta)}(x), \\ \tilde{v}(t,0) = \tilde{v}(t,1) = 0, \\ \tilde{v}(0,x) = \phi(x)u_0(x), \end{cases}$$

where $(t,x) \in (0,T) \times (0,1)$ and $\tilde{h}_1(t,x) := \phi h_1 - \phi_x(av)_x - \phi_{xx}av - \phi_x av_x + b\phi_x v$. (Notice that ϕ_x, ϕ_{xx} are supported in $(\alpha, \alpha+\delta) \subset (\alpha,\beta)$.) Clearly, $\tilde{v}$ satisfies

$$\tilde{v}(T,\cdot) \equiv 0 \text{ on } (0,1).$$

Moreover, using (22) below and the fact that $\tilde{h}_1 \equiv 0$ on $(0,\alpha)$, one has

$$\begin{aligned} \int_0^T\int_0^1 \tilde{h}_1^2 dxdt &\leq K\left(\int_0^T\int_\alpha^1 h_1^2 dxdt + \int_0^T\int_\alpha^1 |v|^2 dxdt \right. \\ &\quad \left. + \int_0^T\int_\alpha^1 |v_x|^2 dxdt\right) \\ &\leq K\left(\|u_0\|^2_{L^2(\alpha,1)} + \int_0^T \|v(t)\|^2_{L^2(\alpha,1)}dt \right. \\ &\quad \left. + \int_0^T \|v_x\|^2_{L^2(\alpha,1)}dt\right) \\ &\leq K_T\|u_0\|^2_{L^2(\alpha,1)}, \end{aligned}$$

where K_T is a positive constant and depends on T.

3) Let z be the solution of

$$\begin{cases} z_t - (a(x)z_x)_x + b(t,x)z_x + c(t,x)z = 0, \\ z(t,0) = z(t,1) = 0, \\ z(0,x) = u_0(x), \end{cases} \tag{23}$$

where $(t,x) \in (0,T) \times (0,1)$. (The well-posedness of (23) follows from Theorem 7.) Then $\tilde{z}(t,x) := \xi(x)z(t,x)$ is the solution of

$$\begin{cases} \tilde{z}_t - (a(x)\tilde{z}_x)_x + b(t,x)\tilde{z}_x + c(t,x)\tilde{z} = \tilde{h}_2(t,x)\chi_{(\alpha,\beta)}(x), \\ \tilde{z}(t,0) = \tilde{z}(t,1) = 0, \\ \tilde{z}(0,x) = \psi(x)u_0(x), \end{cases}$$

where $(t,x) \in (0,T)\times(0,1)$ and $\tilde{h}_2(t,x) := -\xi_x(az)_x - \xi_{xx}az - \xi_x a z_x + b\xi_x z$ (note that ξ_x, ξ_{xx} are supported in $(\alpha, \alpha+\delta) \subset (\alpha, \beta)$). Moreover, $\tilde{z}$ satisfies

$$\tilde{z}(T,\cdot) \equiv 0 \text{ on } (\alpha+\delta, 1),$$

and, proceeding as for $\tilde{h}_1$, one can prove that there exists a positive constant K_T such that

$$\int_0^T \int_0^1 \tilde{h}_2^2 dxdt \le K_T \int_0^1 u_0^2(x)dx.$$

4) Finally, $u := \tilde{v} + \tilde{z}$ is the solution of

$$\begin{cases} u_t - (a(x)u_x)_x + b(t,x)u_x + c(t,x)u = h(t,x)\chi_{(\alpha,\beta)}(x), \\ u(t,0) = u(t,1) = 0, \\ u(0,x) = (\phi + \xi)u_0(x) = u_0(x), \end{cases}$$

where $(t,x) \in (0,T) \times (0,1)$ and $h := \tilde{h}_1 + \tilde{h}_2$. Moreover

$$u(T,\cdot) \equiv 0 \text{ on } (\alpha+\delta, 1),$$

and there exists a positive constant C_T such that

$$\int_0^T \int_0^1 h^2 dxdt \le C_T \int_0^1 u_0^2(x)dx.$$

4.2 Observability property

In this part we prove that Theorem 12.(i) implies the observability property (16). Using (13) and (14), one directly has the next lemma.

LEMMA 11 *Let h be the control given by Theorem 12.(i) and u the corresponding solution of (9). Then*

$$\int_0^{\alpha+\delta} u^2(T,x)dx \le C_T \int_0^1 u_0^2(x)dx,$$

where $C_T := e^{(1+K^2)T}$ and K is as in Hypothesis 1.

Proof of Corollary 9: Let φ in $\mathcal{U}$ be a solution of (15). Let $h \in L^2((0,T)\times(0,1))$ be the control given by Theorem 12.(i) such that

$$\begin{cases} u_t - (a(x)u_x)_x + b(t,x)u_x + c(t,x)u = h(t,x)\chi_{(\alpha,\beta)}(x), \\ u(t,0) = u(t,1) = 0, \\ u(0,x) = \varphi(0,x), \\ u(T,x) = 0, \quad x \in (\alpha+\delta, 1). \end{cases}$$

Multiplying the previous equation by φ and (15) by u, integrating over $(0,1)$ and summing up we obtain

$$\int_0^1 \frac{d}{dt}(u\varphi)dx = \int_0^1 h\chi_{(\alpha,\beta)}\varphi.$$

Here we have used the fact that $|b(t,0)| \leq K\sqrt{a(0)} = 0$. Integrating over $(0,T)$ we have:

$$\int_0^1 u(T,x)\varphi(T,x)dx - \int_0^1 \varphi^2(0,x) = \int_0^T \int_0^1 h\chi_{(\alpha,\beta)}\varphi.$$

Since $u(T,x) = 0$ for all $x \in (\alpha+\delta, 1)$, one has

$$\begin{aligned}\int_0^1 \varphi^2(0,x)dx &= \int_0^{\alpha+\delta} u(T,x)\varphi(T,x) - \int_0^T \int_\alpha^\beta h\varphi dxdt \\ &\leq \epsilon \int_0^{\alpha+\delta} u^2(T,x)dx + C_\epsilon \int_0^{\alpha+\delta} \varphi^2(T,x)dx \\ &+\frac{\epsilon}{2} \int_0^T \int_\alpha^\beta h^2 dxdt + \frac{1}{2\epsilon} \int_0^T \int_\alpha^\beta \varphi^2 dxdt,\end{aligned}$$

where $\epsilon > 0$ will be chosen later. By Lemma 11 it follows that

$$\begin{aligned}\int_0^1 \varphi^2(0,x)dx &\leq \epsilon C_T \int_0^1 \varphi^2(0,x)dx + C_\epsilon \int_0^{\alpha+\delta} \varphi^2(T,x)dx \\ &+\epsilon \int_0^T \int_\alpha^\beta h^2 dxdt + \frac{1}{2\epsilon} \int_0^T \int_\alpha^\beta \varphi^2(t,x)dxdt.\end{aligned}$$

Moreover, (14) implies

$$\begin{aligned}\int_0^1 \varphi^2(0,x)dx &\leq (\epsilon C + \epsilon C_T) \int_0^1 \varphi^2(0,x)dx \\ &+\frac{1}{2\epsilon} \int_0^T \int_\alpha^\beta \varphi^2(t,x)dxdt + C_\epsilon \int_0^{\alpha+\delta} \varphi^2(T,x)dx.\end{aligned}$$

Choosing ϵ such that $1 - \epsilon C - \epsilon C_T > 0$, one has

$$\int_0^1 \varphi^2(0,x)dx \leq K_\epsilon \left(\int_0^T \int_\alpha^\beta \varphi^2(t,x)dxdt + \int_0^{\alpha+\delta} \varphi^2(T,x)dx \right).$$

References

[1] P. Albano, P. Cannarsa. *Lectures on Carleman estimates for elliptic operators and applications.* In preparation.

[2] M. Campiti, G. Metafune, D. Pallara. Degenerate self-adjoint evolution equations on the unit interval. *Semigroup Forum* 57:1-36, 1998.

[3] P. Cannarsa, G. Fragnelli. Controllability of semilinear weakly degenerate parabolic equations. In preparation.

[4] P. Cannarsa, G. Fragnelli, J. Vancostenoble. Regional controllability of semilinear degenerate parabolic equations in bounded domains. To appear in *J. Math. Anal. Appl.*

[5] P. Cannarsa, P. Martinez, J. Vancostenoble. Nulle contrôlabilité régionale pour des équations de la chaleur dégénérées. *Comptes Rendus Mécanique* 330:397-401, 2002.

[6] P. Cannarsa, P. Martinez, J. Vancostenoble. Persistent regional controllability for a class of degenerate parabolic equations, *Commun. Pure Appl. Anal.* 3:607-635, 2004.

[7] K.J. Engel, R. Nagel. *One-parameter Semigroups for Linear Evolution Equations*, Graduate Texts in Mathematics 194, Springer-Verlag, 2000.

[8] A. V. Fursikov, O. Yu. Imanuvilov.*Controllability of evolution equations.* Lecture Notes Series, Research Institute of Mathematics, Global Analysis Research Center, Seoul National University 34, 1996.

[9] I. Lasiecka, R. Triggiani. *Carleman estimates and exact boundary controllability for a system of coupled, non conservative second order hyperbolic equations.* Partial Differential Equations Methods in Control and Shape Analysis, Lectures Notes in Pure and Applied Math. 188, Marcel Dekker, New York, 1994.

[10] G. Lebeau, L. Robbiano. Contrôle exact de l'équation de la chaleur. *Comm. P.D.E.* 20:335-356, 1995.

[11] P. Martinez, J. P. Raymond, J. Vancostenoble. Regional null controllability for a linearized Crocco type equation. *SIAM J. Control Optim.* 42:709–728, 2003.

[12] D. Tataru. Carleman estimates, unique continuation and controllability for anizotropic PDE's.*Contemporary Mathematics* 209:267-279, 1997.

IDENTIFICATION OF AQUIFER TRANSMISSIVITY WITH MULTIPLE SETS OF DATA USING THE DIFFERENTIAL SYSTEM METHOD

M. Giudici,[1] G. A. Meles,[1] G. Parravicini,[2] G. Ponzini,[1] and C. Vassena[1]
[1] *Università degli Studi di Milano, Dipartimento di Scienze della Terra, Sezione di Geofisica, via Cicognara 7, Milano, Italy,* {*Mauro.Giudici, Giansilvio.Ponzini, Chiara.Vassena*}*@unimi.it*
[2] *Università degli Studi di Milano, Dipartimento di Fisica, via Celoria 16, I-20133 Milano, Italy, Guido.Parravicini@unimi.it*

Abstract The mass balance equation for stationary flow in a confined aquifer and the phenomenological Darcy's law lead to a classical elliptic PDE, whose phenomenological coefficient is transmissivity, T, whereas the unknown function is the piezometric head. The differential system method (DSM) allows the computation of T when two "independent" data sets are available, i.e., a couple of piezometric heads and the related source or sink terms corresponding to different flow situations such that the hydraulic gradients are not parallel at any point. The value of T at only one point of the domain, $\mathbf{x}_0$, is required. The T field is obtained at any point by integrating a first order partial differential system in normal form along an arbitrary path starting from $\mathbf{x}_0$. In this presentation the advantages of this method with respect to the classical integration along characteristic lines are discussed and the DSM is modified in order to cope with multiple sets of data. Numerical tests show that the proposed procedure is effective and reduces some drawbacks for the application of the DSM.

keywords: Inverse problems, porous media, multiple data sets

1. Problem definition and classical methods of solution

We consider ground water flow in a confined aquifer, i.e. a permeable porous geological formation with upper and lower impermeable boundaries. The mass balance equation for stationary flow (which means that the fluid density is constant and the porous medium is not deforming), can be written as

$$\partial_x (T\partial_x h) + \partial_y (T\partial_y h) = f\,, \qquad (1)$$

where T is the aquifer transmissivity $[L^2/T]$, h is the piezometric head $[L]$ and f is the source term, i.e. the well discharge rate of abstracted water per unit area of the aquifer $[L/T]$. The development of a forecasting model requires the solution to (1) with respect to h, so that T and f must be known.

Please use the following format when citing this chapter:

Giudici, M., Meles, G.A., Parravicini, G., Ponzini, G., and Vassena, C., 2006, in IFIP International Federation for Information Processing, Volume 202, Systems, Control, Modeling and Optimization, eds. Ceragioli, F., Dontchev, A., Furuta, H., Marti, K., Pandolfi, L., (Boston: Springer), pp. 175-181.

Data on T are usually obtained from the interpretation and processing of well tests, which are very much influenced by the well characteristics (head losses due to screen and drain effect, pump position, etc.) and provide a value which can be representative of a region with a limited radius around the well, say of the same order of magnitude of the screened intervals, which could be of the order of tens of meters. As a consequence, these values are not representative of the flow processes at a regional scale, where flow is modelled in aquifers whose lateral extensions could be as great as tens or hundreds of kilometers and for which the spacing of the numerical grid could be hundreds of meters.

The T field has to be estimated, for example with the solution of an inverse problem for equation (1). This requires the computation of T, given h and f and the least prior knowledge of T.

In the mathematical and geophysical literature this inverse problem has been classically posed as a Cauchy problem, and the solution is found by integration along the flow lines (see, e.g., [8], [9], [10], [2], [3], [13]). For this it is necessary to assign T at a point for each flow line. The application of such an approach to real cases is very difficult, practically impossible. In fact it is difficult to measure T along the inflow or outflow boundary of the domain or wherever at a point along each flow line. It is also difficult to determine the flow lines with enough precision from head data which are available at a limited number of irregularly scattered points. Moreover, since the T field depends upon the hydraulic gradient, grad$h$, the integration of (1) with respect to T along a flow line is intrinsically unstable. Since the integration along each flow path is independent from the integration along the neighbouring flow lines, the instability could lead to results which do not respect any regularity of the T field among nearby flow lines (see [3] for a discussion about practical aspects).

Other approaches, related to non linear least-squares techniques, possibly with regularization, or in the framework of maximum likelihood estimation, assume some knowledge of the unknown parameter [18], [1], e.g., the fact that it is piecewise constant so that the domain can be partitioned into a number of subdomains where T is constant. This approach is known as zonation.

Instead of using additional prior information on T, which always poses problems of data effectiveness, other methods can reduce the above mentioned problem for inversion through the use of data measured at different times and therefore related to different flow situations. See [15], [14], [12], [1], [4], [16], [19], [7] among the others.

The next section describes one of these methods, the Differential System Method (DSM).

2. The Differential System Method

The simplest version of the DSM, see [5] and [11], allows for a solution of the inverse problem when two independent sets of data, $\left\{ \left(h^{(l)}, f^{(l)} \right), l = 1, 2 \right\}$, and the value of T at only one point $\mathbf{x}_0$ of the domain are available. In this case equation (1) can be written for both data sets and leads to a system of first order partial differential equations for T, which can be written in the normal form

$$\text{grad}\, T = -T\mathbf{a} + \mathbf{b}\,, \tag{2}$$

if the following *independence condition* holds:

$$\det A \neq 0\,, \tag{3}$$

where the elements of A are given by the relations

$$A_{i,l} = \partial_i h^{(l)}\,. \tag{4}$$

The T field is obtained at any point $\mathbf{x}$ by integration of the differential equation (2) in the unknown function T along any line connecting $\mathbf{x}$ to $\mathbf{x}_0$, where the value, T_0, of T at $\mathbf{x}_0$ is the initial value for the integration. The integration path, γ, can be chosen according to a *stability condition* that requires that the line integral $\int_\gamma |\mathbf{a}| dl$ be small in order that the error propagation along the integration line γ be small.

The DSM has been tested with stationary [5] and transient [17] synthetic data. A discussion on the discrete stability of the method is given in [5] and numerical experiments are shown in [6].

The numerical tests so far performed show that the stability condition is important also for the choice of the starting point $\mathbf{x}_0$. In fact if $\mathbf{x}_0$ is chosen in an area where $|\mathbf{a}|$ is great, numerical errors prevent the computation of T with a good confidence but for a small neighborhood of $\mathbf{x}_0$. Unfortunately the data on T are usually available where well tests can be performed; as a consequence, $\mathbf{x}_0$ should correspond to the location of an existing well where tests have been performed and nobody can guarantee that $|\mathbf{a}|$ is small there.

Another difficulty for the application of the DSM to real cases is the fact that data sets independent on the whole domain can be obtained for a variation of the physical boundary conditions, which is nevertheless quite rare and above all cannot be controlled. In fact boundary conditions vary as a response to climate change, modification of land use, and so on. On the other hand, a variation of the pumping schedule modifies the flow field in limited regions surrounding the pumping wells only and not throughout the whole aquifer (see, e.g., [16]).

3. The Differential System Method with multiple sets of data

The difficulties discussed at the end of the previous section might be mitigated if the DSM is modified in order to deal with several sets of data, i.e. M pairs $\left(h^{(l)}, f^{(l)}\right)$, $l = 1, \ldots, M$, with $M > 2$. Equation (1) can be written for all the available data sets. The standard version of the DSM can be applied if we locally choose the "best" pair of sets of data to build the matrix A, as defined by (4), and compute the vectors $\mathbf{a}$ and $\mathbf{b}$ to be used in (2).

In particular multiple data sets can be used pairwise to compute the vectors $\mathbf{a}$ and $\mathbf{b}$ in the following way. The domain is subdivided in subregions, where a pair of data sets can be found that best satisfies the following conditions:

1. the independence condition;
2. the stability condition;
3. the smallness of $\mu(A)$, the condition number of A.

In particular, $\mu(A)$ is computed as follows

$$\mu(A) = ||A|| \cdot ||A^{-1}|| = \frac{\sum_{i,j} a_{i,j}^2}{|\det A|}, \tag{5}$$

where the Frobenius norm is used. Once the vectors $\mathbf{a}$ and $\mathbf{b}$ have been computed with the "best" pair of sets, the DSM can be applied with the standard procedure in each subregion.

4. Numerical tests

In this section some results of simple numerical tests are shown. More complex cases have also been analysed, but the results are qualitatively very similar, so that this simple case could be more easy to be analysed and interpreted.

The reference Log(T) field is represented in figure 1, together with the position of the abstraction wells that are used to generate the synthetic head data.

In particular one set of data (set 0) is obtained with no pumping wells and nine data sets (sets 1 to 9) correspond to the cases when one well at time is pumping, with the discharge rates (in L/s) plotted in figure 1. For each data set the noise-free head data are obtained with a finite difference solution of the discrete balance equation; the assigned Dirichlet boundary conditions are linearly varying from left (100 m) to right (80 m). Then the data are corrupted with an uncorrelated noise. Here we show the results when the noise is introduced with a truncation of the piezometric heads at the third decimal digit.

The results of the standard DSM applied to the data sets 0 and 4 are shown in figure 2 when the starting point is at well no. 1 or 9. The differences between the two cases are apparent. In particular when the starting point corresponds to

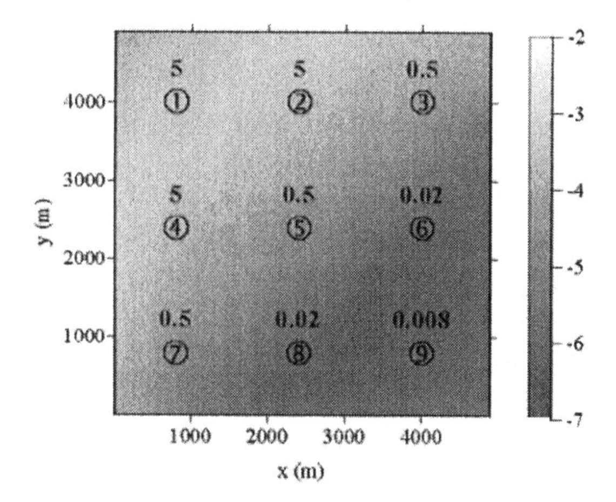

Figure 1. Reference Log(T) field (T in m^2/s). Circled numbers show the positions of the abstraction wells; numbers above the labels show discharge rates in L/s.

well no. 9 negative transmissivities have been identified in a large region (the black area of the bottom plot of figure 2).

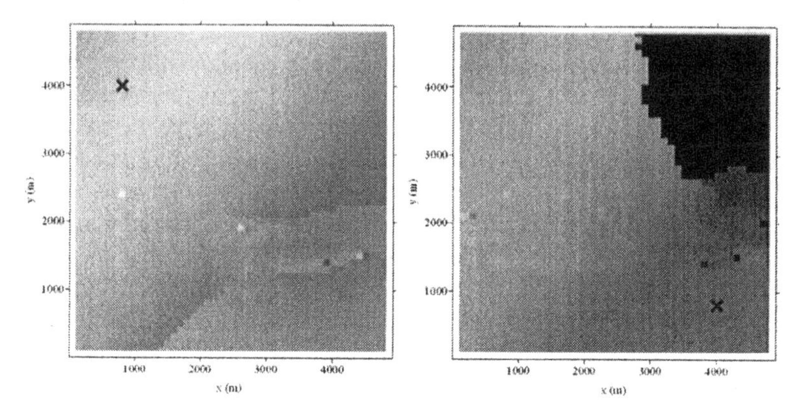

Figure 2. Log(T) field identified with the standard DSM. The crosses denote the positions of the starting point. Gray scale is the same as for figure 1.

The results obtained when all the sets of data are used simultaneously, with the technique described in the previous section, are shown in figure 3, again for the starting point at well no. 1 or 9. These results show that this approach is very useful to reduce the dependence of the final solution on the starting point, which can be chosen almost everywhere without worsening the results of the DSM.

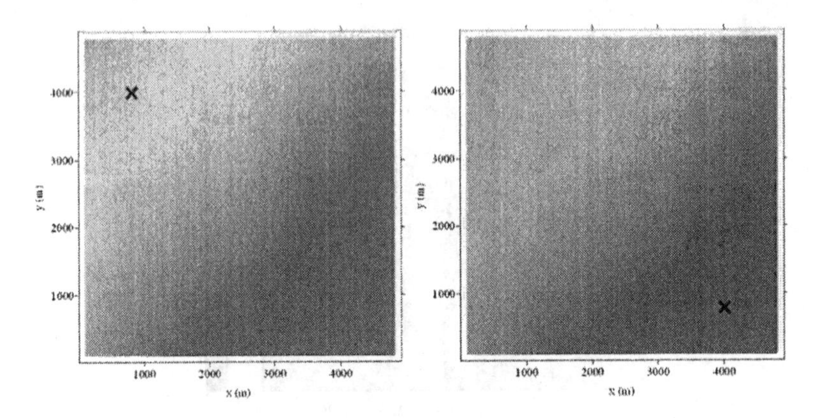

Figure 3. Log(T) field identified with the modified DSM. The crosses denote the positions of the starting point. Gray scale is the same as for figure 1.

5. Conclusions and perspectives

The new features of the DSM are very important, because they permit to limit the problems related to the choice of the starting point and to the availability of data sets which are independent throughout the whole domain.

Several perspectives are still open.

When M sets of data are available, M balance equations like (1) can be written, one for each data set, so that A becomes a rectangular matrix with M rows and two columns. In this case the vectors $\mathbf{a}$ and $\mathbf{b}$ can be computed with the least-squares technique. Numerical tests are going on to evaluate this alternative.

The numerical tests presented here show the importance of considering $\mu(A)$ for the computation of the vectors $\mathbf{a}$ and $\mathbf{b}$, which is the first stage of the DSM. Also the successive stage, the choice of the integration path, can be improved by considering different combinations of the three conditions introduced in this paper: the independence condition, the stability condition and the $\mu(A)$ condition.

Eventually, the method can be applied to transient data; in that case a large number of sets of data, corresponding to measurements at different times, could be available. The criteria for the estimation of the time derivative of the piezometric head and for the choice of the data sets to be used are of paramount importance.

References

[1] J. Carrera, S.P. Neuman. Estimation of aquifer parameters under transient and steady–state conditions: 1, maximum likelihood method incorporating prior information. *Water Resour.*

Res. 22:199-210, 1986.

[2] G. Chavent. Analyse fonctionnelle et identification de coefficients répartis dans les équations aux dérivées partielles. Thése d'etat, Fac. des Science de Paris, 1971.

[3] Y. Emsellem, G. de Marsily. An automatic solution for the inverse problem. *Water Resour. Res.* 7:1264-1283, 1971.

[4] T.R. Ginn, J.H. Cushman, M.H. Houch. A continuous{time inverse operator for groundwater and contaminant transport modeling: deterministic case. *Water Resour. Res.* 26:241-252, 1990.

[5] M. Giudici, G. Morossi, G. Parravicini, G. Ponzini. A new method for the identification of distributed transmissivities. *Water Resour. Res.* 31:1969-1988, 1995.

[6] M. Giudici, F. Delay, G. de Marsily, G. Parravicini, G. Ponzini, A. Rosazza. Discrete stability of the Differential System Method evaluated with geostatistical techniques. *Stochastic Hydrol. and Hydraul.* 12:191-204, 1998.

[7] S. Liu, T.-C. J. Yeh, R. Gardiner. Effectiveness of hydraulic tomography: Sandbox experiments *Water Resour. Res.* doi:10.1029/2001WR000338, 2002.

[8] R.W. Nelson. In place measurement of permeability in heterogeneous media, 1, Theory of a proposed method. *J. Geophys. Res.* 65:1753-1760, 1960.

[9] R.W. Nelson. In place measurement of permeability in heterogeneous media, 2, Experimental and computational considerations. *J. Geophys. Res.* 66:2469-2478, 1961.

[10] R.W. Nelson. Condition for determining areal permeability distribution by calculation. *Soc. Pet. Eng. J.* 2:223-224, 1962.

[11] G. Parravicini, M. Giudici, G. Morossi, G. Ponzini. Minimal a priori assignment in a direct method for determining phenomenological coefficients uniquely. *Inverse Problems* 11:611-629, 1995.

[12] G. Ponzini, A. Lozej. Identification of aquifer transmissivities: the comparison model method. *Water Resour. Res.* 18:597-622, 1982.

[13] G.R. Richter. An inverse problem for the steady state diffusion equation. *SIAM J. Math. Anal.* 41:210-221, 1981.

[14] B. Sagar, S. Yakowitz, L. Duckstein. A direct method for the identification of the parameters of dynamic non-homogeneous aquifers. *Water Resour. Res.* 11:563-570, 1975.

[15] S. Scarascia, G. Ponzini. An approximate solution for the inverse problem in hydraulics. *L'Energia Elettrica* 49:518-531, 1972.

[16] M.F. Snodgrass, P.K. Kitanidis. Transmissivity identification through multi–directional aquifer stimulation. *Stochastic Hydrol. and Hydraul.* 12:299-316, 1998.

[17] R. Vàzquez Gonzàlez, M. Giudici, G. Parravicini, G. Ponzini. The differential system method for the identification of transmissivity and storativity. *Transport in Porous Media* 26:339-371, 1997.

[18] W.-G. W. Yeh. Review of parameter identification procedures in groundwater hydrology: the inverse problem. *Water Resour. Res.* 22:95-108, 1986.

[19] T.-C. J. Yeh, S. Liu. Hydraulic tomography: Development of a new aquifer test method. *Water Resour. Res.* 36:2095-2105, 2000.

CHEAP CONTROL PROBLEM OF LINEAR SYSTEMS WITH DELAYS: A SINGULAR PERTURBATION APPROACH

V. Y. Glizer [1]

[1] *Technion - Israel Institute of Technology, Faculty of Aerospace Engineering, Haifa, 32000, Israel, valery@techunix.technion.ac.il*

Abstract A quadratic cheap control of linear systems with multiple state delays is considered. This optimal control problem is transformed to an optimal control problem of singularly perturbed systems. A composite suboptimal control of the latter is designed based on its asymptotic decomposition into two much simpler parameter-free subproblems, the slow and fast ones. Using this composite control, a suboptimal control of the original cheap control problem is constructed and justified for two classes of the initial function for the state variable. An illustrative example is presented.

keywords: time-delayed system, cheap control, singular perturbation, composite control.

1. Introduction

The cheap control problem, i.e. an optimal control problem with a small control cost (with respect to a state cost) in the cost functional, is of considerable importance in such topics of control theory as: singular (degenerate) optimal control and its regularization, limitations of linear and nonlinear regulators, high gain control, inverse control problems, linear optimal filtering with a small noise in the observation, robust controllability of systems with disturbances, and some others.

The smallness of the control cost yields the singular perturbation in the Hamilton-Jacobi-Bellman equation, as well as in the Hamilton boundary-value problem, associated with the original problem by control optimality conditions.

The cheap control problem for differential equations without delays has been extensively investigated in the open literature (see e.g. [1-3] and references therein), while there are only few works in the open literature devoted to analysis of the cheap control problem with a delayed dynamics (see [4-6]).

Please use the following format when citing this chapter:

Glizer, V.Y., 2006, in IFIP International Federation for Information Processing, Volume 202, Systems, Control, Modeling and Optimization, eds. Ceragioli, F., Dontchev, A., Furuta, H., Marti, K., Pandolfi, L., (Boston: Springer), pp. 183-193.

In [4], a finite horizon linear-quadratic optimal control problem was considered with a small cost of the control in the cost functional, and with single point-wise and distributed state delays in the dynamics. It was assumed that the delay is small of order of the control cost, and that the initial function for the state variable is continuous. The zero-order asymptotic solution to the singularly perturbed set of ordinary and partial functional-differential equations of Riccati type, associated with the original problem by the control optimality conditions, has been derived. Based on this solution, the zero-order asymptotic expansion of the optimal trajectory, as well as the suboptimal feedback control, were obtained. In [5,6], a similar cheap control problem was studied for the case of a nonsmall delay. The zero-order asymptotic solution to the singularly perturbed set of Riccati-type functional-differential equations, associated with the original problem, has been obtained in [5]. Using this asymptotic solution, a suboptimal control was constructed. In [6], based on a proper transformation of the original problem and, then, on an asymptotic decomposition of the resulting one into two much simpler parameter-free subproblems, a suboptimal state-feedback control of the original problem was designed.

In this paper, a finite-horizon linear-quadratic cheap control problem with multiple point-wise and distributed state delays in the dynamics is analyzed. Two classes of the initial function for the state variable are considered: (1) measurable square-integrable functions; (2) measurable essentially bounded functions. A direct method of suboptimal solution of the original cheap control problem is proposed. This method is based on: (i) an equivalent transformation of the original problem to a control problem with singularly perturbed dynamics; (ii) an asymptotic decomposition of the resulting problem into two much simpler parameter-free subproblems, the slow and fast ones. It should be noted that the fast state variable of the control problem, obtained after the transformation, becomes a control in the slow subproblem. Thus, the slow subproblem contains not only the state delays in the dynamics, but also the control ones. In spite of the latter, the set of Riccati-type functional-differential equations, associated with the slow subproblem, is much simpler than the one associated with the original cheap control problem. The fast subproblem does not contain delays, and it is solved analytically. Using the optimal feedback controls of the slow and fast subproblems, a composite control for the transformed problem and, then, a suboptimal state-feedback control for the original problem are designed.

The following main notations are applied in the paper: (1) E^n is the n-dimensional real Euclidean space; (2) $\|\cdot\|$ denotes the Euclidean norm either of a vector or of a matrix; (3) the prime denotes the transposition of a matrix A, (A') or of a vector x, (x'); (4) $L^2[b,c;E^n]$ is the space of n-dimensional vector-valued functions $x(t)$ defined, measurable and square-integrable on the interval $[b,c)$, $\|x(\cdot)\|_{L^2} = \left(\int_b^c x'(t)x(t)dt\right)^{1/2}$ denotes the norm in this space;

(5) $L^\infty[b,c;E^n]$ is the space of n-dimensional vector-valued functions $x(t)$ defined, measurable and essentially bounded on the interval $[b,c)$, $\|x(\cdot)\|_{L^\infty} = \text{ess sup}_{t\in[b,c)}\|x(t)\|$ denotes the norm in this space; (6) I_n is the n-dimensional identity matrix.

2. Problem statement

Consider the controlled system

$$dz(t)/dt = \sum_{i=0}^{N} A_i z(t-h_i) + \int_{-h}^{0} G(\tau)z(t+\tau)d\tau + Bu(t), \quad t \in [0,T], \quad (2.1)$$

where $z(t) \in E^n, u(t) \in E^r, (n \geq r)$, ($u$ is a control); $N > 0$ is an integer; $0 = h_0 < h_1 < ... < h_N = h$ are given constant time-delays; $A_i, (i = 0,1,...,N), G(\tau)$ and B are given time-invariant matrices of corresponding dimensions; B has full rank r; the matrix-valued function $G(\tau)$ is piece-wise continuous for $\tau \in [-h,0]$; $T > 0$ is a given finite duration of the control process.

Using that rank$B = r$ and [7], one can transform (2.1) to an equivalent linear controlled system with state delays, in which the matrix of coefficients for the control has the form $\begin{pmatrix} 0 \\ I_r \end{pmatrix}$. Therefore, in the sequel we assume (without a loss of generality) that B has such a form.

Initial conditions for (2.1) have the form

$$z(\tau) = \varphi(\tau), \quad \tau \in [-h,0); \quad z(0) = \varphi_0, \quad (2.2)$$

where $\varphi(\tau), \tau \in [-h,0)$ and φ_0 are given vector-valued function and vector, respectively. In the sequel, the following two cases are considered: (1) $\varphi(\cdot) \in L^2[-h,0;E^n]$; (2) $\varphi(\cdot) \in L^\infty[-h,0;E^n]$.

Let partition $z(t), A_i, (i = 0,1,...,N)$ and $G(\tau)$ in the accordance with the block-form of B

$$z(t) = \begin{pmatrix} x(t) \\ y(t) \end{pmatrix}, \ A_i = \begin{pmatrix} A_{i1} & A_{i2} \\ A_{i3} & A_{i4} \end{pmatrix}, \ G(\tau) = \begin{pmatrix} G_1(\tau) & G_2(\tau) \\ G_3(\tau) & G_4(\tau) \end{pmatrix}, \quad (2.3)$$

where $x(t) \in E^{n-r}, y(t) \in E^r$; A_{i1} and $G_1(\tau)$ are of the dimension $(n-r) \times (n-r)$, while A_{i4} and $G_4(\tau)$ are of the dimension $r \times r$.

Using (2.3) and the block form of B, one can rewrite (2.1) as follows

$$dx(t)/dt = \sum_{i=0}^{N} [A_{i1}x(t-h_i) + A_{i2}y(t-h_i)]$$

$$+ \int_{-h}^{0} [G_1(\tau)x(t+\tau) + G_2(\tau)y(t+\tau)]\,d\tau, \quad (2.4)$$

$$dy(t)/dt = \sum_{i=0}^{N} [A_{i3}x(t-h_i) + A_{i4}y(t-h_i)]$$

$$+ \int_{-h}^{0} [G_3(\tau)x(t+\tau) + G_4(\tau)y(t+\tau)]\, d\tau + u(t), \ \ t \in [0,T]. \qquad (2.5)$$

For system (2.4)-(2.5) with initial conditions (2.2) the following performance index is considered

$$J_\varepsilon(u) \triangleq \int_0^T \left[x'(t)D_x x(t) + y'(t)D_y y(t) + \varepsilon^2 u'(t)Mu(t)\right] dt \to \min_u, \qquad (2.6)$$

where D_x is a symmetric positive semi-definite matrix, while D_y and M are symmetric positive definite ones; ε is a small positive parameter.

In the sequel, (2.4)-(2.6),(2.2) is called the original problem (OP).

The objective of the paper is to construct a suboptimal state-feedback control for (2.4)-(2.6),(2.2) uniformly valid for all sufficiently small $\varepsilon > 0$.

3. Control optimality conditions for the OP

Using results of [8] yields that, for a given $\varepsilon > 0$, the optimal state-feedback control and the optimal value of the cost functional of the OP have the form

$$u_\varepsilon^*[z(t), z_h(t), t] = -\varepsilon^{-2}M^{-1}B'\left[P(t)z(t) + \int_{-h}^{0} Q(t,\tau)z(t+\tau)d\tau\right], \qquad (3.1)$$

$$J_\varepsilon^* = \varphi_0'P(0)\varphi_0 + 2\varphi_0'\int_{-h}^{0} Q(0,\tau)\varphi(\tau)d\tau + \int_{-h}^{0}\int_{-h}^{0} \varphi'(\tau)R(0,\tau,\theta)\varphi(\theta)d\tau d\theta, \qquad (3.2)$$

where $z_h(t) = \{z(t+\tau) \ \forall \tau \in [-h,0)\}$, and $\{P(t), Q(t,\tau), R(t,\tau,\theta)\}$ is the unique solution of the following set of Riccati-type functional-differential equations in the domain $\Omega = \{(t,\tau,\theta) : t \in [0,T], \tau \in [-h,0], \theta \in [-h,0]\}$

$$dP(t)/dt = -P(t)A_0 - A_0'P(t) - Q(t,0) - Q'(t,0) + P(t)S_\varepsilon P(t) - D, \quad (3.3)$$

$$(\partial/\partial t - \partial/\partial\tau)Q(t,\tau) = -A_0'Q(t,\tau) - P(t)\left[\sum_{i=1}^{N-1} A_i\delta(\tau+h_i) + G(\tau)\right]$$
$$-R(t,0,\tau) + P(t)S_\varepsilon Q(t,\tau), \qquad (3.4)$$

$$(\partial/\partial t - \partial/\partial\tau - \partial/\partial\theta)R(t,\tau,\theta) = -\left[\sum_{i=1}^{N-1} A_i'\delta(\tau+h_i) + G'(\tau)\right] Q(t,\theta)$$

$$-Q'(t,\tau)\left[\sum_{i=1}^{N-1} A_i\delta(\theta+h_i) + G(\theta)\right] + Q'(t,\tau)S_\varepsilon Q(t,\theta), \qquad (3.5)$$

$$P(T) = 0, \quad Q(T,\tau) = 0, \quad R(T,\tau,\theta) = 0, \tag{3.6}$$

$$Q(t,-h) = P(t)A_N, \; R(t,-h,\theta) = A_N^{'}Q(t,\theta), \; R(t,\tau,-h) = Q^{'}(t,\tau)A_N. \tag{3.7}$$

Here, $S_\varepsilon = \varepsilon^{-2}BM^{-1}B^{'}$, $D = \text{diag}(D_x, D_y)$, $\delta(\cdot)$ is the delta-function.

For a given $\varepsilon > 0$, problem (3.3)-(3.7) is very complicated. Moreover, if $\varepsilon \to 0$, it becomes an ill-posed one and, therefore, much more difficult.

In this paper, an approach to constructing a suboptimal state-feedback control for this problem is proposed. This approach directly uses the singular perturbation nature of the OP, and it does not require to solve problem (3.3)-(3.7).

The suboptimal control is constructed in four stages. At the first stage, the OP is transformed equivalently to a control problem with singularly perturbed dynamics and ε-free cost functional. At the second stage, the resulting (transformed) problem is decomposed asymptotically into two much simpler ε-free subproblems, the slow and fast ones. At the third stage, based on the optimal feedback controls of the slow and fast subproblems, a composite state-feedback control for the transformed problem is designed. Finally, at the fourth stage, a suboptimal control for the original problem is obtained based on the composite control of the transformed system.

4. Transformation of the OP

By the control transformation

$$u(t) = (1/\varepsilon)v(t), \tag{4.1}$$

where v is a new control, (2.4)-(2.6) becomes

$$dx(t)/dt = \sum_{i=0}^{N} [A_{i1}x(t-h_i) + A_{i2}y(t-h_i)]$$

$$+ \int_{-h}^{0} [G_1(\tau)x(t+\tau) + G_2(\tau)y(t+\tau)]\, d\tau, \tag{4.2}$$

$$\varepsilon dy(t)/dt = \varepsilon \left\{ \sum_{i=0}^{N} [A_{i3}x(t-h_i) + A_{i4}y(t-h_i)] \right.$$

$$\left. + \int_{-h}^{0} [G_3(\tau)x(t+\tau) + G_4(\tau)y(t+\tau)]\, d\tau \right\} + v(t), \quad t \in [0,T], \tag{4.3}$$

$$J(v) \triangleq \int_0^T \left[x^{'}(t)D_x x(t) + y^{'}(t)D_y y(t) + v^{'}(t)Mv(t) \right] dt \to \min_{v}. \tag{4.4}$$

In the sequel, (4.2)-(4.4),(2.2) is called the transformed problem (TP).

5. Asymptotic decomposition of the TP

5.1 Slow subproblem

The slow subproblem is obtained from the TP by setting there formally $\varepsilon = 0$ and redenoting x, y, v and J by x_s, y_s, v_s and J_s, respectively. Thus, one obtains

$$dx_s(t)/dt = \sum_{i=0}^{N}[A_{i1}x_s(t-h_i) + A_{i2}y_s(t-h_i)]$$

$$+ \int_{-h}^{0}[G_1(\tau)x_s(t+\tau) + G_2(\tau)y_s(t+\tau)]d\tau, \quad t \in [0,T], \tag{5.1}$$

$$v_s(t) = 0, \quad t \in [0,T], \tag{5.2}$$

$$J_s(y_s) = \int_0^T [x_s'(t)D_x x_s(t) + y_s'(t)D_y y_s(t)]dt \to \min_{y_s}, \tag{5.3}$$

$$x_s(\tau) = \varphi_x(\tau), \ \ y_s(\tau) = \varphi_y(\tau), \ \ \tau \in [-h, 0); \ \ x_s(0) = \varphi_{0x}, \tag{5.4}$$

where $\varphi_x(\tau)$ and $\varphi_y(\tau)$ are the upper and lower blocks of $\varphi(\tau)$ of the dimensions $n-r$ and r, respectively; φ_{0x} is the upper block of φ_0 of the dimension $n-r$.

The slow subproblem consists of equation (5.2) and the problem of minimizing the cost functional $J_s(y_s)$ along trajectories of (5.1),(5.4). This problem is called in the sequel the reduced-order problem (ROP). In the ROP, y_s is a control, i.e. this problem has time-delays in state and control.

Due to [8], the optimal feedback control of the ROP exists, is unique and has the form

$$y_s^*[x_s(t), x_{sh}(t), y_{sh}(t), t] = -D_y^{-1}\{[A_{02}'P_s(t) + Q_{s2}'(t,0)]x_s(t)$$

$$+ \int_{-h}^{0}[A_{02}'Q_{s1}(t,\tau) + R_{s1}'(t,\tau,0)]x_s(t+\tau)d\tau$$

$$+ \int_{-h}^{0}[A_{02}'Q_{s2}(t,\tau) + R_{s2}(t,0,\tau)]y_s(t+\tau)d\tau\}, \tag{5.5}$$

where $x_{sh}(t) = \{x_s(t+\tau)\,\forall\tau \in [-h,0)\}$, $y_{sh}(t) = \{y_s(t+\tau)\,\forall\tau \in [-h,0)\}$, and $P_s(t), Q_{sk}(t,\tau), R_{sk}(t,\tau,\theta), (k = 1,2)$ and a matrix $R_{s0}(t,\tau,\theta)$ form the unique solution to the following set of Riccati-type functional-differential equations in Ω

$$dP_s(t)/dt = -P_s(t)A_{01} - A_{01}'P_s(t) - Q_{s1}(t,0) - Q_{s1}'(t,0) - D_x$$

$$+[P_s(t)A_{02} + Q_{s2}(t,0)]D_y^{-1}[P_s(t)A_{02} + Q_{s2}(t,0)]', \tag{5.6}$$

$$(\partial/\partial t - \partial/\partial\tau)Q_{s1}(t,\tau) = -A_{01}'Q_{s1}(t,\tau) - P_s\left[\sum_{i=1}^{N-1} A_{i1}\delta(\tau+h_i) + G_1(\tau)\right]$$

$$-R_{s0}(t,0,\tau)+[P_s(t)A_{02}+Q_{s2}(t,0)]D_y^{-1}[A_{02}'Q_{s1}(t,\tau)+R_{s1}'(t,\tau,0)], \tag{5.7}$$

$$(\partial/\partial t-\partial/\partial\tau)Q_{s2}(t,\tau)=-A_{01}'Q_{s2}(t,\tau)-P_s\left[\sum_{i=1}^{N-1}A_{i2}\delta(\tau+h_i)+G_2(\tau)\right]$$

$$-R_{s1}(t,0.\tau)+[P_s(t)A_{02}+Q_{s2}(t,0)]D_y^{-1}[A_{02}'Q_{s2}(t,\tau)+R_{s2}(t,0,\tau)], \tag{5.8}$$

$$(\partial/\partial t-\partial/\partial\tau-\partial/\partial\theta)R_{s0}(t,\tau,\theta)=-\left[\sum_{i=1}^{N-1}A_{i1}'\delta(\tau+h_i)+G_1'(\tau)\right]Q_{s1}(t,\theta)$$

$$-Q_{s1}'(t,\tau)\left[\sum_{i=1}^{N-1}A_{i1}\delta(\theta+h_i)+G_1(\theta)\right]$$

$$+[Q_{s1}'(t,\tau)A_{02}+R_{s1}(t,\tau,0)]D_y^{-1}[A_{02}'Q_{s1}(t,\theta)+R_{s1}'(t,\theta,0)], \tag{5.9}$$

$$(\partial/\partial t-\partial/\partial\tau-\partial/\partial\theta)R_{s1}(t,\tau,\theta)=-\left[\sum_{i=1}^{N-1}A_{i1}'\delta(\tau+h_i)+G_1'(\tau)\right]Q_{s2}(t,\theta)$$

$$-Q_{s1}'(t,\tau)\left[\sum_{i=1}^{N-1}A_{i2}\delta(\theta+h_i)+G_2(\theta)\right]$$

$$+[Q_{s1}'(t,\tau)A_{02}+R_{s1}(t,\tau,0)]D_y^{-1}[A_{02}'Q_{s2}(t,\theta)+R_{s2}(t,0,\theta)], \tag{5.10}$$

$$(\partial/\partial t-\partial/\partial\tau-\partial/\partial\theta)R_{s2}(t,\tau,\theta)=-\left[\sum_{i=1}^{N-1}A_{i2}'\delta(\tau+h_i)+G_2'(\tau)\right]Q_{s2}(t,\theta)$$

$$-Q_{s2}'(t,\tau)\left[\sum_{i=1}^{N-1}A_{i2}\delta(\theta+h_i)+G_2(\theta)\right]$$

$$+[Q_{s2}'(t,\tau)A_{02}+R_{s2}(t,\tau,0)]D_y^{-1}[A_{02}'Q_{s2}(t,\theta)+R_{s2}(t,0,\theta)], \tag{5.11}$$

$$P_s(T)=0,\ \ Q_{sk}(T,\tau)=0,\ \ R_{sj}(T,\tau,\theta),\ \ (k=1,2;\ j=0,1,2), \tag{5.12}$$

$$Q_{sk}(t,-h)=P_s(t)A_{Nk},\ \ (k=1,2), \tag{5.13}$$

$$R_{s0}(t,-h,\tau)=A_{N1}'Q_{s1}(t,\tau),\ \ R_{s0}(t,\tau,-h)=Q_{s1}'(t,\tau)A_{N1}, \tag{5.14}$$

$$R_{s1}(t,-h,\tau)=A_{N1}'Q_{s2}(t,\tau),\ \ R_{s1}(t,\tau,-h)=Q_{s1}'(t,\tau)A_{N2}, \tag{5.15}$$

$$R_{s2}(t,-h,\tau)=A_{N2}'Q_{s2}(t,\tau),\ \ R_{s2}(t,\tau,-h)=Q_{s2}'(t,\tau)A_{N2}. \tag{5.16}$$

REMARK 1 *Although set of equations (5.6)-(5.16), associated with the ROP, does not look like a simple one, it is essentially simpler than set (3.3)-(3.7), associated with the OP. The number of unknown scalar functions in the first set* $[2n^2+n-2nr+r(r-1)/2]$ *is much less that the one* $(2n^2+n)$ *in the second set. Moreover, the first set is* ε*-free and, therefore, it is well-posed.*

5.2 Fast subproblem

The fast subproblem is obtained as follows: (1) the slow variable $x(\cdot)$ is removed from equation (4.3) and performance index (4.4) of the TP; (2) the transformation of variables $t = \varepsilon\xi, y(\varepsilon\xi) = y_f(\xi), v(\varepsilon\xi) = v_f(\xi), J(v(\varepsilon\xi)) = \varepsilon J_f(v_f(\xi))$ is made in the resulting problem, where ξ, y_f, v_f and J_f are new independent variable, state, control and cost functional, respectively. As a result, one obtains the problem

$$dy_f(\xi)/d\xi = \varepsilon\left[\sum_{i=0}^{N} A_{i4}y_f(\xi - h_i/\varepsilon) + \int_{-h}^{0} G_4(\tau)y_f(\xi + \tau/\varepsilon)d\tau\right] + v_f(\xi), \tag{5.17}$$

$$J_f(v_f) = \int_0^{T/\varepsilon} [y_f'(\xi)D_y y_f(\xi) + v_f'(\xi)Mv_f(\xi)]d\xi \to \min_{v_f}. \tag{5.18}$$

Now, neglecting formally the term with the multiplier ε in (5.17) and replacing T/ε by $+\infty$ in (5.18) yield the fast subproblem

$$dy_f(\xi)/d\xi = v_f(\xi), \tag{5.19}$$

$$J_f(v_f) = \int_0^{+\infty} [y_f'(\xi)D_y y_f(\xi) + v_f'(\xi)Mv_f(\xi)]d\xi \to \min_{v_f}. \tag{5.20}$$

Due to [9], the fast subproblem with a given initial value $y_f(0)$ of the state variable has the unique optimal state-feedback control

$$v_f^*[y_f(\xi)] = -M^{-1}P_f y_f(\xi), \tag{5.21}$$

where P_f is the unique symmetric positive definite solution of the algebraic Riccati equation

$$P_f M^{-1} P_f - D_y = 0. \tag{5.22}$$

Moreover, the optimal trajectory $y_f(\xi)$ satisfies the inequality

$$\|y_f(\xi)\| \leq a\exp(-\beta\xi)\|y_f(0)\|, \quad \xi \geq 0, \tag{5.23}$$

where $a > 0$ and $\beta > 0$ are some constants.

6. Composite control for the TP

The design of composite control for problem (4.2)-(4.4),(2.2) consists of two stages. At the first stage, the auxiliary control is constructed

$$v_a[x(t), y(t), x_h(t), y_h(t), t] = v_s(t) + v_f^*[\tilde{y}(t/\varepsilon)], \tag{6.1}$$

where $\tilde{y}(t/\varepsilon)$ is defined as follows

$$\tilde{y}(t/\varepsilon) \triangleq y(t) - y_s^*[x(t), x_h(t), \bar{y}_h(t), t], \quad \bar{y}_h(t) \triangleq y_h(t) - y_{fh}(t/\varepsilon), \tag{6.2}$$

$$y_{fh}(t/\varepsilon) \triangleq \{y_f((t+\tau)/\varepsilon)\ \forall\tau\in[-h,0)\},\ \ y_f(\xi)=0\ \forall\xi<0. \qquad (6.3)$$

By substituting (6.2)-(6.3) into (6.1), one has after some rearrangement

$$v_a[x(t),y(t),x_h(t),y_h(t),t] = -M^{-1}P_f\{y(t)$$
$$+D_y^{-1}[(A_{02}^{'}P_s(t)+Q_{s2}^{'}(t,0))x(t)$$
$$+\int_{-h}^{0}(A_{02}^{'}Q_{s1}(t,\tau)+R_{s1}^{'}(t,\tau,0))x(t+\tau)d\tau$$
$$+\int_{-h}^{0}(A_{02}^{'}Q_{s2}(t,\tau)+R_{s2}(t,0,\tau))(y(t+\tau)-y_f((t+\tau)/\varepsilon))d\tau]\}. \qquad (6.4)$$

Expression (6.4) for $v_a[\cdot]$ contains $y_f(\cdot)$. At the second stage, using (5.23) and the limit process for $\varepsilon\to 0$, we eliminate the term depending on $y_f(\cdot)$ from (6.4). Thus, we obtain the composite state-feedback control for system (4.2)-(4.3),(2.2)

$$v_c[x(t),y(t),x_h(t),y_h(t),t] = -M^{-1}P_f\{y(t)$$
$$+D_y^{-1}[(A_{02}^{'}P_s(t)+Q_{s2}^{'}(t,0))x(t)$$
$$+\int_{-h}^{0}(A_{02}^{'}Q_{s1}(t,\tau)+R_{s1}^{'}(t,\tau,0))x(t+\tau)d\tau$$
$$+\int_{-h}^{0}(A_{02}^{'}Q_{s2}(t,\tau)+R_{s2}(t,0,\tau))y(t+\tau)d\tau]\}. \qquad (6.5)$$

7. Suboptimal control for the OP

By using (4.1) and (6.5), one obtains the state-feedback control for (2.4)-(2.6),(2.2)

$$u_{so}[x(t),y(t),x_h(t),y_h(t),t] = \varepsilon^{-1}v_c[x(t),y(t),x_h(t),y_h(t),t]. \qquad (7.1)$$

Let $J_{\varepsilon 1}^{so}$ be the value of the cost functional in (2.6) obtained by employing the control $u_{so}[\cdot]$ in system (2.4)-(2.5),(2.2) in the case $\varphi(\cdot)\in L^2[-h,0;E^n]$, while $J_{\varepsilon 2}^{so}$ be such a value in the case $\varphi(\cdot)\in L^\infty[-h,0;E^n]$. Let $J_{\varepsilon 1}^{*}$ and $J_{\varepsilon 2}^{*}$ be the optimal values of the cost functional in the OP corresponding to the first and second cases of $\varphi(\cdot)$, respectively.

THEOREM 2 *There exists a positive number ε_0, such that, for all $\varepsilon\in(0,\varepsilon_0]$, the following inequalities are satisfied:*
$0<J_{\varepsilon 1}^{so}-J_{\varepsilon 1}^{*}\le a\varepsilon^{3/2}[\|\varphi_0\|+\|\varphi(\cdot)\|_{L^2}]^2$,
$0<J_{\varepsilon 2}^{so}-J_{\varepsilon 2}^{*}\le a\varepsilon^{2}[\|\varphi_0\|+\|\varphi(\cdot)\|_{L^\infty}]^2$,
where $a>0$ is some constant independent of ε.

8. Example

Consider the scalar cheap control problem

$$dy(t)/dt = y(t) + 2y(t-0.5) + u(t), \tag{8.1}$$

$$J_\varepsilon(u) \triangleq \int_0^2 [y^2(t) + \varepsilon^2 u^2(t)]dt \to \min_u \tag{8.2}$$

subject to each of the following versions of the initial conditions

$$y(\tau) = \varphi_1(\tau) = |\tau + 0.251|^{-1/4}, \quad \tau \in [-0.5, 0); \;\; y(0) = \varphi_0 = 1.5, \tag{8.3}$$

$$y(\tau) = \varphi_2(\tau) = \tau + 1.41, \quad \tau \in [-0.5, 0); \;\; y(0) = \varphi_0 = 1.5. \tag{8.4}$$

It is seen that $\varphi_1(\cdot) \in L^2[-0.5, 0; E^1]$, while $\varphi_2(\cdot) \in L^\infty[-0.5, 0; E^1]$.

Due to Section 3, the optimal control for (8.1)-(8.2) with (8.3) (or (8.4)) is

$$u_\varepsilon^*[y(t), y_{0.5}(t), t] = -\varepsilon^{-2}[P(t)y(t) + \int_{-0.5}^0 Q(t,\tau)y(t+\tau)dt], \tag{8.5}$$

where $y_{0.5}(t) = \{y(t+\tau) \; \forall \tau \in [-0.5, 0)\}$, and $P(t), Q(t,\tau)$ along with a function $R(t,\tau,\theta)$ form the unique solution of the following set of Riccati-type functional-differential equations in the domain $\Omega = \{(t,\tau,\theta) : t \in [0,2], \tau \in [-0.5, 0], \theta \in [-0.5, 0]\}$

$$dP(t)/dt = -2P(t) - 2Q(t,0) + \varepsilon^{-2}P^2(t) - 1, \;\; P(2) = 0, \tag{8.6}$$

$$(\partial/\partial t - \partial/\partial \tau)Q(t,\tau) = [\varepsilon^{-2}P(t) - 1]Q(t,\tau) - R(t,0,\tau), \;\; Q(2,\tau) = 0, \tag{8.7}$$

$$(\partial/\partial t - \partial/\partial \tau - \partial/\partial \theta)R(t,\tau,\theta) = \varepsilon^{-2}Q(t,\tau)Q(t,\theta), \;\; R(2,\tau,\theta) = 0, \tag{8.8}$$

$$Q(t,-0.5) = 2P(t), \quad R(t,-0.5,\tau) = R(t,\tau,-0.5) = 2Q(t,\tau). \tag{8.9}$$

Although all equations of this set are scalar, its solution is a complicated task, especially, for $\varepsilon \to 0$.

Let construct the suboptimal control for (8.1)-(8.2). By control transformation (4.1) in (8.1)-(8.2) and asymptotic decomposition of the resulting (transformed) problem, one obtains the slow subproblem, which consists of the equation $v_s(t) = 0, t \in [0,2]$ and the ROP $\int_0^2 y_s^2(t)dt \to \min_{y_s}$ yielding the solution $y_s^*(t) = 0, t \in [0,2]$. The fast subproblem is the scalar version of (5.19)-(5.20) with $D_y = M = 1$ yielding the optimal control $v_f^*[y_f(\xi)] = -y_f(\xi)$. Using these solutions of the slow and fast subproblems, and (6.5) and (7.1), we obtain the simple analytical memoryless expression for the suboptimal control $u_{so}[y(t), y_{0.5}(t), t] = -\varepsilon^{-1}y(t)$.

Let $J_{\varepsilon,k}^*$, $(k = 1,2)$ be the optimal values of the cost functional in (8.1)-(8.2) with initial conditions (8.3) and (8.4), respectively. Let $J_{\varepsilon,k}^{so}$, $(k = 1,2)$ be the values of the cost functional in (8.2) obtained by employing $u_{so}[\cdot]$ in (8.1) with initial conditions (8.3) and (8.4), respectively. In the following table, these values are presented for some values of ε.

Table 1. Values $J^*_{\varepsilon 1}$, $J^{so}_{\varepsilon 1}$ and $J^*_{\varepsilon 2}$, $J^{so}_{\varepsilon 2}$

ε	0.1	0.08	0.06	0.04	0.02
$J^*_{\varepsilon 1}$	0.434	0.311	0.202	0.122	0.054
$J^{so}_{\varepsilon 1}$	0.562	0.383	0.243	0.136	0.057
$J^*_{\varepsilon 2}$	0.346	0.256	0.177	0.109	0.050
$J^{so}_{\varepsilon 2}$	0.404	0.287	0.192	0.114	0.051

By using Table 1, the following inequalities are obtained:
$J^{so}_{\varepsilon 1} - J^*_{\varepsilon 1} \leq 0.48\varepsilon^{3/2}[\|\varphi_0\| + \|\varphi_1\|_{L^2}]^2$, $J^{so}_{\varepsilon 2} - J^*_{\varepsilon 2} \leq 0.68\varepsilon^2[\|\varphi_0\| + \|\varphi_2\|_{L^\infty}]^2$.

References

[1] P.V. Kokotovic, H.K. Khalil, J. O'Reilly. *Singular perturbation methods in control: analysis and design*. Academic Press, London, 1986.

[2] J.H. Braslavsky, R.H. Middleton, J.S. Frendenberg. Cheap control performance of a class of nonright-invertible nonlinear systems. *IEEE Trans. Automat. Control*. 47:1314-1319, 2002.

[3] E.N. Smetannikova, V.A. Sobolev. Regularization of cheap periodic control problems. *Automat. Remote Control*. 66:903-916, 2005.

[4] V.Y. Glizer. Asymptotic solution of a linear quadratic optimal control problem with delay in state and cheap control. Research Report, Technion-Israel Institute of Technology, Faculty of Aerospace Engineering, N. 787, 1996.

[5] V.Y. Glizer. Asymptotic solution of a cheap control problem with state delay. *Dynam. Control*. 9:339-357, 1999.

[6] V.Y. Glizer, L.M. Fridman, V. Turetsky. Cheap suboptimal control of uncertain systems with state delays: an integral sliding mode approach. Research Report, Technion-Israel Institute of Technology, Faculty of Aerospace Engineering, N. 954, 2005.

[7] H. Kwakernaak, R. Sivan. *Linear optimal control systems*. John Wiley & Sons, New York, 1972.

[8] A. Ichikawa. Quadratic control of evolution equations with delays in control. *SIAM J. Control Optimiz.* 20:645-668, 1982.

[9] R.E. Kalman. Contributions to the theory of optimal control. *Bol. Soc. Mat. Mexicana*. 5:102-119, 1960.

DIFFERENTIABLE LOCAL BARRIER-PENALTY PATHS

C. Grossmann [1]
[1] *TU Dresden, Institut für Numerische Mathematik, D-01062 Dresden, Germany, grossm@math.tu-dresden.de**

Abstract Perturbations of Karush-Kuhn-Tucker conditions play an important role for primal-dual interior point methods. Beside the usual logarithmic barrier various further techniques of sequential unconstrained minimization are well known. However other than logarithmic embeddings are rarely studied in connection with Newton path-following methods. A key property that allows to extend the class of methods is the existence of a locally Lipschitz continuous path leading to a primal-dual solution of the KKT-system. In this paper a rather general class of barrier/penalty functions is studied. In particular, under LICQ regularity and strict complementarity assumptions the differentiability of the path generated by any choice of barrier/penalty functions from this class is shown. This way equality as well as inequality constraints can be treated directly without additional transformations. Further, it will be sketched how local convergence of the related Newton path-following methods can be proved without direct applications of self-concordance properties.

keywords: Perturbed KKT-systems, general barrier-penalty embedding, differentiable path, path-following methods, interior point methods

1. Barrier/penalty functions and primal-dual paths

Barrier/penalty methods and its path-following variants form an important class of numerical methods for constrained optimization problems via a family of unconstrained ones (cf. [2], [8], [9]). While large classes of classical barrier/penalty methods are well studied already in [3] path-following methods mainly restrict to log-barrier terms. The aim of the present paper is to provide a convergence concept for a wide range of path-following Newton methods under strong regularity assumptions. The concerning results are derived in detail in [5]. In addition, following [6] for the log-barrier method we discuss the relaxation of the LICQ regularity assumption by MFCQ. In this case we show

*Paper written with financial support of DFG grant GR 1777/2-2.

Please use the following format when citing this chapter:

Grossmann, C., 2006, in IFIP International Federation for Information Processing, Volume 202, Systems, Control, Modeling and Optimization, eds. Ceragioli, F., Dontchev, A., Furuta, H., Marti, K., Pandolfi, L., (Boston: Springer), pp. 195-204.

that the study of the behavior of the log-barrier method applied to a locally linearized problem provides full information upon the convergence properties of the approximated duals of the original nonlinear problem.

Considered are nonlinear programming problems

$$\begin{aligned} & f(x) \to \min ! \\ \text{s.t.} \quad & x \in G := \{ x \in R^n : g_i(x) = 0, \; i \in I^g, \quad g_i(x) \le 0, \; i \in I^u \}, \end{aligned} \tag{1}$$

where

$$I^g := \{1, \ldots, q\}, \quad I^u := \{q+1, \ldots, m\}, \quad I := I^g \cup I^u$$

and $f,\, g_i : R^n \to R,\; i \in I$ denote twice Lipschitz continuously differentiable functions. Let abbreviate $g : R^n \to R^m$ with $g = (g_1, \ldots, g_m)^T$ and

$$G_g := \{x \in R^n : g_i(x) = 0, \, i \in I^g\}, \quad G_u^0 := \{x \in R^n : g_i(x) < 0, \, i \in I^u\}.$$

Problem (1) is supposed to possess some local solution $x^* \in G$ that satisfies the linear independence constraint qualification (LICQ). In particular, this implies that a uniquely defined multiplier vector $y^* \in R^m$ exists such that the KKT-conditions

$$\begin{aligned} & \nabla_x L(x^*, y^*) = 0, \qquad g_i(x^*) = 0, \; i \in I^g, \\ & g_i(x^*) \le 0, \quad y_i^* \ge 0, \; i \in I^u, \quad {y^*}^T g(x^*) = 0. \end{aligned} \tag{2}$$

hold. In addition to LICQ we assume strict complementarity, i.e. $y_i^* \ne 0 \;\Leftrightarrow\; i \in I_0$ and that x^* satisfies the well-known second order sufficiency condition. Here denotes $I_0 := I_0(x^*) := \{ i \in I : g_i(x^*) = 0 \}$. Further, let $\nabla_x L,\ \nabla^2_{xx} L$ be the partial gradient and Hessian, respectively, of the Lagrangian L. Taking into account strict complementarity second order sufficiency condition simplifies to (2) and

$$v^T \nabla^2_{xx} L(x^*, y^*) v > 0 \quad \forall v \in R^n, \quad \nabla g_i(x^*)^T v = 0, \; i \in I_0, \; v \ne 0. \tag{3}$$

Further, we notice that LICQ also implies $G_g \cap G_u^0 \ne \emptyset$ which allows to apply classical barrier methods locally to all the inequality constraints of the optimization problem (1).

In barrier/penalty-methods the constraints of the original nonlinear programming problem (1) are incorporated into the objective in such a way that violations of the constraints are asymptotically avoided by extra costs. Instead of problem (1) we consider the related unconstrained auxiliary problems

$$\begin{aligned} & F(x, s) := f(x) + \sum_{i \in I} \phi_i(g_i(x), s) \to \min ! \\ \text{s.t.} \quad & x \in B_s := \{ x \in R^n \mid \; \phi_i(g_i(x), s) < +\infty, \; i \in I\}. \end{aligned} \tag{4}$$

Here $\phi_i(\cdot,\cdot) : R \times R_{++} \to \overline{R}$, $i \in I$, denote barrier/penalty-functions which depend upon the barrier/penalty-parameter $s > 0$. Let

$$R_+ := \{t \in R : t \geq 0\}, \qquad R_{++} := \{t \in R : t > 0\}, \qquad \overline{R} := R \cup \{+\infty\}.$$

To ensure differentiability of the local path throughout this paper we assume that for any $s > 0$ the barrier/penalty-functions $\phi_i(\cdot, s) : R \to \overline{R}$ are differentiable in $\mathrm{dom}\, \phi_i(\cdot, s)$ and satisfy

$$\frac{\partial}{\partial t} \phi_i(t,s) = \psi_i\left(\frac{t}{s}\right) \qquad \forall\, t \in \mathrm{dom}\, \phi_i(\cdot, s), \quad s > 0, \tag{5}$$

with some $\psi_i : R \to R$, $i \in I^g$, $\psi_i : R \to \overline{R}$, $i \in I^u$, $\psi_i \not\equiv 0$, $i \in I$. The functions ψ_i we call the generating functions for the barrier/penalty-method. The relation (5) between barrier/penalty-functions and their generating functions was proposed in [4] for path-following algorithms applied to inequality constrained problems.

REMARK 1 *The same structural assumption (5) was considered by Auslender et. al. [1] for saddle point problems. The following supposed properties, however, differ from those made in [1] due to our goal to establish convergence of path-following Newton methods.*

Assumed properties for $\psi_i,\ i \in I^u$:

U1: $\mathrm{dom}\, \psi_i = (-\infty, d_i)$ with some $d_i \in \overline{R}$ and $\lim\limits_{r \to d_i -} \psi_i(r) = +\infty$.

U2: $\psi_i : R \to \overline{R}$ convex, differentiable in $\mathrm{dom}\, \psi_i$ with ψ_i' locally Lipschitz

$$|\psi_i'(\rho_1) - \psi_i'(\rho_2)| \leq L_1(r)|\rho_1 - \rho_2| \quad \forall\, \rho_1, \rho_2 \leq r < d_i$$

and

$$|\psi_i'(\rho_1) - \psi_i'(\rho_2)| \leq L_2(r)\left|\frac{1}{\rho_1} - \frac{1}{\rho_2}\right| \quad \forall\, \rho_1, \rho_2 \leq r < \min\{0, d_i\}.$$

with some nondecreasing $L_1(\cdot), L_2(\cdot) : R_{++} \to R_{++}$.

U3: $\psi_i'(r) \geq 0 \quad \forall\, r \in \mathrm{dom}\, \psi_i, \quad \lim\limits_{r \to -\infty} \psi_i(r) = 0, \quad \lim\limits_{r \to -\infty} r^2\, \psi_i'(r)$ exists and is finite.

Assumed properties for $\psi_i,\ i \in I^g$:

G1: $\mathrm{dom}\, \psi_i = R, \quad R = \psi_i(\mathrm{dom}\, \psi_i)$.

G2: ψ_i differentiable with ψ_i' locally Lipschitz continuous

$$|\psi_i'(\rho_1) - \psi_i'(\rho_2)| \leq L_3(r)\, |\rho_1 - \rho_2| \qquad \forall\, |\rho_1|, |\rho_2| \leq r$$

with some nondecreasing $L_3(\cdot) : R_{++} \to R_{++}$.

G3: $\psi_i'(r) \geq 0, \quad \forall r \in R \quad$ and $\quad \psi_i(r) \neq 0 \implies \psi_i'(r) > 0$.

Examples for generating functions are:

- $\psi_i(r) := \begin{cases} |d_i - r|^{-p} & , \text{ if } r < d_i, \\ +\infty & , \text{ if } r \geq d_i, \end{cases}$
  in case $d_i = 0$: $p = 1$ log-barrier, $p = 2$ Fiacco/McCormick's SUMT; otherwise shifted version.

- $\psi_i(r) := \max^{p-1}\{0, r\},\ r \in R,$
  $p \geq 2$ corresponds to p-th order penalty function.

- $\psi_i(r) = \exp(r),\ r \in R$
  exponential penalty.

- $\psi_i(r) = \operatorname{sign}(r)|r|^{p-1},\ r \in R$
  with fixed $p > 1$, (penalty for equality constraints).

We notice that strict complementarity, LICQ and the second order sufficiency conditions guarantee that the wanted minimizers of $F(\cdot, s)$ can be characterized by the necessary and sufficient local optimality condition

$$x(s) \in B_s : \qquad \nabla f(x(s)) + \sum_{i \in I} y_i(s) \nabla g_i(x(s)) = 0$$

with the so-called barrier/penalty multipliers $y_i(s) := \psi_i(\frac{g_i(x)}{s})$. The main result concerning stability behavior of the specific perturbation of the KKT-system is

THEOREM 2 *Under the made assumptions, there exist some* $\bar{s} > 0$, $\delta > 0$ *such that for any* $s \in (0, \bar{s}]$ *the parametric system*

$$\begin{aligned} \nabla f(x(s)) + \sum_{i \in I} y_i(s)\, \nabla g_i(x(s)) &= 0 \\ y_i(s) - \psi_i(g_i(x(s))/s) &= 0, \quad i \in I \end{aligned} \tag{6}$$

possesses a unique solution $(x(s), y(s))$ *with* $x(s) \in B_s \cap U_\delta(x^*)$, *and we have*

$$\lim_{s \to 0+} (x(s), y(s)) = (x^*, y^*).$$

With

$$x(0) := x^*, \quad y(0) := y^*, \tag{7}$$

the functions $x(\cdot)$, $y(\cdot)$ *are continuously differentiable in* $(0, \bar{s}]$, *possess right sided derivatives at* $s = 0$ *and these derivatives are bounded for* $s \to 0+$.

The proof of this theorem essentially rests on the implicit function theorem applied to the following perturbed KKT-system

$$\begin{aligned} \nabla f(x(s,r)) + \sum_{i \in I_0} y_i(s,r)\, \nabla g_i(x(s,r)) &= r, \\ s\, \psi_i^{-1}(y_i(s,r)) &= g_i(x(s,r)), \quad i \in I_0. \end{aligned}$$

For this the main property (see [5]) is the regularity of the matrix

$$H(s) := \begin{pmatrix} \nabla_{xx}L(x(s),y(s)) & \nabla g_1(x(s)) & \cdot & \cdot & \nabla g_m(x(s)) \\ \psi_1'\left(\frac{g_1(x(s))}{s}\right)\nabla g_1(x(s))^T & -s & 0 & \cdot & 0 \\ \psi_2'\left(\frac{g_2(x(s))}{s}\right)\nabla g_2(x(s))^T & 0 & -s & \cdot & 0 \\ \cdot\cdot & \cdot & \cdot & \cdot & \cdot \\ \psi_m'\left(\frac{g_m(x(s))}{s}\right)\nabla g_m(x(s))^T & 0 & 0 & \cdot & -s \end{pmatrix}$$

for sufficiently small $s > 0$.

As a direct consequence of Theorem 2 holds

COROLLARY 3 *Under the given assumptions there exist some constants* $s_0 \in (0, \bar{s}]$ *and* $c_L > 0$ *such that*

$$\left.\begin{array}{l} \|x(s) - x(t)\| \le c_L\,|s-t| \\ \|y(s) - y(t)\| \le c_L\,|s-t| \end{array}\right\} \qquad \forall\, s,\, t \in [0, s_0]. \tag{8}$$

2. Log-Barriers Under Weaker Assumptions

In this section we follow widely [6] and restrict us to inequality constrained optimization problems, i.e. to

$$f(x) \to \min! \quad \text{s.t.} \quad x \in G = \{\, x \in R^n \,:\, g_i(x) \le 0,\ i \in I^u \,\}. \tag{9}$$

To this problem we apply log-barrier embedding and obtain the auxiliary problem

$$F(x,s) := f(x) - s \sum_{i \in I^u} \ln(-g_i(x)) \to \min! \quad \text{s. t.} \quad x \in G^0. \tag{10}$$

In contrast to the first part of the paper, now the regularity assumptions are relaxed as follows:

(A1) x^* is some local minimizer of (9).

(A2) MFCQ is satisfied at x^*, i.e.,

$$U^0 := \{u \in R^n : \nabla g_i(x^*)^T u < 0 \quad \forall i \in I_0\} \neq \emptyset.$$

(A3) the *strict complementarity condition w.r. to* Y^* holds, i.e.,

$$\exists\, y^* \in Y^* \quad \text{with} \quad y_i^* > 0\ \forall i \in I_0. \tag{11}$$

For a general study of regularity conditions and stability in nonlinear programming we refer to [7].

Next we introduce a locally linearized problem and show that log-barrier methods applied to it behave asymptotically like log-barrier methods applied to the original problem. Let denote

$$A := (\dots \nabla g_i(x^*) \dots)_{i \in I_0} \qquad \text{(column-wise)}$$

Related to x^* with $d = x - x^*$ we study the locally linearized problem

$$\nabla f(x^*)^T d \to \min ! \qquad \text{s.t.} \qquad d \in \mathcal{R}(A), \quad A^T d \leq 0. \tag{12}$$

Here $\mathcal{R}(A)$ stands for the range of A. Notice that the point $d^* = 0$ forms the unique solution of problem (12). Log-barriers applied to (12) yield the auxiliary problems

$$\begin{aligned} \varphi_s(d) = \nabla f(x^*)^T d - s \sum_{i \in I_0} \ln(-\nabla g_i(x^*)^T d) &\to \min ! \\ \text{s. t.} \qquad d \in D^0 = \{ d \in \mathcal{R}(A) \, : \, A^T d < 0 \}. \end{aligned} \tag{13}$$

For these we have

LEMMA 4 *For any $s > 0$ problem (13) possesses a unique solution $\tilde{d}(s)$. Further, there is a unique solution d^* of the problem*

$$\prod_{i \in I_0} \frac{-\nabla g_i(x^*)^T d}{\nabla f(x^*)^T d} \to \textit{max !} \qquad \textit{s.t.} \quad d \in D^0, \quad \|d\| = 1, \tag{14}$$

and it holds $\tilde{d}(s) = t_s d^*$ with some $t_s > 0$ for all $s > 0$ as well as $\|\tilde{d}(s)\| = O(s)$.

THEOREM 5 *The log-barrier method (13) yields for the barrier multipliers $\tilde{y}(s)$ related to the solutions $\tilde{x}(s)$ that.*

$$\tilde{y}_i(s) := \frac{-s}{\tilde{g}_i(\tilde{x}(s))} \equiv \mu_i, \quad i \in I_0, \tag{15}$$

where

$$\mu_i := -\frac{1}{m_0} \frac{\nabla f(x^*)^T d^*}{\nabla g_i(x^*)^T d^*}, \; i \in I_0 \;\; \textit{and } d^* \textit{ solves (14).} \tag{16}$$

Setting $\mu_i = 0 \; \forall i \in I_1$, μ is a multiplier of the original problem (9).

Next we study the nonlinear problem

$$F(x, s) = f(x) - s \sum_{i=1}^{m} \ln(-g_i(x)) \to \min !$$

$$\text{s. t.} \quad x \in G^\varepsilon := \{ x \in R^n \, : \, g_i(x) < 0, \; i = 1, \dots, m, \; \|x - x^*\| < \varepsilon \},$$

In addition to (A1), (A2), (A3) assume:

(A4) The second-order optimality condition holds:

$$u^T \nabla^2_{xx} L(x^*, y) u > 0 \text{ for all } y \in Y^* \text{ and all } u \in U^*, u \neq 0,$$

where

$$U^* := \{u \,:\, \nabla f(x^*)^T u = 0,\ \nabla g_i(x^*)^T u \leq 0 \,\forall\, i \in I_0\}$$

is the critical cone for x^*.

THEOREM 6 *There are $\bar{s} > 0$ and $\varepsilon > 0$ such that for all $s \in (0, \bar{s})$, the function $F(\cdot, s)$ on G^ε has a global minimizer $x(s)$ which is the unique stationary point of $F(\cdot, s)$ on G^ε. The associated multipliers $y(s)$ converge to μ given in Theorem 5 where*

$$\text{dist}((x(s), y(s)), (x^*, Y^*)) \;\leq\; C^* s \quad \textit{with some} \quad C^* > 0,$$

the Hessian $\nabla^2 F(x(s), s)$ is uniformly positive definite and $x(\cdot)$ is continuously differentiable on $(0, \bar{s})$.

3. Path-Following Primal-Dual Methods

We consider the convergence of Newton's method applied to the complete primal-dual system (6). Unlike in primal methods its first part stabilizes the approximation of the duals. However, as in the primal approach system, (6) also becomes increasingly ill-conditioned as $s \to 0+$.

Let denote

$$z := \begin{pmatrix} x \\ y \end{pmatrix} \in Z := R^n \times R^m$$

the vector of all primal and dual components. Further, let $T \,:\, Z \to Z$ denote the mapping

$$T(z, s) := \begin{pmatrix} T_1(z) \\ T_2(z, s) \end{pmatrix}$$

with

$$T_1(z) := \nabla_x L(x, y), \qquad T_2(z, s) := \psi\left(\frac{g(x)}{s}\right) - y.$$

With these notations (6) can be written as the following parametric system of nonlinear equations

$$T(z, s) \;=\; 0. \tag{17}$$

For a fixed barrier/penalty-parameter $s > 0$ a single Newton-step maps an old guess $z \in Z$ to a new approximate $\hat{z} \in Z$ of the solution $z(s)$ of (17) by

$$T'(z, s)(\hat{z} - z) + T(z, s) = 0. \tag{18}$$

Taking into account the structure of (6), the Jacobian has the form

$$T'(z,s) := \begin{pmatrix} Q & B \\ DB^T & -I \end{pmatrix}$$

with

$$\begin{aligned} Q &:= Q(x,y) := \nabla^2_{xx} L(x,y), \\ B &:= B(x) := (\nabla g_1(x), \ldots, \nabla g_m(x)), \\ D &:= D(x) := \operatorname{diag}\left\{ \tfrac{1}{s}\, \psi_i'\left(\tfrac{g_i(x)}{s}\right)\right\}_{i \in I}. \end{aligned}$$

Since system (6) is increasingly ill-conditioned for $s \to 0+$ an adapted analysis for the Newton system is required to obtain sharp error bounds. In connection with log-barrier interior point methods, self-concordance (cf. [8], [9]) forms a common tool. We apply a different approach (cf. [4]) that analyzes such ill-posed systems directly in the Euclidean norm.

Let remark that in case of log-barriers we have $\Psi(r) = 1/r$ which allows to rewrite the second part as

$$g_i(x)\, y_i \; - \; s \; = \; 0, \quad i \in I.$$

A similar transformation is possible recommended if ψ_i is strictly monotone in domψ_i. This transformation stabilizes the numerical process, but does not remove the generic asymptotic singular behavior of the system (17) for $s \to 0+$.

In path-following Newton methods for a fixed barrier/penalty-parameter $s_k > 0$, and known $z^k \in Z$, we define the new iterate $z^{k+1} \in Z$, by only one Newton-step, i.e.

$$T'(z^k, s_k)(z^{k+1} - z^k) + T(z^k, s_k) = 0 \tag{19}$$

and update the parameter by $s_{k+1} = \gamma\, s_k$ with some $\gamma \in (0,1)$. This yields the long-step version of a

Path-Following Algorithm

Step 1: Select parameters ε, c, $s_0 > 0$, and $\nu \in (0,1)$.
Find $x^0 \in B_{s_0}$ such that

$$\|z^0 - z(s_0)\| \le c\, s_0. \tag{20}$$

Set $k := 0$.

Step 2: Determine $z^{k+1} \in Z$ via the linear system

$$\begin{aligned} T'(z^k, s_k)\, d^k &= -T(z^k, s_k)) \\ z^{k+1} &:= z^k + d^k \end{aligned} \tag{21}$$

Step 3: If $s_k \leq \varepsilon$ then stop. Otherwise set $s_{k+1} := \nu\, s_k$ and go to Step 2 with $k := k+1$.

THEOREM 7 *For sufficiently small $s_0 > 0$ and $c > 0$ there exists some parameter $\nu \in (0,1)$ such that then the given path-following algorithm is well defined and generates iterates $z_k \in Z$ that satisfy*

$$\|z^k - z(s_k)\| \leq c\, s_k, \qquad k = 0, 1, \ldots \,. \tag{22}$$

Furthermore, the algorithm terminates after at most $k^* := \lceil \ln(\varepsilon/s_0)/\ln(\nu) \rceil$ steps and the estimate

$$\|z^{k^*} - z^*\| \leq (c_L + c)\, \varepsilon \tag{23}$$

holds, where c_L denotes the Lipschitz constant from Corollary 3.

For the proof as well as for further details we refer to [5].

To ensure a larger range of convergence the given path-following algorithm has to be endowed with an additional step size procedure in step 2, i.e. we apply

$$z^{k+1} := z^k + \alpha_k\, d^k$$

with some $\alpha_k > 0$ appropriately defined, e.g. by Armijo's rule.

An additional stabilization can be obtained by the use of the available approximations of the Lagrangian multipliers. The basic idea rests on

$$\psi_i\left(\frac{g_i(x(s)}{s}\right) \approx y_i(s) \rightarrow y_i^*, \quad i \in I.$$

Taking into account $g_i(x^*) = 0$, $i \in I_0$ the generating function is modified by shifts to satisfy

$$\psi_i(0) = y_i(s), \quad i \in I_0. \tag{24}$$

Standard IP-methods like log-barrier do not allow this, but shifted methods that are also covered by the assumed properties of ψ_i do.

Consider shifted log-barrier, i.e.

$$\psi_i(r) := \begin{cases} 1/(d_i - r) & , \text{ if } r < d_i, \\ +\infty & , \text{ if } r \geq d_i, \end{cases}$$

In this case this leads to the update

$$\hat{d}_i = 1/y_i(s), \quad i \in I_0\,, \tag{25}$$

with I_0 approximately identified via the magnitude of $y_i(s)$.

The case of shifted quadratic loss penalties

$$\psi_i(r) = \max\{0, d_i + r\}$$

leads to

$$\hat{d}_i = y_i(s), \quad i \in I_0\,. \tag{26}$$

This is directly related to augmented Lagrangian techniques.

References

[1] A. Auslender, R. Cominetti, M. Haddou. Asymptotic analysis for penalty and barrier methods in convex and linear programming. *Math. Oper. Res.* 22:43-62, 1997.

[2] A. Forsgren, P.E. Gill, M.H. Wright. Interior methods for nonlinear optimization. *SIAM Review* 44:525-597, 2002.

[3] A.V. Fiacco, G.P. McCormick. *Nonlinear programming: Sequential unconstrained minimization techniques*, Wiley, New York, 1968.

[4] C. Grossmann, Penalty/Barrier path-following in linearly constrained optimization. *Discussiones Mathematicae, Differential Inclusions, Control and Optimization* 20:7-26, 2000.

[5] C. Grossmann, M. Zadlo. General primal-dual barrier-penalty path-following Newton methods for nonlinear programming. TU Dresden, Preprint TUD-NM-10-03, 2003 (to appear in Optimization)

[6] C. Grossmann, D. Klatte, B. Kummer. Convergence of primal-dual solutions for the nonconvex log-barrier method without LICQ. *Kybernetika* 40: 571-584, 2004.

[7] D. Klatte, B. Kummer. *Nonsmooth equations in optimization - regularity, calculus, methods and applications.* Kluwer, Dordrecht, 2002.

[8] Y. Nesterov, A. Nemirovskii. *Interior-point polynomial algorithms in convex programming*, SIAM Publications, Philadelphia, 1994.

[9] S.J. Wright. *Primal-Dual Interior-Point Methods.* SIAM Publications, Philadelphia, 1997.

STATIONARITY AND REGULARITY CONCEPTS FOR SET SYSTEMS

A. Kruger[1]

[1] *University of Ballarat, School of Information Technology and Mathematical Sciences, Centre of Information and Applied Optimization, Ballarat, Australia, a.kruger@ballarat.edu.au*

Abstract The paper investigates stationarity and regularity concepts for set systems in a normed space. Several primal and dual constants characterizing these properties are introduced and the relations between the constants are established. The equivalence between the regularity property and the strong metric inequality is established. The extended extremal principle is formulated.

keywords: nonsmooth analysis, normal cone, optimality, extremality, stationarity, regularity, set-valued mapping, Asplund space

1. Introduction

Starting with the pioneering work by Dubovitskii and Milyutin [2] it is quite natural when dealing with optimality conditions to reformulate optimality in the original optimization problem as a (some kind of) extremal behaviour of a certain system of sets. An easy example is a problem of unconditional minimization of a real-valued function $\varphi : X \to R$. If $x^\circ \in X$ one can consider the sets $\Omega_1 = \text{epi}\,\varphi = \{(x,\mu) \in X \times R : \varphi(x) \leq \mu\}$ (the epigraph of φ) and $\Omega_2 = X \times \{\mu : \mu \leq \varphi(x^\circ)\}$ (the lower halfspace). The local optimality of x° is then equivalent to the condition $\Omega_1 \cap \text{int}\,\Omega_2 \cap B_\rho(x^\circ) = \emptyset$ for some $\rho > 0$.

Besides extremality, stationarity and regularity concepts for set systems can be defined in a natural way. Regularity properties of set systems are closely related to similar properties of multifunctions. They can play the role of constraint qualifications in optimization problems.

The paper is organized as follows. Several primal constants characterizing the mutual arrangement of sets in a normed space are introduced in Section 2. Based on these constants the extremality, stationarity and regularity properties for the set system are defined. Two special cases are considered in Section 3: a system of convex sets and a system of (not necessarily convex) cones. In Section 4 two more primal constants based on comparing point-to-set distances are introduced. They give rise to another two regularity properties: the *metric*

Please use the following format when citing this chapter:

Kruger, A., 2006, in IFIP International Federation for Information Processing, Volume 202, Systems, Control, Modeling and Optimization, eds. Ceragioli, F., Dontchev, A., Furuta, H., Marti, K., Pandolfi, L., (Boston: Springer), pp. 205-214.

inequality and the *strong metric inequality*. The latter one appears to be equivalent to the regularity property defined in Section 2. Section 5 is devoted to the dual constants and dual criteria of stationarity and regularity. The *extended extremal principle* is formulated.

Mainly standard notations are used throughout the paper. The ball of radios ρ centered at x in a normed space is denoted $B_\rho(x)$. We write B_ρ if $x = 0$, and simply B if $x = 0$ and $\rho = 1$. If Ω is a set then $\operatorname{int}\Omega$ denotes its interior.

2. Definitions

Let us consider a system of closed sets $\Omega_1, \Omega_2, \ldots, \Omega_n$ $(n > 1)$ in a normed space X with $x^\circ \in \cap_{i=1}^n \Omega_i$.

The following constant can be used for characterizing the mutual arrangement of sets $\Omega_1, \Omega_2, \ldots, \Omega_n$ near x° ([9, 10]):

$$\theta_\rho[\Omega_1, \ldots, \Omega_n](x^\circ) = \sup\{r \geq 0 : \Big(\bigcap_{i=1}^n (\Omega_i - a_i)\Big) \bigcap B_\rho(x^\circ) \neq \emptyset,\ \forall a_i \in B_r\}. \quad (1)$$

It shows how far the sets can be "pushed apart" while still intersecting in a neighborhood of x°. Evidently $\theta_\rho[\Omega_1, \ldots, \Omega_n](x^\circ)$ is nonnegative (and can be equal to $+\infty$) and nondecreasing as a function of ρ.

A slightly more general form of (0) can be of interest ([11]):

$$\theta_\rho[\Omega_1, \ldots, \Omega_n](\omega_1, \ldots, \omega_n) = \sup\{r \geq 0 : \Big(\bigcap_{i=1}^n (\Omega_i - \omega_i - a_i)\Big) \bigcap B_\rho \neq \emptyset,\ \forall a_i \in B_r\}. \quad (2)$$

This constant corresponds to the case when instead of the common point $x^\circ \in \cap_{i=1}^n \Omega_i$ each of the sets Ω_i is considered near its own point $\omega_i \in \Omega_i$, $i = 1, 2, \ldots, n$. The sets do not need to be intersecting. It is equivalent to considering the system of translated sets $\Omega_1 - \omega_1$, $\Omega_2 - \omega_2$, $\ldots$, $\Omega_n - \omega_n$ near 0:

$$\theta_\rho[\Omega_1, \ldots, \Omega_n](\omega_1, \ldots, \omega_n) = \theta_\rho[\Omega_1 - \omega_1, \ldots, \Omega_n - \omega_n](0).$$

If $\omega_1 = \omega_2 = \ldots = \omega_n = x^\circ$ then, of course,

$$\theta_\rho[\Omega_1, \ldots, \Omega_n](\omega_1, \ldots, \omega_n) = \theta_\rho[\Omega_1, \ldots, \Omega_n](x^\circ).$$

If (0) or (1) is positive more precise estimates of regularity/stationarity can be obtained based on using the "linearized" constants:

$$\theta[\Omega_1, \ldots, \Omega_n](x^\circ) = \liminf_{\rho \to +0} \theta_\rho[\Omega_1, \ldots, \Omega_n](x^\circ)/\rho, \quad (3)$$

$$\theta[\Omega_1, \ldots, \Omega_n](\omega_1, \ldots, \omega_n) = = \liminf_{\rho \to +0} \theta_\rho[\Omega_1, \ldots, \Omega_n](\omega_1, \ldots, \omega_n)/\rho. \quad (4)$$

Finally, one can define one more limiting constant based on (4):

$$\hat{\theta}[\Omega_1,\ldots,\Omega_n](x^\circ) = \liminf_{\omega_i \xrightarrow{\Omega_i} x^\circ} \theta[\Omega_1,\ldots,\Omega_n](\omega_1,\ldots,\omega_n). \tag{5}$$

The notation $\omega \xrightarrow{\Omega} x$ in (5) means that $\omega \to x$ with $\omega \in \Omega$.

The constants (3)–(5) are in a sense derivative-like objects. (3) and (4) can be considered as analogs of the usual derivative, while (5) has some properties of the *strict derivative*: it accumulates information about local properties of the sets not only at a given point but also at all nearby points.

All the constants (0)–(5) are nonnegative. When investigating extremality-stationarity-regularity properties of the set system one needs to check whether the corresponding constant is zero or strictly positive.

DEFINITION 1 *The system of sets $\Omega_1, \Omega_2, \ldots, \Omega_n$ is*

(i) extremal *at x° if $\theta_\rho[\Omega_1,\ldots,\Omega_n](x^\circ) = 0$ for all $\rho > 0$.*

(ii) locally extremal *at x° if $\theta_\rho[\Omega_1,\ldots,\Omega_n](x^\circ) = 0$ for some $\rho > 0$.*

(iii) stationary *at x° if $\theta[\Omega_1,\ldots,\Omega_n](x^\circ) = 0$.*

(iv) weakly stationary *at x° if $\hat{\theta}[\Omega_1,\ldots,\Omega_n](x^\circ) = 0$.*

(v) regular *at x° if $\hat{\theta}[\Omega_1,\ldots,\Omega_n](x^\circ) > 0$.*

PROPOSITION 2 *(i) $\Rightarrow$ (ii) $\Rightarrow$ (iii) $\Rightarrow$ (iv) in Definition 1.*

Opposite implications are not true in general.

The notion of (local) extremality of the set system was introduced (in a different but equivalent way) in [12], where dual necessary conditions were formulated. This result currently known as the *extremal principle* has had numerous applications to different optimization problems (see [13]).

Conditions (iii) and (IV) give natural extensions of the notion of local extremality. Condition (iii) corresponds to the traditional concept of stationarity in optimization theory, while (iv) means that arbitrarily close to x° there exist points whose properties are arbitrarily close to the traditional stationarity property. The first version of the weak stationarity property was defined (under a different name) in [6] (see also [7, 8]).

Stationarity and regularity properties of set systems were considered in [10, 11].

Regularity of the set system is a natural counterpart of the weak stationarity property. It is closely related to the *metric regularity* of multifunctions [4, 5] and can be used e.g. when formulating *constraint qualifications* in mathematical programming.

The condition $\theta[\Omega_1,\ldots,\Omega_n](x^\circ) > 0$ also defines a kind of regularity which is weaker than the one defined in part (iv) of Definition 1. It can be referred to as *weak regularity*. We will not use this concept in the current paper.

The next proposition gives an equivalent definition of regularity.

PROPOSITION 3 *The system of sets Ω_1, Ω_2, ..., Ω_n is regular at x° if and only if there exists an $\alpha > 0$ and a $\delta > 0$ such that*

$$\Big(\bigcap_{i=1}^{n}(\Omega_i - \omega_i - a_i)\Big)\bigcap B_\rho \neq \emptyset \tag{6}$$

for all $\rho \in (0,\delta]$, $\omega_i \in \Omega_i \cap B_\delta(x^\circ)$, $a_i \in B_{\alpha\rho}$, $i = 1,2,\ldots,n$.

$\hat{\theta}[\Omega_1,\ldots,\Omega_n](x^\circ)$ equals to the exact upper bound of all such α.

3. Stationarity and regularity of convex set and cone systems

In the convex case, as one could expect, the concepts of extremality and local extremality coincide and appear to be equivalent to both stationarity and weak stationarity.

PROPOSITION 4 (SEE [10]) *Let Ω_1, Ω_2, ..., Ω_n be convex.*

(i) If $\theta_\rho[\Omega_1,\ldots,\Omega_n](x^\circ) > 0$ for some $\rho > 0$ then $\theta_\rho[\Omega_1,\ldots,\Omega_n](x^\circ) > 0$ for all $\rho > 0$.

(ii) The function $\rho \to \theta_\rho[\Omega_1,\ldots,\Omega_n](x^\circ)/\rho$, considered on the set of positive numbers, is nonincreasing.

(iii) $\theta[\Omega_1,\ldots,\Omega_n](x^\circ) = \sup_{\rho>0}\theta_\rho[\Omega_1,\ldots,\Omega_n](x^\circ)/\rho$.

(iv) $\hat{\theta}[\Omega_1,\ldots,\Omega_n](x^\circ) = \theta[\Omega_1,\ldots,\Omega_n](x^\circ)$.

(v) (i) ⇔ (ii) ⇔ (iii) ⇔ (iv) in Definition 1.

(vi) If $\operatorname{int}\Omega_i \neq \emptyset$, $i = 1,2,\ldots,n-1$, then the first four conditions in Definition 1 are equivalent to

$$\bigcap_{i=1}^{n-1}\operatorname{int}\Omega_i\bigcap\Omega_n = \emptyset, \tag{7}$$

while condition (v) is equivalent to

$$\bigcap_{i=1}^{n-1}\operatorname{int}\Omega_i\bigcap\Omega_n \neq \emptyset.$$

As it follows from part (v) of Proposition 4, under the assumption that all but one sets have nonempty interior, all defined above extremality and stationarity notions reduce in the convex case to the traditional condition (7). Note that the initial definitions make sense for convex sets even without the assumption that the sets have nonempty interior.

PROPOSITION 5 (SEE [10]) *Let $\Omega_1, \Omega_2, \ldots, \Omega_n$ be cones.*

(i) *If $\omega_i \in \Omega_i$, $i = 1, 2, \ldots, n$, and $\rho > 0$ then*

$$\theta_\rho[\Omega_1, \ldots, \Omega_n](\omega_1, \ldots, \omega_n) = \rho\theta_1[\Omega_1, \ldots, \Omega_n](\omega_1/\rho, \ldots, \omega_n/\rho).$$

In particular, $\rho \to \theta_\rho[\Omega_1, \ldots, \Omega_n](0)$ is positively homogeneous:

$$\theta_\rho[\Omega_1, \ldots, \Omega_n](0) = \rho\theta_1[\Omega_1, \ldots, \Omega_n](0).$$

(ii) *If $\omega_i \in \Omega_i$, $i = 1, 2, \ldots, n$, then*

$$\theta[\Omega_1, \ldots, \Omega_n](\omega_1, \ldots, \omega_n) = \liminf_{t \to \infty} \theta_1[\Omega_1, \ldots, \Omega_n](t\omega_1, \ldots, t\omega_n).$$

In particular, $\theta[\Omega_1, \ldots, \Omega_n](0) = \theta_1[\Omega_1, \ldots, \Omega_n](0)$.

(iii) $\hat{\theta}[\Omega_1, \ldots, \Omega_n](0) = \inf_{\omega_i \in \Omega_i} \theta_1[\Omega_1, \ldots, \Omega_n](\omega_1, \ldots, \omega_n)$.

(iv) *If $x^\circ = 0$ then (i) $\Leftrightarrow$ (ii) $\Leftrightarrow$ (iii) in Definition 1 and these conditions are equivalent to*

$$\theta_1[\Omega_1, \ldots, \Omega_n](0) = 0.$$

(v) *The system of sets $\Omega_1, \Omega_2, \ldots, \Omega_n$ is weakly stationary at 0 if and only if*

$$\inf_{\omega_i \in \Omega_i} \theta_1[\Omega_1, \ldots, \Omega_n](\omega_1, \ldots, \omega_n) = 0.$$

(vi) *The system of sets $\Omega_1, \Omega_2, \ldots, \Omega_n$ is regular at 0 if and only if there exists $\alpha > 0$ such that*

$$\theta_1[\Omega_1, \ldots, \Omega_n](\omega_1, \ldots, \omega_n) \geq \alpha \quad \forall \omega_i \in \Omega_i.$$

4. Metric inequality

Some other approaches based on comparing distances can be used for characterizing stationarity/regularity properties of set systems. Let $d(\cdot, \cdot)$ be the distance function in X associated with the norm. We will keep the same notation for point-to-set distances. Thus, $d(x, \Omega) = \inf_{\omega \in \Omega} \|x - \omega\|$ is the distance from a point x to a set Ω and $d(x, \emptyset) = \infty$. The following constant can be useful:

$$\vartheta[\Omega_1, \ldots, \Omega_n](x^\circ) = \limsup_{x \to x^\circ} \left[d(x, \bigcap_{i=1}^n \Omega_i) \,/\, \max_{1 \leq i \leq n} d(x, \Omega_i) \right]_\circ. \tag{8}$$

The "extended" division operation $(\cdot/\cdot)_\circ$ is used in (8) to simplify the definition. It makes division by zero legal. The formal rules are as follows:

1 $(\alpha/\beta)_\circ = \alpha/\beta$, if $\beta \neq 0$;

2 $(\alpha/0)_\circ = +\infty$, if $\alpha > 0$;

3 $(\alpha/0)_\circ = -\infty$, if $\alpha < 0$;

4 $(0/0)_\circ = 0$.

The fourth rule is the most important one here. In the case $x^\circ \in \operatorname{int} \cap_{i=1}^n \Omega_i$ it automatically leads to $\vartheta[\Omega_1, \ldots, \Omega_n](x^\circ) = 0$. Otherwise, all the points $x \in \cap_{i=1}^n \Omega_i$ can be ignored when calculating the value of the upper limit in (8).

The "strict" version of (8) looks a little more complicated: small perturbations (shifts) are applied to the sets.

$$\hat{\vartheta}[\Omega_1, \ldots, \Omega_n](x^\circ) = \limsup_{\substack{x \to x^\circ \\ x_i \to 0}} \left[d(x, \bigcap_{i=1}^n (\Omega_i - x_i)) \,/\, \max_{1 \le i \le n} d(x + x_i, \Omega_i) \right]_\circ . \tag{9}$$

When investigating the properties of set systems it can be important to know whether the corresponding constant (8) or (9) is finite.

PROPOSITION 6 *The following assertions hold:*

(i) $\vartheta[\Omega_1, \ldots, \Omega_n](x^\circ) < \infty$ *if and only if there exists a* $\beta > 0$ *and a* $\delta > 0$ *such that*

$$d(x, \bigcap_{i=1}^n \Omega_i) \le \beta \max_{1 \le i \le n} d(x, \Omega_i) \tag{10}$$

for all $x \in B_\delta(x^\circ)$.

$\vartheta[\Omega_1, \ldots, \Omega_n](x^\circ)$ *coincides with the exact lower bound of all such* β.

(ii) $\hat{\vartheta}[\Omega_1, \ldots, \Omega_n](x^\circ) < \infty$ *if and only if there exists a* $\beta > 0$ *and a* $\delta > 0$ *such that*

$$d(x, \bigcap_{i=1}^n (\Omega_i - x_i)) \le \beta \max_{1 \le i \le n} d(x + x_i, \Omega_i) \tag{11}$$

for all $x \in B_\delta(x^\circ)$, $x_i \in B_\delta$, $i = 1, 2, \ldots, n$.

$\hat{\vartheta}[\Omega_1, \ldots, \Omega_n](x^\circ)$ *coincides with the exact lower bound of all such* β.

The condition formulated in part (i) of Proposition 6 is equivalent to the regularity condition known as the *metric inequality* [3, 4, 16] (some authors consider the sum of distances instead of the maximum in the right-hand side of (10)). The condition in part (ii) can be considered as the *strong metric inequality*. If (10) is valid for all x then the system of sets is said to be *linear regular* [1, 15]. This property is important when investigating convex optimization problems. One can consider some other regularity properties of set systems with interesting relations to linear regularity (see [15]).

(11) is certainly stronger than (10) even in the convex case. Take for instance $\Omega_1 = \Omega_2 = \{(x, y) \in R^2 : y = 0\}$. Then (10) holds true for all x (with $\beta = 1$) while (11) does not.

The next theorem proved in [10] gives the relation between (9) and (5). It allows to use (9) for characterizing stationarity and regularity properties of set systems.

THEOREM 7 $\hat{\vartheta}[\Omega_1, \ldots, \Omega_n](x^\circ) = 1/\hat{\theta}[\Omega_1, \ldots, \Omega_n](x^\circ)$.

COROLLARY 8 *The system of sets* $\Omega_1, \Omega_2, \ldots, \Omega_n$ *is regular at* x° *if and only if* $\hat{\vartheta}[\Omega_1, \ldots, \Omega_n](x^\circ) < \infty$.

It follows from Corollary 8 that regularity of a set system implies the metric inequality.

5. Dual criteria

The stationarity and regularity properties of set systems were defined above in terms of primal space elements. When the sets are closed these properties admit some dual characterizations in terms of "normal" elements.

Let X^* denote the space (topologically) dual to X and $\langle \cdot, \cdot \rangle$ be the bilinear form defining duality between X and X^*. Recall that the *(Fréchet) normal cone* to a set Ω at $x^\circ \in \Omega$ is defined as

$$N(x^\circ|\Omega) = \left\{ x^* \in X^* : \limsup_{x \xrightarrow{\Omega} x^\circ} \frac{\langle x^*, x - x^\circ \rangle}{\|x - x^\circ\|} \leq 0 \right\}. \tag{12}$$

In the rest of the section the sets $\Omega_1, \Omega_2, \ldots, \Omega_n$ are assumed closed. Define a "dual" constant:

$$\eta[\Omega_1, \ldots, \Omega_n](x^\circ) = \lim_{\delta \to +0} \inf \left\{ \left[\|\sum_{i=1}^n x_i^*\| \Big/ \sum_{i=1}^n \|x_i^*\| \right]_\infty : \right.$$
$$\left. x_i^* \in N(x_i|\Omega_i),\ x_i \in \Omega_i \cap B_\delta(x^\circ), i = 1, \ldots, n \right\}. \tag{13}$$

Another "extended" division operation $(\cdot, \cdot)_\infty$ is used here. It differs from the $(\cdot, \cdot)_\circ$ operation, which was used in (8), (9), in the fourth rule definition:

4. $(0/0)_\infty = \infty$.

This allows one to exclude the case $x_1^* = x_2^* = \cdots = x_n^* = 0$ when calculating the exact lower bound in (12). If this is the only case ($x^\circ \in \operatorname{int} \bigcap_{i=1}^n \Omega_i$) one automatically gets $\eta[\Omega_1, \ldots, \Omega_n](x^\circ) = \infty$.

Using (12) one can define (a kind of) stationarity for the set system $\Omega_1, \Omega_2, \ldots,$ $\ldots, \Omega_n$ by the condition $\eta[\Omega_1, \ldots, \Omega_n](x^\circ) = 0$ while the inequality $\eta[\Omega_1, \ldots$ $\ldots, \Omega_n](x^\circ) > 0$ can be considered as a regularity condition. As it follows from the next proposition the dual stationarity condition can be considered as some generalization of the separation property (for nonconvex set systems).

PROPOSITION 9 *(i)* $\eta[\Omega_1, \ldots, \Omega_n](x^\circ) = 0$ *if and only if for any* $\delta > 0$ *there exist elements*

$$\omega_i \in \Omega_i \cap B_\delta(x^\circ),\ x_i^* \in N(\omega_i|\Omega_i),\ i = 1, 2, \ldots, n,$$

such that

$$\sum_{i=1}^n \|x_i^*\| = 1,\ \|\sum_{i=1}^n x_i^*\| < \delta.$$

(ii) $\eta[\Omega_1, \ldots, \Omega_n](x^\circ) > 0$ *if and only if there exists a* $\gamma > 0$ *and a* $\delta > 0$, *such that*

$$\|\sum_{i=1}^n x_i^*\| \geq \gamma \sum_{i=1}^n \|x_i^*\|$$

for all $x_i^* \in N(x_i|\Omega_i),\ x_i \in \Omega_i \cap B_\delta(x^\circ),\ i = 1, \ldots, n.$

$\eta[\Omega_1, \ldots, \Omega_n](x^\circ)$ *equals to the exact upper bound of all such* γ.

The relation between primal and dual stationarity/regularity conditions is given by the next theorem.

THEOREM 10 (SEE [9]) *(i)* $\hat{\theta}[\Omega_1, \ldots, \Omega_n](x^\circ) \leq \eta[\Omega_1, \ldots, \Omega_n](x^\circ)$.

(ii) If X is Asplund and $\hat{\theta}[\Omega_1, \ldots, \Omega_n](x^\circ) < 1$ *then*

$$\eta[\Omega_1, \ldots, \Omega_n](x^\circ) \leq \frac{\hat{\theta}[\Omega_1, \ldots, \Omega_n](x^\circ)}{1 - \hat{\theta}[\Omega_1, \ldots, \Omega_n](x^\circ)}. \tag{14}$$

COROLLARY 11 $\eta[\Omega_1, \ldots, \Omega_n](x^\circ) = 0$ *then the system of sets* $\Omega_1, \Omega_2, \ldots,$ Ω_n *is weakly stationary at* x°.

(ii) If X is Asplund then the Extended extremal principle *is valid:*

The system $\Omega_1, \Omega_2, \ldots, \Omega_n$ *is weakly stationary at* x° *if and only if* $\eta[\Omega_1, \ldots,$ $\ldots, \Omega_n](x^\circ) = 0$.

Due to Proposition 2 it follows from the second part of Corollary 11 that in the Asplund space setting the equality $\eta[\Omega_1, \ldots, \Omega_n](x^\circ) = 0$ is a necessary condition of local extremality of the set system. This result first proved in [12] for spaces admitting an equivalent Fréchet differentiable norm and then extended in [14] to general Asplund spaces, is currently known as the *Extremal principle* [13, 14] and is one of the main tools for deducing necessary optimality conditions in nonsmooth and nonconvex problems.

Taking into account the extremal characterizations of Asplund spaces in [14] one can conclude that asplundity of the space is not only sufficient but also necessary for the Extended extremal principle to be valid. This gives another proof of the well known fact that, being a rather rich subclass of general Banach spaces (see [17]), Asplund spaces provide the appropriate framework for using Fréchet normals and subdifferentials.

THEOREM 12 *The following assertions are equivalent:*

(i) X is an Asplund space.

(ii) The Extremal principle is valid in X.

(iii) The Extended extremal principle is valid in X.

References

[1] H.H. Bauschke, J.M. Borwein, W. Li. Strong conical hull intersection property, bounded linear regularity, Jameson's property (G), and error bounds in convex optimization. *Math. Program.* 86:135-160, 1999.

[2] A.Y. Dubovitskii, A.A. Milyutin. Extremum problems in the presence of restrictions. *U.S.S.R. Comp. Maths. Math. Phys.* 5:1-80, 1965.

[3] A.D. Ioffe. Approximate subdifferentials and applications. 3: the metric theory. *Mathematika* 36:1-38, 1989.

[4] A.D. Ioffe. Metric regularity and subdifferential calculus. *Russian Math. Surveys* 55:501-558, 2000.

[5] D. Klatte, B. Kummer. *Nonsmooth Equations in Optimization: Regularity, Calculus, Methods and Applications*. Kluwer Academic Publishers, Dordrecht, 2002.

[6] A.Y. Kruger. On extremality of set systems. *Dokl. Nats. Akad. Nauk Belarusi* 42:24-28, 1998 (in Russian).

[7] A.Y. Kruger. Strict (ε, δ)-semidifferentials and extremality conditions. *Optimization* 51:539-554, 2002.

[8] A.Y. Kruger. On Fréchet subdifferentials. *J. Math. Sci. (N. Y.)*, 116:3325-3358, 2003.

[9] A.Y. Kruger. Weak stationarity: eliminating the gap between necessary and sufficient conditions. *Optimization* 53:147-164, 2004.

[10] A.Y. Kruger. Stationarity and regularity of set systems. *Pacif. J. Optimiz.* 1:101-126, 2005.

[11] A.Y. Kruger. About regularity of set systems. Research Report, University of Ballarat, School of Information Technology and Mathematical Sciences, N. 3, 2005.

[12] A.Y. Kruger, B.S. Mordukhovich. Extremal points and the Euler equation in nonsmooth optimization. *Dokl. Akad. Nauk BSSR*, 24(8):684-687, 1980 (in Russian).

[13] B.S. Mordukhovich. The extremal principle and its applications to optimization and economics. In *Optimization and Related Topics*, A. Rubinov and B. Glover, eds., Applied Optimization, Vol. 47, Kluwer Academic Publishers, Dordrecht, 2001, 343–369.

[14] B.S. Mordukhovich, Y. Shao. Extremal characterizations of Asplund spaces. *Proc. Amer. Math. Soc.* 124:197-205, 1996.

[15] K.F. Ng, W.H. Yang. Regularities and their relations to error bounds. *Math. Program., Ser. A*, 99:521–538, 2004.

[16] H.V. Ngai, M. Théra. Metric inequality, subdifferential calculus and applications. *Set-Valued Anal.* 9:187-216, 2001.

[17] R.R. Phelps. *Convex Functions, Monotone Operators and Differentiability, 2nd edition*, Lecture Notes in Mathematics, Vol. 1364, Springer-Verlag, New York, 1993.

INTRINSIC MODELING OF LINEAR THERMO-DYNAMIC THIN SHELLS

C. Lebiedzik [1]

[1] *Wayne State University, Department of Mathematics, Detroit MI USA, kate@math.wayne.edu**

Abstract We consider the problem of modeling dynamic thin shells with thermal effects based on the intrinsic geometry methods of Michel Delfour and Jean-Paul Zolésio. This model relies on the oriented distance function which describes the geometry. Here we further develop the Kirchhoff-based shell model introduced in our previous work by subjecting the elastically and thermally isotropic shell to an unknown temperature distribution. This yields a fully-coupled system of four linear equations whose variables are the displacement of the shell mid-surface and the thermal stress resultants.

keywords: Intrinsic shell model, dynamic thermoelasticity

1. Introduction

In this paper we continue the development of a Kirchhoff-based shell model using the intrinsic-geometric methods introduced by Michel Delfour and Jean-Paul Zolésio [6, 5]. The aim of this method is to produce a coordinate-free version of the shell equations, in contrast to the classical equations which require explicit representation of the nonconstant coefficients. With the intrinsic approach, one can exploit the underlying geometry of the shell to derive equations in which the nonconstant coefficients are written in the form of tangential operators. This enables us to better modify and apply known techniques that were developed for use in the constant-coefficient case (flat plate models).

In our previous work [2–4] we have developed a linear dynamic model of the thin shell and shown several interesting stability/controllability results. However, as thermal effects are very important in many applications of engineering, we wish to include them in our shell model. We proceed in the development of a (linear) thermoelastic shell model based essentially on similar assumptions to those which are used in the derivation of classical linear thermoelastic plate models (see, *e.g.* [7]). As such, we subject the elastically and thermally

*Paper written with financial support of the National Science Foundation under Grant DMS-0408565.

Please use the following format when citing this chapter:

Lebiedzik, C., 2006, in IFIP International Federation for Information Processing, Volume 202, Systems, Control, Modeling and Optimization, eds. Ceragioli, F., Dontchev, A., Furuta, H., Marti, K., Pandolfi, L., (Boston: Springer), pp. 215-225.

isotropic shell to an unknown temperature distribution. Eventually this yields a fully-coupled system of four linear equations whose variables are the displacement of the shell mid-surface and the thermal stress resultants. The form of these equations is familiar – in fact it looks very similar to a 'linear' version of the well-known Von Kármán system [7]. However, it must be noted that all the operators are tangential operators and thus the curvature of the shell is very much in evidence.

2. Preliminary Considerations

In this section we present a brief overview of the oriented distance function and the intrinsic tangential calculus that forms the basis of our shell model. In addition, we introduce the set of hypotheses on the shell that will be in force for the rest of this paper.

2.1 Overview of the intrinsic geometry

In order to improve readability we here include a brief discussion of the oriented distance function and the intrinsic geometric methods of Delfour and Zolésio. Since by necessity this overview will lack detail, the reader is referred to [5, 6] for a definitive exposition on this topic.

Consider a domain $\mathcal{O} \subset R^3$ whose nonempty boundary $\partial\mathcal{O}$ is a C^1 two-dimensional submanifold of R^3. Define the oriented (or signed) distance function to $\mathcal{O}$ as $b(x) = d_{\mathcal{O}}(x) - d_{R^3 \setminus \mathcal{O}}(x)$ where d is the Euclidean distance from the point x to the domain $\mathcal{O}$. In other words, $b(x)$ is simply the positive or negative distance to the boundary $\partial\mathcal{O}$. It can be shown that for every $x \in \partial\mathcal{O}$, there exists a neighborhood where the function $\nabla b = \nu$, the unit outward external normal to $\partial\mathcal{O}$ [6].

Consider a subset $\Gamma \subseteq \partial\mathcal{O}$ which will eventually become the mid-surface of our shell. We define the projection $p(x)$ of a point x onto Γ as $p(x) = x - b(x)\nabla b(x)$. Then, we define a shell S_h of thickness h as

$$S_h(\Gamma) \equiv \left\{x \in R^3 : p(x) \in \Gamma,\ |b(x)| < h/2\right\} \quad . \tag{1}$$

When $\Gamma \neq \partial\mathcal{O}$, the shell S_h has a lateral boundary $\Sigma_h(\Gamma) \equiv \{x \in R^3 : p(x) \in \Upsilon,\ |b(x)| < h/2\}$ where $\Upsilon \equiv \partial\Gamma$ denotes the boundary of Γ. A natural curvilinear coordinate system (X, z) is thus induced on the shell S_h, where the coordinate vector X gives the position of a point on the mid-surface Γ, and $z \in \left(-\frac{h}{2}, \frac{h}{2}\right)$ gives the vertical (normal) distance from the mid-surface. We also define the "flow mapping" $T_z(X)$ as $T_z(X) = X + z\nabla b(X)$ for all X and z in S_h. The curvatures of the shell will be denoted H and K. These can be reconstructed from the boundary distance function $b(x)$ by noting that at any point (X, z), the matrix $D^2 b$ has eigenvalues $0,\ \lambda_1,\ \lambda_2$. The curvatures are then given by $\mathrm{tr}(D^2 b) = 2H = \lambda_1 + \lambda_2$ and $K = \lambda_1 \lambda_2$.

Next, we mention briefly some useful aspects of the tangential differential calculus. Given $f \in C^1(\Gamma)$, we define the tangential gradient ∇_Γ of the scalar function f by means of the projection as

$$\nabla_\Gamma f \equiv \nabla(f \circ p)(x)|_\Gamma \quad . \tag{2}$$

This notion of the tangential gradient is equivalent to the classical definition using an extension F of f in the neighborhood of Γ, *i.e.* $\nabla_\Gamma f = \nabla F|_\Gamma - \frac{\partial F}{\partial \nu}\nu$ [6]. Following the same idea we can define the tangential Jacobian matrix of a vector function $v \in C^1(\Gamma)^3$ as $D_\Gamma v \equiv D(v \circ p)|_\Gamma$ or $(D_\Gamma v)_{ij} = (\nabla_\Gamma v_i)_j$, the tangential divergence as $\mathrm{div}_\Gamma v \equiv \mathrm{div}(v \circ p)|_\Gamma$, the Hessian $D^2_\Gamma f$ of $f \in C^2(\Gamma)$ as $D^2_\Gamma f = D_\Gamma(\nabla_\Gamma f)$, the Laplace-Beltrami operator of $f \in C^2(\Gamma)$ as $\Delta_\Gamma f \equiv \mathrm{div}_\Gamma(\nabla_\Gamma f) = \Delta(f \circ p)|_\Gamma$, the tangential linear strain tensor of elasticity as $\varepsilon_\Gamma(v) \equiv \frac{1}{2}(D_\Gamma v + {}^*D_\Gamma v) = \varepsilon(v \circ p)|_\Gamma$, and the tangential vectorial divergence of a second-order tensor A as $\mathrm{div}_\Gamma A \equiv \mathrm{div}(A \circ p)|_\Gamma = \mathrm{div}_\Gamma A_i$. Using these definitions one can derive Green's formula in the tangential calculus [6]:

$$\int_\Gamma f \mathrm{div}_\Gamma v \, d\Gamma + \int_\Gamma \langle \nabla_\Gamma f, v \rangle \, d\Gamma = \int_\Upsilon \langle f v, \nu \rangle \, d\Upsilon + 2 \int_\Gamma f \, H \langle v, \nabla b \rangle \, d\Gamma \tag{3}$$

where ν is the outward unit normal to the curve Υ. From [6, 5] we have that $\langle \nabla_\Gamma w, \nabla b \rangle = 0$ and $D_\Gamma v \nabla b = 0$ by definition for any scalar w and vector v. In addition, if we consider a purely tangent vector $v = v_\Gamma$, *i.e.* $\langle v_\Gamma, \nabla b \rangle = 0$, we can take the tangential gradient of both sides of this expression and derive that $D^2 b \, v_\Gamma + {}^*D_\Gamma v_\Gamma \nabla b = 0$. Finally, throughout this paper we will use $\langle \cdot, \cdot \rangle$ to denote the scalar product of two vectors and $A..B$ to denote the double contraction of two matrices – *i.e.* $A..B = tr(AB)$.

2.2 Model hypotheses

ASSUMPTION 1 *We impose the following assumptions on the shell.*

(i) The shell is assumed to be made of an isotropic and homogeneous material, so that the Lamé coefficients $\lambda > 0$ *and* $\mu > 0$ *are constant.*

(ii) The thickness h *of the shell is small enough to accommodate the curvatures* H *and* K*, i.e. the product of the thickness by the curvatures is small as compared to 1. As a consequence we shall drop terms of order equal or greater than 2 in the series expansions.*

(iii) (Kirchhoff Hypothesis) Let T *be a transformation of the shell* S_h*, and let* $\mathbf{e} = (e_\Gamma, w)$ *be the corresponding transformation of the mid-surface. In the classical thin plate theory named after Kirchhoff, the displacement vectors* T *and* $e \circ p$ *are related by the hypothesis that the filaments of the plate initially perpendicular to the middle surface remain straight and perpendicular*

to the deformed surface, and undergo neither contraction nor extension. In the intrinsic geometry we have $T = \mathbf{e} \circ p - b(\,{}^*D_{\Gamma_0}\mathbf{e}\,\nabla b) \circ p$.

(iv) We will assume the boundary Υ *consists of two open connected regions* Υ_0 *and* Υ_1*, with* $\Upsilon = \overline{\Upsilon_0 \cup \Upsilon_1}$ *and* $\emptyset = \Upsilon_0 \cap \Upsilon_1$*. We will clamp the shell on* Υ_0*, and allow* Υ_1 *to be free.*

(v) The shell is assumed to be subject to an unknown temperature distribution $\tau(\mathbf{x}, t)$ *which is measured from a reference temperature. The shell is assumed to be thermally isotropic, the change in* τ *is small compared to the reference temperature* τ_0 *of the shell, and the thermal strain is assumed to be linear. Thus the thermal strains of the shell are given by* $\varepsilon^\tau(T) = \bar{\alpha}\tau I$*, where* $\bar{\alpha}$ *is the coefficient of thermal expansion.*

We denote by $\mathbf{e}$ the transformation of the shell mid-surface and by e_Γ and e_n the tangential and normal components of $\mathbf{e}$ in local coordinates. We define w to be the magnitude of the normal displacement. As such, we have that

$$w = \langle \mathbf{e}, \nabla b \rangle, \qquad e_n = w\,\nabla b, \qquad e_\Gamma = \mathbf{e} - e_n\,. \tag{4}$$

The variable τ denotes the temperature in the shell body, as measured from a reference temperature τ_0, taken to be the absolute temperature of the body. Because of the assumptions of thinness of the shell and linearity of the thermal strains (Hypothesis 1 (ii) and (v)), it is reasonable to suppose that the temperature varies linearly with respect the thickness of the shell,

$$\tau = \tau_1 \circ p + b\,\tau_2 \circ p\,, \tag{5}$$

with τ_1, τ_2 variables defined on the mid-surface of the shell Γ. Note that τ_1 corresponds physically to the thermal energy of stretching (membrane energy), whereas τ_2 corresponds to the thermal effect of shell bending. The final form of the equations of the shell will not involve τ_i, but instead will naturally involve the thermal stress resultants θ and φ, defined as

$$\varphi = \bar{\alpha}\,\tau_1, \qquad \theta = \bar{\alpha}\,\tau_2\,, \tag{6}$$

where $\bar{\alpha}$ is the coefficient of thermal expansion.

Here we list the following definitions and properties derived in [2]:

LEMMA 2 *The following strain-displacement relation holds for a shell modeled in the intrinsic geometry under Hypothesis 1 (i)-(iii).*

$$\begin{aligned} \varepsilon(T) = {} & (\varepsilon_\Gamma(e_\Gamma) + wD^2b + V_\Gamma e_\Gamma) \circ p \\ & -b\,(-\varepsilon_\Gamma(D^2b\,e_\Gamma) + C_\Gamma e_\Gamma + S_\Gamma w + G_\Gamma w + w(D^2b)^2) \circ p\,, \end{aligned} \tag{7}$$

where ε_Γ is the tangential linear strain tensor of elasticity and

$$C_\Gamma u \;=\; \frac{1}{2}(D^2b\;{}^*D_\Gamma u + D_\Gamma u\; D^2b)$$

$$\begin{aligned} V_\Gamma u &= \frac{1}{2}((D^2b\,u)\otimes\nabla b+\nabla b\otimes(D^2b\,u)) \\ G_\Gamma w &= \frac{1}{2}((\nabla b\otimes\nabla_\Gamma w)D^2b+D^2b(\nabla_\Gamma w\otimes\nabla b)) \\ S_\Gamma w &= \frac{1}{2}(D_\Gamma^2 w+{}^*D_\Gamma^2 w)\,. \end{aligned} \tag{8}$$

C_Γ and V_Γ are 1st-order and 0-order operators, respectively, that in practice operate on a tangential vector u. G_Γ is a 1st-order operator, and S_Γ is the symmetrization of the Hessian matrix of a scalar function w (the Hessian matrix is not symmetric in the tangential calculus [6]). Define the space V

$$V=\left\{\mathbf{e}\in\left[H^1(\Gamma)\right]^2\times H^2(\Gamma)\,\middle|\, e_\Gamma=w=\frac{\partial}{\partial\nu}w=0\text{ on }\Upsilon_0\right\}. \tag{9}$$

3. Thermoelastic shell model

THEOREM 3 *Define the following operator C acting on a matrix A:*

$$C(A)=\lambda\mathrm{tr}(A)\,I+2\mu\,A\,, \tag{10}$$

the expression $\tilde{\chi}$

$$\tilde{\chi}=C_\Gamma e_\Gamma-\varepsilon_\Gamma(D^2b\,e_\Gamma)\,, \tag{11}$$

the parameters $\beta=(\lambda+2\mu)^{-1}$, $\zeta=\beta(\frac{2}{3}\mu+\lambda)$, $\kappa=\frac{\lambda_0}{\rho c\bar{\alpha}\beta}$, and $\eta=\frac{\zeta\bar{\alpha}\tau_0}{\lambda_0}$. Here ρ is shell density, c is specific heat, and λ_0 is thermal conductivity.

Then, the displacement $\mathbf{e}\in C([0,\infty);V)$ and thermal variables θ, $\varphi\in C([0,\infty);L_2(\Gamma))$ satisfy the following system of shell equations which holds on $\Gamma\times(0,\infty)$:

$$\begin{aligned} &\partial_{tt}w-\gamma\Delta_\Gamma\partial_{tt}w+\Delta_\Gamma^2 w+\frac{\gamma}{2}\mathrm{div}_\Gamma(D^2b\,\partial_{tt}e_\Gamma) \\ &\qquad -\zeta\Delta_\Gamma\theta-\zeta(4H^2-2K)\theta-2\zeta\gamma^{-1}H\varphi+P_1(e_\Gamma)+Q_1(w)=0 \end{aligned} \tag{12}$$

$$\begin{aligned} &(I+\gamma(D^2b)^2)\partial_{tt}e_\Gamma-\beta[\gamma^{-1}\mathrm{div}_\Gamma C(\varepsilon_\Gamma(e_\Gamma))+D^2b\,\mathrm{div}_\Gamma C(\tilde{\chi}) \\ &\qquad -\mathrm{div}_\Gamma(D^2b\,C(\tilde{\chi}))]+\zeta(\mathrm{div}_\Gamma(D^2b\,\theta)-D^2b\,\nabla_\Gamma\theta)+\zeta\gamma^{-1}\nabla_\Gamma\varphi \\ &\qquad -\frac{\gamma}{2}D^2b\nabla_\Gamma\partial_{tt}w+P_2(w)+Q_2(e_\Gamma)=0 \end{aligned} \tag{13}$$

where P_1 denotes coupling terms and Q_1 denotes lower order terms in the plate equation; and P_2, Q_2 in the wave equation:

$$\begin{aligned} P_1(e_\Gamma) &= \beta[2\lambda H\gamma^{-1}\mathrm{div}_\Gamma e_\Gamma+2\mu\gamma^{-1}\mathrm{tr}(D^2b\,\varepsilon_\Gamma(e_\Gamma)) \\ &\quad -2\mu\langle(D^2b)^2..D^3b,e_\Gamma\rangle+2\mu\mathrm{div}_\Gamma\mathrm{div}_\Gamma(\tilde{\chi})+4\mu H\mathrm{tr}(D^3be_\Gamma\,D^2b) \\ &\quad -\lambda\Delta_\Gamma\langle 2\nabla_\Gamma H,e_\Gamma\rangle-\lambda(4H^2-2K)\langle 2\nabla_\Gamma H,e_\Gamma\rangle \end{aligned}$$

$$\begin{aligned}
Q_1(w) &= \beta[k_\gamma w + 4\mu \mathrm{div}_\Gamma((D^2b)^2\nabla_\Gamma w) + \lambda\Delta_\Gamma((4H^2-2K)w) \\
&\quad +2\mu\mathrm{div}_\Gamma\mathrm{div}_\Gamma((D^2b)^2w)+2\mu\mathrm{div}_\Gamma(K\nabla_\Gamma w)+\lambda(4H^2-2K)\Delta_\Gamma w] \\
&\qquad +2\mu\mathrm{tr}(S_\Gamma w(D^2b)^2) + 4\mu H\mathrm{tr}((D^2b)^3w) \\
P_2(w) &= \beta[-2\lambda\gamma^{-1}\nabla_\Gamma(Hw) + 2\mu\gamma^{-1}\mathrm{div}_\Gamma(w\,D^2b) \\
&\qquad +\lambda 2\nabla_\Gamma H(\Delta_\Gamma w - (4H^2-2K)w) - 2\mu(D^2b)^2..D^3bw] \\
&\qquad -2\mu\mathrm{div}_\Gamma(D^2b\,S_\Gamma w) + 2\mu D^2b\,\mathrm{div}_\Gamma(S_\Gamma w) \\
Q_2(e_\Gamma) &= -2\beta\mu\gamma^{-1}(K\,e_\Gamma + 2(D^2b)^2e_\Gamma)
\end{aligned}$$

The thermal variables φ and θ satisfy the following coupled heat-like equations:

$$\frac{1}{\kappa}\partial_t\varphi - \Delta_\Gamma\varphi - 2H\theta + \eta(2H\partial_t w + \mathrm{div}_\Gamma\partial_t e_\Gamma) = f_1 \tag{14}$$

$$\begin{aligned}
\frac{1}{\kappa}\partial_t\theta - \Delta_\Gamma\theta - \eta(\Delta_\Gamma\partial_t w + \mathrm{tr}(C_\Gamma\partial_t e_\Gamma) - \mathrm{div}_\Gamma(D^2b\partial_t e_\Gamma) \\
+(4H^2-2K)\partial_t w) = f_2
\end{aligned} \tag{15}$$

where f_1 and f_2 represent heat sources or sinks. We have the following free boundary conditions on $\Upsilon_1 \times (0,\infty)$:

$$\begin{aligned}
&\Big(\mathcal{C}(w\,D^2b + \varepsilon_\Gamma(e_\Gamma)) - \zeta\varphi I\Big)\cdot\nu = 0 \\
&\langle(\lambda\beta\nabla_\Gamma, \mathrm{tr}(D^3b\,e_\Gamma)\rangle + 4\mu\beta(D^2b)^2\nabla_\Gamma w + 2\mu\beta K\nabla_\Gamma w - \nabla_\Gamma(\Delta_\Gamma w)\nu \\
&+\langle\gamma(D^2b\partial_{tt}e_\Gamma - 2\nabla_\Gamma\partial_{tt}w) + 2\mu\beta B^3_\Gamma(C_\Gamma e_\Gamma - \varepsilon_\Gamma(D^2b\,e_\Gamma) + (D^2b)^2\,w), \nu\rangle \\
&\qquad -\langle\beta\lambda\nabla_\Gamma((4H^2-2K)w) - \zeta\nabla_\Gamma\theta, \nu\rangle + 2\mu B^2_\Gamma w = 0 \\
&\lambda\beta\mathrm{tr}(D^3b\,e_\Gamma) + \lambda\beta(4H^2-2K)w - \Delta_\Gamma w - \zeta\theta \\
&\qquad +2\mu\beta\langle(C_\Gamma e_\Gamma - \varepsilon_\Gamma(D^2b\,e_\Gamma) + (D^2b)^2)\nu, \nu\rangle + 2\mu\beta B^1_\Gamma w = 0 \\
&\langle\theta,\nu\rangle = -\lambda_2(\theta - \tilde\theta), \qquad \langle\varphi,\nu\rangle = \lambda_2(\varphi - \tilde\varphi)
\end{aligned}$$

clamped boundary conditions on $\Upsilon_0 \times (0,\infty)$:

$$w = \frac{\partial}{\partial\nu}w = 0, \quad e_\Gamma = 0, \quad \theta = \varphi = 0 \tag{16}$$

Here λ_2 is the coefficient from Newton's law of cooling, and $\tilde\theta = \bar\alpha\tilde\tau_1$, $\tilde\varphi = \bar\alpha\tilde\tau_2$, where $\tilde\tau$ is the temperature of the external medium of the shell.

Proof: *Elastic Equations.* We begin with the stress-strain relationships for a general shell body. Let T be a transformation of the body S_h. We have that the stress

$$\sigma = \mathcal{C}\varepsilon(T) - \mathcal{C}\varepsilon^\tau(T)\,, \tag{17}$$

where ε^τ denotes the thermal strain. By Hypothesis 1 (v), we can write that $\varepsilon^\tau = \bar\alpha\tau I$, where τ is the temperature and $\bar\alpha$ is the coefficient of thermal expansion. Next, we note that the body is assumed to be isotropic, so that $\mathcal{C}(A) = \lambda \mathrm{tr}(A) + 2\mu A$ where λ and μ are the Lamé coefficients. Finally, we use all this information to compute the potential energy

$$\mathcal{E}_p = \frac{1}{2}\int_{S_h} \varepsilon..\sigma = \frac{1}{2}\Big[\lambda \int_{S_h} (\mathrm{tr}\varepsilon(T))^2 + 2\mu \int_{S_h} \mathrm{tr}(\varepsilon(T)^2)$$
$$-\bar{\nu}\int_{S_h} \mathrm{tr}(\varepsilon(T))\bar{\alpha}\tau\Big], \qquad \text{where } \lambda + \tfrac{2}{3}\mu = \bar{\nu}.$$

REMARK 4 *At this point in the computation of the elastic energy, it is customary to impose the hypothesis of plane stresses: $\sigma..(\nabla b \otimes \nabla b) = 0$ (which in local coordinates is denoted $\sigma_{33} = 0$). As is well understood (see, e.g. [1]), this assumption implies a change of Lamé coefficient λ to $\frac{E\nu}{1-\nu^2}$, while μ remains unchanged. The same situation arises in the case of plates, we refer to [7] for further details. This modified expression for λ is more in line with both experimental evidence and asymptotic models. This does not affect any of the mathematical arguments to follow, so the imposition of this hypothesis is left to the discretion of the reader.*

Let us denote $\mathcal{E}_p = \mathcal{E}_{p,e} + \mathcal{E}_{p,t}$, with the elastic contribution to the potential energy $\mathcal{E}_{p,e}$ given as calculated in [2] using Lemma 2 as

$$\begin{aligned}\mathcal{E}_{p,e} \;=\; & h\frac{\lambda}{2}\,|2H\,w + \mathrm{div}_\Gamma e_\Gamma|^2_{L_2(\Gamma)} + h\mu\int_\Gamma \mathrm{tr}\left[(\varepsilon_\Gamma(e_\Gamma) + D^2b\,w + V_\Gamma e_\Gamma)^2\right] \\ & + h\frac{\lambda\gamma}{2}\left|\Delta_\Gamma w + \mathrm{tr}(C_\Gamma e_\Gamma) - \mathrm{div}_\Gamma(D^2b\,e_\Gamma) + (4H^2 - 2K)w\right|^2_{L_2(\Gamma)} \\ & + \mu\gamma h\int_\Gamma \mathrm{tr}\left[\left(S_\Gamma w + C_\Gamma e_\Gamma - \varepsilon_\Gamma(D^2b\,e_\Gamma) + G_\Gamma w + w(D^2b)^2\right)^2\right] d\Gamma\end{aligned}$$

where $\gamma = \frac{h^2}{12}$. We compute the thermal contribution $\mathcal{E}_{p,t}$ explicitly from Lemma 2, the expansion of τ (5), and the definition of φ and θ (6) :

$$\mathcal{E}_{p,t} = \frac{\bar{\nu}}{2}\int_{S_h} \mathrm{tr}(\varepsilon(T))\bar{\alpha}\tau = \frac{\bar{\nu}}{2}\int_{-\frac{h}{2}}^{\frac{h}{2}}\int_{\Gamma^z} \mathrm{tr}(\varepsilon(T))\bar{\alpha}(\tau_1\circ p + z\tau_2\circ p) \tag{18}$$
$$= \frac{\bar{\nu}}{2}\int_{-\frac{h}{2}}^{\frac{h}{2}}\int_\Gamma ((\mathrm{tr}(\varepsilon(T))\theta_1\circ p)\circ T_z)j(z) + ((z\mathrm{tr}(\varepsilon(T))\theta_2\circ p)\circ T_z)j(z).$$

after using the Federer decomposition and a change of variable. Now, Hypothesis 1 (ii) allows us to say $j(z) \approx 1$; and noting that the functions p and T_z are inverses by definition and evaluating the z-integral, we have

$$\begin{aligned}\mathcal{E}_{p,t} = & \frac{h\bar{\nu}}{2}\int_\Gamma (\mathrm{div}_\Gamma e_\Gamma + 2Hw)\varphi \qquad\qquad (19)\\ & + \frac{h^3\bar{\nu}}{24}\int_\Gamma (\Delta_\Gamma w + \mathrm{tr}(C_\Gamma e_\Gamma) - \mathrm{div}_\Gamma(D^2b\,e_\Gamma + (4H^2 - 2K)w)\theta.\end{aligned}$$

Thus, collecting (17) and (19) gives the desired expression for the potential energy of the shell. For the kinetic energy of the shell we have, from [2] and the Kirchhoff hypothesis:

$$\begin{aligned}\mathcal{E}_k &= \frac{\rho h}{2}\int_\Gamma |\partial_t e_\Gamma|^2 + |\partial_t w|^2 \\ &\quad + \frac{\rho h\gamma}{2}\int_\Gamma |D^2 b\,\partial_t e_\Gamma|^2 + |\nabla_\Gamma \partial_t w|^2 + |D^2 b\,\partial_t e_\Gamma - \nabla_\Gamma\partial_t w|^2.\end{aligned} \tag{20}$$

The kinetic energy of the thermal variables will be discussed later.

Among all kinematically admissible displacements, the actual motion of the shell will make stationary the Lagrangian

$$\mathcal{L}(\mathbf{e}) = \int_0^t \mathcal{E}_k(\mathbf{e}) - \mathcal{E}_{p,e}(\mathbf{e}) + \mathcal{E}_{p,t}(\mathbf{e})\,.$$

Note that we take the variation with respect to $\mathbf{e}$ only: $\frac{\partial}{\partial\psi}\mathcal{L}(\mathbf{e}+\psi\,\hat{\mathbf{e}})\Big|_{\psi=0} = 0$. This results in the following weak form of the equations:

$$\begin{aligned}
&\int_0^t \Big[-\rho[2(\partial_t e_\Gamma, \partial_t \hat{e_\Gamma})_\Gamma + 2\gamma((D^2b)\partial_t e_\Gamma, (D^2b)\partial_t\hat{e_\Gamma})_\Gamma \\
&-\gamma(\nabla_\Gamma\partial_t w, (D^2b)\partial_t\hat{e_\Gamma})_\Gamma \\
&-\gamma((D^2b)\partial_t e_\Gamma, \nabla_\Gamma\partial_t\hat{w})_\Gamma + 2(\partial_t w, \partial_t\hat{w})_\Gamma + 2\gamma(\nabla_\Gamma\partial_t w, \nabla_\Gamma\partial_t\hat{w})_\Gamma] \\
&2\lambda\gamma(\Delta_\Gamma w, \Delta_\Gamma\hat{w})_\Gamma + 4\mu\gamma\int_\Gamma \mathrm{tr}((S_\Gamma w + G_\Gamma w)(S_\Gamma\hat{w} + G_\Gamma\hat{w})) \\
&+2\lambda(\mathrm{div}_\Gamma e_\Gamma, \mathrm{div}_\Gamma\hat{e_\Gamma})_\Gamma + 4\mu\int_\Gamma \mathrm{tr}(\varepsilon_\Gamma(e_\Gamma)\varepsilon_\Gamma(\hat{e_\Gamma})) \\
&-2\mu(D^2b\,e_\Gamma, D^2b\,\hat{e_\Gamma})_\Gamma + 4\lambda(Hw, \mathrm{div}_\Gamma\hat{e_\Gamma})_\Gamma \\
&+4\lambda(\mathrm{div}_\Gamma e_\Gamma, H\hat{w})_\Gamma + 4\mu(w, \mathrm{tr}(\varepsilon_\Gamma(\hat{e_\Gamma})\,D^2b))_\Gamma + 4\mu(\mathrm{tr}(\varepsilon_\Gamma(e_\Gamma)\,D^2b), \hat{w})_\Gamma \\
&+2(\sqrt{k_\gamma}w, \sqrt{k_\gamma}\hat{w})_\Gamma + 2\lambda\gamma(\mathrm{div}_\Gamma(D^2b\,e_\Gamma), \mathrm{div}_\Gamma(D^2b\,\hat{e_\Gamma}))_\Gamma \\
&+4\mu\gamma\int_\Gamma \mathrm{tr}(\varepsilon_\Gamma(D^2b\,e_\Gamma)\varepsilon_\Gamma(D^2b\,\hat{e_\Gamma})) - 2\lambda\gamma(\mathrm{div}_\Gamma(D^2b\,e_\Gamma), \mathrm{tr}(C_\Gamma\hat{e_\Gamma}))_\Gamma \\
&-2\lambda\gamma(\mathrm{tr}(C_\Gamma e_\Gamma), \mathrm{div}_\Gamma(D^2b\,\hat{e_\Gamma}))_\Gamma + 2\lambda\gamma(\mathrm{tr}(C_\Gamma e_\Gamma), \mathrm{tr}(C_\Gamma\hat{e_\Gamma}))_\Gamma \\
&+4\mu\gamma\int_\Gamma \mathrm{tr}(C_\Gamma e_\Gamma\,C_\Gamma\hat{e_\Gamma}) - 4\mu\gamma\int_\Gamma \mathrm{tr}(C_\Gamma e_\Gamma\,\varepsilon_\Gamma(D^2b\,\hat{e_\Gamma})) \\
&-4\mu\gamma\int_\Gamma \mathrm{tr}(\varepsilon_\Gamma(D^2b\,e_\Gamma)\,C_\Gamma\hat{e_\Gamma}) - 2\lambda\gamma(\mathrm{tr}(D^3b\,e_\Gamma), \Delta_\Gamma\hat{w})_\Gamma \\
&-2\lambda\gamma(\Delta_\Gamma w, \mathrm{tr}(D^3b\hat{e_\Gamma}))_\Gamma \qquad (21) \\
&+4\mu\gamma\int_\Gamma \mathrm{tr}(C_\Gamma e_\Gamma\,S_\Gamma\hat{w}) + 4\mu\gamma\int_\Gamma \mathrm{tr}(S_\Gamma w\,C_\Gamma\hat{e_\Gamma}) \\
&-4\mu\gamma\int_\Gamma \mathrm{tr}(S_\Gamma w\,\varepsilon_\Gamma(D^2b\,\hat{e_\Gamma})) - 4\mu\gamma\int_\Gamma \mathrm{tr}(\varepsilon_\Gamma(D^2b\,e_\Gamma)\,S_\Gamma\hat{w})
\end{aligned}$$

$$
\begin{aligned}
&-2\lambda\gamma((4H^2-2K)w, \mathrm{tr}(D^3b\,\hat{e_\Gamma}))_\Gamma \\
&-2\lambda\gamma(\mathrm{tr}(D^3b\,e_\Gamma), (4H^2-2K)\hat{w})_\Gamma \\
&-4\mu\gamma(\mathrm{tr}(D^3b\,e_\Gamma(D^2b)^2), \hat{w})_\Gamma - 4\mu\gamma(w, \mathrm{tr}(D^3b\,\hat{e_\Gamma}(D^2b)^2))_\Gamma \\
&+2\lambda\gamma((4H^2-2K)w, \Delta_\Gamma\hat{w})_\Gamma + 2\lambda\gamma(\Delta_\Gamma w, (4H^2-2K)\hat{w})_\Gamma \\
&+4\mu\gamma(w, \mathrm{tr}(S_\Gamma\hat{w}(D^2b)^2))_\Gamma + 4\mu\gamma(\mathrm{tr}(S_\Gamma w(D^2b)^2), \hat{w})_\Gamma \\
&+2\bar{\nu}(\varphi, \mathrm{div}_\Gamma\hat{e_\Gamma})_\Gamma + 4\bar{\nu}(\varphi, H\,\hat{w})_\Gamma + 2\bar{\nu}\gamma(\theta, \Delta_\Gamma\hat{w})_\Gamma \\
&+2\bar{\nu}\gamma(\theta, \mathrm{tr}(C_\Gamma\hat{e_\Gamma})_\Gamma - 2\bar{\nu}\gamma(\theta, \mathrm{div}_\Gamma(D^2b\,\hat{e_\Gamma}))_\Gamma \\
&+2\bar{\nu}\gamma((4H^2-2K)\theta, \hat{w})_\Gamma\Big]\,dt = 0
\end{aligned}
$$

We integrate the expression (20) in order to derive the equations (12) and (13) of Theorem 3. These calculations are presented explicitly in [3], so we omit the details. We note that the regularity of the weak solution **e** is high enough to permit the necessary integration by parts to derive the strong form – this is proved in Proposition 4.2 of [3]. After this, inspection of (20) reveals that there are two fourth-order terms in w. The first, $(\Delta_\Gamma w, \Delta_\Gamma\hat{w})$, will yield the required tangential biharmonic operator Δ_Γ^2 in the strong form. However, the next term is also fourth-order, and we would like to combine the two in analogy to the case of plates, where the second term becomes the biharmonic plus a boundary integral. In fact, in [3] we show that

$$
\begin{aligned}
\int_\Gamma \mathrm{tr}((S_\Gamma w + G_\Gamma w)(S_\Gamma\hat{w} + G_\Gamma\hat{w}))\,d\Gamma &= \int_\Gamma \Delta_\Gamma w\Delta_\Gamma\hat{w}\,d\Gamma \\
&+ \int_\Gamma \langle K\nabla_\Gamma w, \nabla_\Gamma\hat{w}\rangle\,d\Gamma + 2\int_\Gamma \langle D^2b\nabla_\Gamma w, D^2b\nabla_\Gamma\hat{w}\rangle\,d\Gamma \\
&+ \int_\Upsilon (B_\Gamma^1 w\frac{\partial}{\partial\nu}\hat{w} - B_\Gamma^2 w\hat{w})\,d\Upsilon
\end{aligned}
$$

with B_Γ^1 and B_Γ^2 being defined as

$$
\begin{aligned}
B_\Gamma^1 w &= -(\tau\otimes\tau)..(S_\Gamma w + G_\Gamma w) \\
B_\Gamma^2 w &= \langle\nabla_\Gamma((\tau\otimes\nu)..(S_\Gamma w + G_\Gamma w)), \tau\rangle\,.
\end{aligned} \tag{22}
$$

The operators B_Γ^1 and B_Γ^2 are simply the tangential versions of the same operators which appear in the modeling of Kirchhoff plates [7]. One can show this explicitly by choosing a local basis $\nu = (\nu_1, \nu_2)$; $\tau = (-\nu_2, \nu_1)$ and substituting appropriately. The additional boundary operator B_Γ^3 which appears in the free boundary conditions of Theorem 3 is given by

$$
B_\Gamma^3 A = \partial_t\langle\tau, A\nu\rangle + \langle\mathrm{div}_\Gamma A, \nu\rangle = \langle\nabla_\Gamma(\tau\otimes\nu..A), \tau\rangle + \langle\mathrm{div}_\Gamma A, \nu\rangle\,. \tag{23}
$$

This operator comes from integration of cross-terms involving $S_\Gamma\hat{w}$.

Thermal Equations. Next, we must obtain the equations of motion for the thermal variables φ and θ. Recall that τ is the temperature of the body, measured from a reference temperature τ_0. By combining Fourier's law of heat conduction, the entropy balance law, the second law of thermodynamics for irreversible processes, and using the fact that the change of temperature is small to linearize, we have the following equation for heat transfer in a three-dimensional isotropic, elastic body (see [7], p. 29, and [8], Chapter 1):

$$\Delta\tau - \frac{1}{\kappa}\partial_t\tau - \frac{\eta}{\bar{\alpha}}\partial_t\left(\varepsilon(T)..I\right) = -\frac{\mathcal{H}}{\lambda_0} \tag{24}$$

with

$$\kappa = \frac{\lambda_0}{c\rho}, \qquad \eta = \left(\lambda + \frac{2}{3}\mu\right)\frac{\bar{\alpha}^2\tau_0}{\lambda_0}, \tag{25}$$

where $\lambda_0 > 0$ is the coefficient of thermal conductivity (assumed to be constant), c is the specific heat, ρ is the density of the material, and $\mathcal{H}$ are heat sources and sinks inside the body.

Recalling the definition (5), the equality (2), re-writing the thermal loads as $\mathcal{H} = \mathcal{H}_1 \circ p + b\mathcal{H}_2 \circ p$ (justified again by the assumption that the change in temperature is small), and substituting gives

$$\begin{aligned}
&\Delta(\tau_1 \circ p) + \Delta(b\tau_2 \circ p) - \frac{1}{\kappa}\partial_t(\tau_1 \circ p + b\tau_2 \circ p)\\
&\quad -\frac{\eta}{\alpha}\partial_t(\mathrm{div}_\Gamma e_\Gamma + 2Hw)\circ p + \frac{\eta}{\bar{\alpha}}\partial_t(b\,(\Delta_\Gamma w + \mathrm{tr}(C_\Gamma e_\Gamma)\\
&\qquad -\mathrm{div}_\Gamma(D^2 b\, e_\Gamma) + (4H^2 - 2K)w)\circ p) = -\frac{\mathcal{H}_1 \circ p}{\lambda_0} - b\frac{\mathcal{H}_2 \circ p}{\lambda_0}.
\end{aligned}$$

Expanding $\Delta(b\tau_2 \circ p) = b\,\Delta(\tau_2 \circ p) + \tau_2 \circ p\Delta b + 2\langle\nabla b, \nabla(\tau_2 \circ p)\rangle$ and multiplying by $\bar{\alpha}$ gives

$$\begin{aligned}
&\Delta(\varphi \circ p) + b\,\Delta(\theta \circ p) + 2H\,\theta \circ p + 2\langle\nabla b, \nabla\theta \circ p\rangle\\
&-\frac{1}{\kappa}\partial_t(\varphi \circ p + b\theta \circ p) - \eta\partial_t(\mathrm{div}_\Gamma e_\Gamma + 2Hw)\circ p + \eta\partial_t(b\,(\Delta_\Gamma w\\
&+\mathrm{tr}(C_\Gamma e_\Gamma) - \mathrm{div}_\Gamma(D^2 b\, e_\Gamma) + (4H^2 - 2K)w)\circ p) = f_1 \circ p + b\, f_2 \circ p
\end{aligned} \tag{26}$$

after defining $f_i = -\frac{\bar{\alpha}\mathcal{H}_i}{\lambda_0}$. Notice that equation (26) is of the form $A_F \circ p + bA_B \circ p = f_1 \circ p + b\, f_2 \circ p$ where A_F denotes the thermal change due to the flexure of the shell, and A_B the change due to bending of the shell. This gives us two coupled equations on the three-dimensional body:

$$\begin{aligned}
&[(\Delta - \frac{1}{\kappa}\partial_t)b\varphi + 2H\,\theta]\circ p + 2\langle\nabla b, \nabla\theta \circ p\rangle\\
&-\eta\partial_t(\mathrm{div}_\Gamma e_\Gamma + 2Hw)\circ p = f_1 \circ p
\end{aligned}$$

$$(\Delta - \frac{1}{\kappa}\partial_t)\theta \circ p + \eta\partial_t((\Delta_\Gamma w + \mathrm{tr}(C_\Gamma e_\Gamma) \\ -\mathrm{div}_\Gamma(D^2 b\, e_\Gamma) + (4H^2 - 2K)w) \circ p) = f_2 \circ p .$$

Restricting these to the midsurface gives immediately that

$$\Delta_\Gamma \varphi + 2H\theta - \frac{1}{\kappa}\partial_t \varphi - \eta\partial_t(\mathrm{div}_\Gamma e_\Gamma + 2Hw) = f_1$$

$$\Delta_\Gamma \theta - \frac{1}{\kappa}\partial_t \theta - \eta\partial_t(\Delta_\Gamma w + \mathrm{tr}(C_\Gamma e_\Gamma) - \mathrm{div}_\Gamma(D^2 b\, e_\Gamma) + (4H^2 - 2K)w) = f_2$$

as desired, since $\langle \nabla b, \nabla_\Gamma \theta \rangle = 0$. Finally, the boundary conditions on φ and θ are given by Newton's law of cooling.

THEOREM 5 (WELL-POSEDNESS) *The thermoelastic shell model presented in Theorem 3 generates a C_0 semigroup of contractions $\left\{e^{\mathbf{A}t}\right\}_{t\geq 0}$ on the space*

$$\mathcal{H} = H^2(\Gamma) \times H^1_\gamma(\Gamma) \times [H^1(\Gamma)]^2 \times [L_2(\Gamma)]^2 \times L_2(\Gamma) \times L_2(\Gamma)$$

Therefore for initial data $\mathbf{x}^0 = [w^0, w^1, e^0_\Gamma, e^1_\Gamma, \theta^0, \phi^0] \in \mathcal{H}$, the solution $\mathbf{x}(t) = [w, \partial_t w, e_\Gamma, \partial_t e_\Gamma, \theta, \phi]$ is given by $\mathbf{x}(t) = e^{\mathbf{A}t}\mathbf{x}^0$.

Proof: Straightforward calculations show that $\mathbf{A}$ is maximal dissipative – that is, $\langle \mathbf{A}X, X\rangle_\mathcal{H} \leq 0$ and $\langle \mathbf{A}^*X, X\rangle_\mathcal{H} \leq 0$ for all $X \in \mathcal{H}$. Thus, by the Lumer-Phillips theorem, the system of equations (12)-(16) is well-posed.

References

[1] M. Bernadou. *Finite Element Methods for Thin Shell Problems.* J. Wiley and Sons, 1996.

[2] J. Cagnol, I. Lasiecka, C. Lebiedzik, and J.-P. Zolésio. Uniform Stability in Structural Acoustic Models with Flexible Curved Walls *J. Diff. Eqns.* 186:88-121, 2003.

[3] J. Cagnol and C. Lebiedzik. On the free boundary conditions for a shell model based on intrinsic differential geometry. *Applicable Analysis.* 83:607-633, 2004.

[4] J. Cagnol and C. Lebiedzik. Optimal Control of a Structural Acoustic Model with Flexible Curved Walls. In *Control and boundary analysis*, Lect. Notes Pure Appl. Math., 240, Chapman & Hall/CRC, Boca Raton, FL, 2005.

[5] M. C. Delfour and J.-P. Zolésio. Differential equations for linear shells: comparison between intrinsic and classical models In *Advances in mathematical sciences: CRM's 25 years (Montreal, PQ, 1994), CRM Proc. Lecture Notes* Amer. Math. Soc. Providence, RI, 1997.

[6] M. C. Delfour and J.-P. Zolésio. *Intrinsic differential geometry and theory of thin shells.* To appear, 2006.

[7] J. Lagnese and J.-L. Lions. *Modelling, Analysis and Control of Thin Plates.* Masson, Paris, 1988.

[8] W. Nowacki. *Thermoelasticity, 2nd Edition.* Pergamon Press, New York, 1986.

DISCONTINUOUS CONTROL IN BANACH SPACES

L. Levaggi[1]
[1]*Department of Mathematics, University of Genova, Italy, levaggi@dima.unige.it**

Abstract The application of state-discontinuous feedback laws to infinite-dimensional control systems, with particular reference to sliding motions, is discussed for linear systems with distributed control. Using differential inclusions a definition of generalized solutions for the discontinuous closed loop system is introduced. Sliding modes can both be defined as viable generalized solutions or by extending the equivalent control method to infinite dimensional systems. Regularity properties of the sliding manifold under which the two methods are equivalent are investigated. Then, a comparison between classical results obtained for finite dimensional spaces and properties of infinite dimensional sliding modes is made.

keywords: Variable Structure Systems; Infinite Dimensional Systems; Sliding Mode Control.

1. Introduction

Variable structure control methods and in particular sliding mode controls, are by now recognised as classical tools for the regulation of systems governed by ordinary differential equations in a finite dimensional setting. For an overview of the finite-dimensional theory see [21]. While being easy to design, they possess attractive properties of robustness and insensitivity with respect to disturbances and unmodeled dynamics. These characteristics are all the more important when dealing with infinite-dimensional systems. Recent research has been devoted to the extension of sliding mode control and therefore the use of discontinuous feedback laws, to the infinite-dimensional setting. The early works [14, 15, 17] were confined to some special classes of systems, but at present both theory and application of sliding mode control have been extended to a rather general setting [18, 16, 19]. In particular in [18] the key concept of

*Paper written with financial support from MIUR cofinanced project "Control, optimization and stability of non-linear systems: geometrical and analytical methods".

Please use the following format when citing this chapter:

Levaggi, L., 2006, in IFIP International Federation for Information Processing, Volume 202, Systems, Control, Modeling and Optimization, eds. Ceragioli, F., Dontchev, A., Furuta, H., Marti, K., Pandolfi, L., (Boston: Springer), pp. 227-236.

equivalent control is extended to evolution equations governed by unbounded linear operators that generate C_0-semigroups.

The application of a state-discontinuous feedback law brings about the question of how to define what is the meaning of solution for the resulting closed loop. This issue becomes crucial for sliding mode control, since one seeks to constrain the evolution of the system to belong to the feedback discontinuity manifold. For ordinary differential equations the problem is solved by introducing Filippov solutions [4]. In the Banach space setting a generalised solution concept has been proposed in [10, 9] and a relationship between the equivalent control method and generalised solutions of infinite-dimensional systems with discontinuous right-hand side has been established, under some regularity assumptions. In Section 3 these results are extended to a more general setting by requiring less stringent hypotheses on the interaction between the evolution operator and the sliding surface. This allows for more flexibility in the construction of the sliding manifold and this is of primary importance for application purposes.

2. Generalized solutions for affine discontinuous control systems

The setting of the paper is the following: we consider controlled differential equations of the form

$$\begin{cases} \dot{x}(t) = Ax(t) + Bu(x(t)) \\ x(0) = x_0, \end{cases} \tag{1}$$

where x is the state variable and u is the control variable.

ASSUMPTION 1 *The following conditions are assumed to hold:*

(i) $A : \mathcal{D}(A) \subset X \rightarrow X$ *is the infinitesimal generator of a* C_0*-semigroup* $K(t)$, $t \geq 0$, *on the reflexive Banach space* X*;*

(ii) U *is a Banach space and* $B : U \rightarrow X$ *is a continuous linear operator;*

(iii) $u : \mathcal{D}(u) \subset X \rightarrow U$ *is a densely defined function that satisfies the growth condition*

$$\|u(x)\| \leq M\|x\| + N, \qquad \forall x \in \mathcal{D}(u) \tag{2}$$

for some positive constants M *and* N*.*

Following [10, 9] we introduce the following multivalued function F

$$F(x) = \bigcap_{\varepsilon>0} \overline{\text{co}}\, Bu(\overline{\mathcal{B}}(x,\varepsilon) \cap \mathcal{D}(u)), \quad x \in X, \tag{3}$$

where $\overline{\mathcal{B}}(x,r)$ is the closed ball of center x and radius r. We call *generalized solution* of (1) a mild solution of the differential inclusion

$$\begin{cases} \dot{x}(t) - Ax(t) \in F(x(t)) \\ x(0) = x_0, \end{cases} \tag{4}$$

A continuous function $x : [0,T] \to X$ is called a mild solution of (4) if there exists $g \in L^1(0,T;X)$ with $g(s) \in F(x(s))$ for almost all $s \in [0,T]$ such that

$$x(t) = K(t)x_0 + \int_0^t K(t-s)g(s)\,ds, \qquad t \in [0,T]$$

(see i.e. [22, 2] and references therein for a discussion about mild solutions and existence theorems).

THEOREM 2 *[10] If u satisfies (2), $F(x)$ is a non void, closed, convex and bounded subset of X for all $x \in X$. Moreover, F is strongly-weakly upper semi-continuous and locally bounded and therefore there always exist mild solutions of (4).*

In what follows, we will be particularly interested in the following class of solutions: if S is a subset of X and $x_0 \in S$ a mild solution of inclusion (4) that satisfies $x(t) \in S$ for all $t > 0$ is called *viable* on S. S is a *viable domain* for (4) if for any $x_0 \in S$ there exists a viable solution of the differential inclusion starting from x_0. A *generalized viable solution* of (1) is a viable solution of (4). The results by Cârjă and Vrabie in [2, 3] can be applied to our differential inclusion, so that we have necessary and sufficient conditions for the existence of viable generalized solutions.

3. Sliding modes on linear sliding manifolds

From now on the attention is restricted to a particular class of control functions u.

ASSUMPTION 3 *Let Y be a Banach space, $C : X \to Y$ a continuous linear operator, $C \neq 0$ and $\mathcal{D}(u) = X \setminus S$, $S = \ker C$.*

Thus S is a proper linear subspace of X, with void interior and $\mathcal{D}(u)$ is dense. A *sliding mode* is attained when, upon reaching the surface S, the state is henceforth constrained to remain (slide) on it. From the control view point the choice of C has to be done in such a way that, once the evolution is constrained on the sliding surface, the control goal is fulfilled. Let us suppose that S has been selected and the existence of the sliding mode has been proved (this can generally be done using Lyapunov-like techniques). Mimicking the finite dimensional case, an equivalent control can be defined as a feedback law that selects a constrained motion on S from those allowed by the system (1) with

$x_0 \in S$. As in the classical theory existence and uniqueness of such a control law is necessary for well-posedness (in some sense), we require it also in this context. Therefore it is assumed that

ASSUMPTION 4 *The operator $CB : U \to CB(U) \subset Y$ is continuously invertible and $X = S \oplus B(U)$*

(observe that, since C is not given by the problem, but is a control tool this just poses restrictions on the construction of S, not on the class of control systems under consideration). Then define

$$u_{\text{eq}}(x) := -(CB)^{-1}CAx, \qquad \forall x \in S, \tag{5}$$

and call $Q = B(CB)^{-1}C$, $P = I - Q$ the projections on $B(U)$ along S and vice-versa respectively. The projected equation obtained by substituting u_{eq} in (1) is

$$\begin{cases} \dot{x} = (A - QA)x \\ x(0) = x_0 \in S. \end{cases} \tag{6}$$

The above differential equation is well-posed in a "classical sense" (see [5] for a discussion about this issue) whenever the operator $A - QA$ generates a strongly continuous semigroup $\tilde{K}(t)$, $t \geq 0$ on S. The following result gives a condition under which the equivalent control method just described is meaningful and relate it to the generalized solutions of Section 2.

THEOREM 5 *Let $S \cap \mathcal{D}(A)$ be dense in S and suppose that Assumptions 1, 3 and 4 hold. Suppose moreover that QA is a perturbation of Miyadera-Voigt type, i.e. that there exist $t_0 > 0$ and $q < 1$ such that*

$$\int_0^{t_0} \|QAK(t)x\|\, dt \leq q\|x\|, \quad \forall x \in \mathcal{D}(A). \tag{7}$$

Then $\tilde{A} = A - QA$ generates a C_0-semigroup $\tilde{K}(t)$, $t \geq 0$ on S.
Moreover the trajectory on S obtained through the equivalent control method is a generalized solution of (1) viable on S if and only if $Bu_{\text{eq}}(x) \in F(x)$ for all $x \in S \cap \mathcal{D}(A)$.

PROOF. Condition (7) assures that $A - QA$ generates a C_0-semigroup $H(t)$, $t \geq 0$ on X by the perturbation theorem of Miyadera and Voigt (for a proof see for example [5], Section III.3.c). It is easy to prove that S is $H(t)$-invariant, so that the restriction $\tilde{K}(t)$ of $H(t)$ on S is a semigroup on S generated by $\tilde{A}$. The invariance of $H(\cdot)$ is equivalent to this property: there exists $\omega \in \mathbb{R}$ such that for any $\lambda > \omega$ one has $\mathcal{R}(\lambda; A - QA)S \subset S$ ([20] Theorem 5.1 p. 121). If $y = \mathcal{R}(\lambda; A - QA)x$ for some $x \in S$, then $\lambda y - Ay + QAy = x$. Applying C we get $\lambda Cy = 0$, therefore $y \in S$ if $\lambda \neq 0$ and thus $A - QA$ is a generator on S.

For the second part of the proof, we need the following results:

LEMMA 6 (COROLLARY 3.16 IN [5]) *Let A generate the C_0-semigroup $K(t)$, $t \geq 0$ on X and Q be a continuous linear operator such that QA satisfies condition (7) for some $t_0 > 0$ and $q \in [0, 1)$. Then the semigroup $H(t)$, $t \geq 0$ generated by $A - QA$ satisfies*

$$H(t)x = K(t)x + \int_0^t K(t-s)QAH(s)x\,ds \tag{8}$$

$$\int_0^{t_0} \|QAH(s)x\|\,ds \leq \frac{q}{1-q}\|x\|, \tag{9}$$

for all $x \in \mathcal{D}(A)$, and any $t \geq 0$.

LEMMA 7 (THEOREM 4.8.3 AND COROLLARY 4.8.1 IN [1]) *Suppose X is a reflexive Banach space and $f : [0, T] \to X$ is in $L^1(0, T; X)$. Then there exists a unique function $x : [0, T] \to X$ which is weakly continuous and such that for each $y \in \mathcal{D}(A^*)$ one has*

$$\langle x(t), y\rangle = \langle x_0, y\rangle + \int_0^t \langle x(s), A^*y\rangle\,ds + \int_0^t \langle f(s), y\rangle\,ds, \qquad 0 \leq t \leq T \tag{10}$$

and this function is given by

$$x(t) = K(t)x_0 + \int_0^t K(t-s)f(s)\,ds.$$

Let us go back to the proof of the theorem. By the density assumption there exist a sequence $\{x_n\}$ in $\mathcal{D}(A) \cap S$ such that $x_n \to x$. Setting $z_n(t) = \tilde{K}(t)x_n$, by (8) and (10) one has

$$\langle z_n(t), y\rangle = \langle x_n, y\rangle + \int_0^t \langle z_n(s), A^*y\rangle\,ds - \int_0^t \langle QAz_n(s), y\rangle\,ds,$$

for all $t \geq 0$ and $y \in \mathcal{D}(A^*)$. As $\tilde{K}(t)$, $t \geq 0$ is a C_0-semigroup it follows that $\|z_n(t) - z(t)\| \to 0$ uniformly on compact subsets of $[0, +\infty)$, therefore

$$\lim_{n\to+\infty} \int_0^t \langle QAz_n(s), y\rangle\,ds = -\langle z(t), y\rangle + \langle x, y\rangle + \int_0^t \langle z(s), A^*y\rangle\,ds. \tag{11}$$

For any $t \geq 0$ and any x the vector $\int_0^t \tilde{K}(s)x\,ds$ is in $\mathcal{D}(\tilde{A}) \subset \mathcal{D}(A)$ and $\tilde{K}(t)x - x = \tilde{A}\int_0^t \tilde{K}(s)x\,ds$, thus

$$\begin{aligned}\int_0^t \langle z(s), A^*y\rangle\,ds &= \langle(\tilde{A} + QA)\int_0^t z(s)\,ds, y\rangle \\ &= \langle z(t), y\rangle - \langle x, y\rangle + \langle QA\int_0^t z(s)\,ds, y\rangle.\end{aligned}$$

Combining the above results, by the density of $\mathcal{D}(A^*)$ it follows that

$$\int_0^t QAz_n(s)\,ds \rightharpoonup QA\int_0^t z(s)\,ds, \qquad \text{for all } t \geq 0. \tag{12}$$

Note that this just depends on the fact that $\tilde{A}$ is a generator and a perturbation of A. Condition (9) will now be exploited to show that the above convergence holds also in the abstract Sobolev space $W^{1,1}(0,T;X)$, thus proving the thesis. To simplify notations let $f_n(t) = \int_0^t QAz_n(s)\,ds$ and $f(t) = QA\int_0^t z(s)\,ds$ for $t \geq 0$. Obviously $f_n \in AC(0,T;X)$ for any n and $T > 0$ by the absolute continuity of the Bochner integral. Moreover by (9), for $T \leq t_0$ and any n, m

$$\|f_n' - f_m'\|_{L^1(0,T;X)} = \int_0^T \|QA\tilde{K}(s)(x_n - x_m)\|\,ds \leq \frac{q}{1-q}\|x_n - x_m\|.$$

Therefore $\{f_n'\}$ is a Cauchy sequence in $L^1(0,T;X)$ and since this space is complete, there exists $h \in L^1(0,T;X)$ such that $f_n' \to h$ in $L^1(0,T;X)$. Using the same arguments it is easy to see that $\{f_n\}$ is convergent in $L^1(0,T;X)$ and by (12) the limit has to be f. The only thing to prove now is that in fact f is absolutely continuous and $h = f'$ almost everywhere. This can be done by a standard argument involving derivatives in the distribution sense, applied to the abstract setting. In fact let $\mathcal{D}'(0,T;X)$ be the space of X-valued distributions on $(0,T)$, i.e. $\mathcal{D}'(0,T;X) = \mathcal{L}(\mathcal{D}(0,T),X)$. The derivative of a distribution in $\mathcal{D}'(0,T;X)$ is defined in the usual way and for $f \in L^1(0,T;X)$, $\varphi \in \mathcal{D}(0,T)$ it gives

$$\left(\frac{d}{dt}f\right)(\varphi) := -f(\varphi') := -\int_0^T f(s)\varphi'(s)\,ds.$$

Therefore for any $\varphi \in \mathcal{D}(0,T)$

$$\begin{aligned} -\int_0^T f(s)\varphi'(s)\,ds &= \lim_{n\to+\infty} -\int_0^T f_n(s)\varphi'(s)\,ds \\ &= \lim_{n\to+\infty} \int_0^T f_n'(s)\varphi(s)\,ds \\ &= \int_0^T h(s)\varphi(s)\,ds, \end{aligned}$$

that is $f' = h$ in $L^1(0,T;X)$.
From (11) it then follows that

$$\langle z(t), y\rangle = \langle x, y\rangle + \int_0^t \langle z(s), A^*y\rangle\,ds - \int_0^t \langle h(s), y\rangle\,ds,$$

where $h \in L^1(0,T;X)$

$$h(s) = \frac{d}{ds}QA\int_0^s z(r)\,dr, \qquad \text{a. e. } s \in [0,T]$$

and the thesis follows from Lemma 7. △

REMARK 8 *Observe that the condition on the equivalent control stated in the above result is also necessary in the finite dimensional setting in order that a sliding mode on S is feasible once the control law u has been chosen.*

Theorem 5 extends similar results in [10, 9], considerably enlarging the class of control systems for which the stated equivalence is valid. In [10] the operator A was assumed to generate a compact semigroup, while in [9] the requirement was the extendibility on S of the operator QA. Suppose that U is finite-dimensional, or for simplicity that the control is scalar. Then $Cx = \langle \gamma, x \rangle$ for some $\gamma \in X^*$ and QA admits an extension iff $\gamma \in \mathcal{D}(A^*)$, while this condition is not required for (7) to be verified. For example let $X = L^2(0,1)$ and A be the unbounded operator associated to the heat equation with Dirichlet boundary conditions, i.e. $Ax = x''$ with $\mathcal{D}(A) = H^2(0,1) \cap H_0^1(0,1)$. The input operator is $Bu = ub$ with $u \in \mathbb{R}$, $b \in X$ while C is chosen as above; to simplify matters suppose that $\gamma \in L^2(0,1)$ is such that $(\gamma, b) = 1$, where $(\cdot, \cdot)$ is the usual scalar product in X. Therefore $Qx = b(\gamma, x)$ and we have

$$||QAK(t)x||^2 = \int_0^1 b^2(\xi)(\gamma, AK(t)x)^2 d\xi = ||b||^2(\gamma, AK(t)x)^2$$

so that

$$\int_0^h ||QAK(t)x|| = ||b|| \int_0^h |(\gamma, AK(t)x)|dt.$$

Recall now that A is the generator of an analytic semigroup, therefore fractional powers of A are well defined. Let $\gamma \in \mathcal{D}(A^{1/2})$. Then $(\gamma, AK(t)x) = (A^{1/2}\gamma, A^{1/2}K(t)x)$ and exploiting classical results about fractional powers of operators (see i.e. [20]) we get

$$\int_0^h |(\gamma, AK(t)x)|dt \leq ||A^{1/2}\gamma|| \int_0^h ||A^{1/2}K(t)x||dt \leq C||x|| \int_0^h t^{-1/2}dt$$

for small h. Since the right-hand side tends to zero for h tending to zero, there exist t_0 and q so that condition (7) is satisfied. Note that in this case a continuous extension of QA would require $\gamma \in \mathcal{D}(A) \subset \mathcal{D}(A^{1/2})$.
Observe also that in the finite dimensional case the equivalent control is defined everywhere on S and is continuous, therefore although the chosen feedback law is discontinuous, its effect in sliding motion is equivalent to the enforcement of a continuous control $u_{\text{eq}}(x)$. This is no more true for general Banach spaces. Now u_{eq} is only densely defined and A-bounded. The equation of motion on S is regulated by $\dot{x} = -Ax + QAx$ and QA is an unbounded perturbation. The control giving the constrained motion is not continuous, thus the application of a discontinuous control law in (1) results in an evolution obtained through an

unbounded perturbation of the generator A. In the next example we show now how it is possible to interpret the equivalent control as a boundary feedback. Let $H = L^2(\Omega)$ with $\Omega \subset I\!R^N$ open, bounded with "smooth" boundary Γ. Consider the differential problem

$$\begin{cases} z_t(t,\xi) - A(\xi,\partial)z(t,\xi) = 0 & \text{in } (0,T] \times \Omega \\ z(0,\xi) = z_0(\xi), & \text{in } \Omega \\ z(t,\sigma) = (z(t,\cdot),\omega)g(\sigma) & \text{in } (0,T] \times \Gamma \end{cases}$$

where $A(\xi,\partial)$ is a second order elliptic differential operator, $z_0 \in H$ and $g \in L^2(\Gamma)$. Let γ be the trace operator of restriction on Γ and $\mathcal{D}(A) = H^2(\Omega) \cap \ker \gamma$, $(Ax)(\xi) = A(\xi,\partial)x(\xi)$ on $\mathcal{D}(A)$. Then $-A$ generates a C_0-semigroup on H. Let

$$D : L^2(\Gamma) \to \mathcal{D}(A^{1/4-\varepsilon}) = H^{1/2-2\varepsilon}$$

be the Dirichlet map $v = Dg$ iff $A(\xi,\partial)v = 0$, $\gamma v = g$.
The differential problem can be reformulated in semigroup form (see results in [6] for the parabolic case and [7] for hyperbolic systems)

$$\dot{z} = A_F z, \qquad z(0) = z_0$$

where

$$\begin{aligned} A_F z &= -A[I - Dg(z,\omega)] \\ \mathcal{D}(A_F) &= \{z \in H : z - Dg(z,\omega) \in \mathcal{D}(A)\}. \end{aligned}$$

Now as $A = A^*$

$$A_F^* = -Ax + (Ax, Dg)w, \quad \mathcal{D}(A_F^*) = \mathcal{D}(A^*).$$

Let us now consider a scalar sliding mode control for the abstract system

$$\dot{x} + Ax = bu, \qquad u \in I\!R,\ b \in H$$

with sliding surface $S = \{x \in H : (x,c) = 0\}$ with $(b,c) = 1$. From the application of the equivalent control method the evolution in sliding is governed by the projected differential equation

$$\dot{x} = -Ax + (Ax,c)b.$$

The relation with the Dirichlet boundary feedback problem above is now straightforward by considering as ω the function b corresponding to the input operator B and choosing c as the relevation Dg of an $L^2(\Gamma)$ function.

REMARK 9 *Note that in the above case, using the stated generation result of the feedback operator A_F, we can require weaker regularity on the function c in order that the equation of the sliding mode is well-posed, as opposed to the application of Theorem 5. Consider, however, that the above improvement of the result is strongly related to the analyticity properties of the semigroup that governs the evolution, while Theorem 5 is valid for a larger class of systems.*

References

[1] A. V. Balakrishnan. *Applied functional analysis*. Applications of Mathematics, No. 3, Springer-Verlag, 1976.

[2] O. Cârjă, I. I. Vrabie. Some new viability results for semilinear differential inclusions. *NoDEA Nonlinear Differential Equations Appl.* 4:401–424, 1997.

[3] O. Cârjă, I. I. Vrabie. Viability for semilinear differential inclusions via the weak sequential tangency condition. *J. Math. Anal. Appl.* 262:24–38, 2001.

[4] A. F. Filippov. *Differential equations with discontinuous righthand sides*. Mathematics and its Applications (Soviet Series), Kluwer Academic Publishers Group, Dordrecht, 1988.

[5] K. J. Engel, R. Nagel. *One-parameter semigroups for linear evolution equations*, Graduate Texts in Mathematics, vol. 194, Springer-Verlag, 2000.

[6] I. Lasiecka, R. Triggiani. Feedback semigroups and cosine operators for boundary feedback parabolic and hyperbolic equations. *J. Differ. Equ.* 47:246–272, 1983.

[7] I. Lasiecka, R. Triggiani. Finite rank, relatively bounded perturbations of semigroups generators. I. Well-posedness and boundary feedback hyperbolic dynamics. *Ann. Scuola Norm. Sup. Pisa Cl. Sci. (4)*, 12:641–668, 1985.

[8] I. Lasiecka, R. Triggiani. Finite rank, relatively bounded perturbations of semigroups generators. II. Spectrum and Riesz basis assignment with applications to feedback systems. *Ann. Mat. Pura Appl. (4)*, 143:47–100,1986.

[9] L. Levaggi. Infinite dimensional systems' sliding motions. *Eur. J. Control* 8:508–516, 2002.

[10] L. Levaggi. Sliding modes in Banach spaces. *Differ. Integral Equ.* 15:167–189, 2002.

[11] J.-L. Lions. *Quelques méthodes de résolution des problèmes aux limites non linéaires*. Dunod, Paris, 1969.

[12] J.-L. Lions. *Optimal control of systems governed by partial differential equations*. Die Grundlehren der mathematischen Wissenschaften, Band 170, Springer-Verlag, New York, 1971.

[13] J.-L. Lions, E. Magenes. *Non-homogeneous boundary value problems and applications. vol. i*. Die Grundlehren der mathematischen Wissenschaften, Band 181, Springer-Verlag, New York, 1972.

[14] Yu. V. Orlov, V. I. Utkin. Use of sliding modes in distributed system control problems. *Automat. Remote Control* 43:1127–1135, 1982.

[15] Yu. V. Orlov, V. I. Utkin. Sliding mode control in indefinite-dimensional systems. *Automatica J. IFAC* 23:753–757, 1987.

[16] Y. Orlov, D. Dochain. Discontinuous feedback stabilization of minimum-phase semilinear infinite-dimensional systems with application to chemical tubular reactor. *IEEE Trans. Automat. Control* 47:1293–1304, 2002.

[17] Y. Orlov, V. I. Utkin. Unit sliding mode control in infinite-dimensional systems. *Appl. Math. Comput. Sci.* 8:7–20, 1998.

[18] Y. V. Orlov. Discontinuous unit feedback control of uncertain infinite-dimensional systems. *IEEE Trans. Automat. Control* 45:834–843, 2000.

[19] Y. Orlov, Y. Lou, P. D. Christofides. Robust stabilization of infinite-dimensional systems using sliding-mode output feedback control. *Internat. J. Control* 77:1115–1136, 2004.

[20] A. Pazy. *Semigroups of linear operators and applications to partial differential equations.* Applied Mathematical Sciences, vol. 44, Springer-Verlag, New York, 1983.

[21] V. I. Utkin. *Sliding modes in control and optimization.* Communications and Control Engineering Series, Springer-Verlag, Berlin, 1992.

[22] I. I. Vrabie. *Compactness methods for nonlinear evolutions.* Pitman Monographs and Surveys in Pure and Applied Mathematics, vol. 32, Longman Scientific & Technical, Harlow, 1987.

[23] Tullio Zolezzi. Variable structure control of semilinear evolution equations, in *Partial differential equations and the calculus of variations - Vol. II*, Progr. Nonlinear Differential Equations Appl., Birkhäuser, Boston, MA, 1989, pp. 997–1018.

RAZUMIKHIN-TYPE THEOREMS OF INFINITE DIMENSIONAL STOCHASTIC FUNCTIONAL DIFFERENTIAL EQUATIONS

Kai Liu [1] and Yufeng Shi [2]

[1]*Department of Mathematical Sciences, The University of Liverpool, Liverpool, L69 7ZL, U. K.** [2]*School of Mathematics and System Sciences, Shandong University, Jinan 250100, China, yfshi@sdu.edu.cn* †

Abstract The argument of Razumikhin-type has been well developed and showed significant advantage for the stability of stochastic functional differential equations in finite dimensions. However, so far there have been almost no results of Razumikhin-type on the stability of mild solutions of stochastic functional differential equations in infinite dimensions. The main aim of this paper is to establish Razumikhin-type stability theorems for stochastic functional differential equations in infinite dimensions. By virtue of these new criteria, we can establish the exponential stability of stochastic delay differential equations and stochastic delay partial differential equations.

Key words: Lyapunov function; Razumikhin-type theorem; Stochastic functional differential equations in infinite dimensions.

1. Introduction

Stochastic functional differential equations in infinite dimensional spaces are motivated by the development of analysis and the theory of stochastic processes itself such as stochastic partial differential equations with some hereditary characteristics on the one hand, and by such topics as wave propagation in random media, turbulence, population biology and stochastic control in applications on

*This work was supported by EPSRC Grant No. GR/R37227.

†Partially supported by Foundation for University Key Teacher by Ministry of Education of China, National Natural Science Foundation of China grant 10201018 and Doctor Promotional Foundation of Shandong grant 02BS127.

Please use the following format when citing this chapter:

Liu, K., and Shi, Y., 2006, in IFIP International Federation for Information Processing, Volume 202, Systems, Control, Modeling and Optimization, eds. Ceragioli, F., Dontchev, A., Furuta, H., Marti, K., Pandolfi, L., (Boston: Springer), pp. 237-247.

the other. The analysis and control of such systems then involve investigating their stability, which is often regarded as the first characteristic of dynamical systems (or models) studied.

The purpose of this paper is to investigate stability of mild solutions for certain infinite dimensional stochastic functional differential equations. Roughly speaking, we shall consider the following stochastic functional differential equations in a certain Hilbert space H with norm $\|\cdot\|_H$:

$$\begin{aligned} du(t) &= Au(t)\,dt + F(t, u(t), u_t)\,dt + G(t, u(t), u_t)\,dW(t), \quad t \geq 0, \\ u_0 &= \xi \in C^b_{\mathcal{F}_0}([-r, 0]; H), \end{aligned} \tag{1}$$

where A is the infinitesimal generator of a certain C_0-semigroup $\{T(t), t \geq 0\}$ of bounded linear operators on H and $F : \mathbf{R}^+ \times H \times C([-r, 0]; H) \to H$ and $G : \mathbf{R}^+ \times H \times C([-r, 0]; H) \to \mathcal{L}(K, H)$ are two measurable nonlinear mappings. Here K is some real separable Hilbert space and $u_t = \{u(t+\theta) : -r \leq \theta \leq 0\}$ is regarded as a $C([-r, 0]; H)$-valued stochastic process. The family of all bounded, $\mathcal{F}_0$-measurable, $C([-r, 0]; H)$-valued random variables is denoted $C^b_{\mathcal{F}_0}([-r, 0]; H)$. The process $\{W(t)\}_{t \geq 0}$ is some given K-valued, Q-Wiener process with $tr(Q) < \infty$ and $\xi(t) : \Omega \times [-r, 0] \to H$, $r \geq 0$, is a given initial datum such that $\xi(t)$ is $\mathcal{F}_0$-measurable and $\sup_{-r \leq t \leq 0} E\|\xi(t)\|^2_H < \infty$.

Stochastic evolution equations in Hilbert spaces have been studied by many authors over the last several years. For instance, Da Prato and Zabczyk [3] and Pardoux [4] (amongst others) have established results on the existence and uniqueness of solutions for a certain class of infinite dimensional stochastic evolution equations. For variable delay case, the similar problems have been studied by Real [5] for stochastic linear evolution equations and by Caraballo and Liu [6] and Caraballo, Liu and Truman [7] for nonlinear cases. On the other hand, under various circumstances there exists an extensive literature on stability of infinite dimensional stochastic differential equations. In particular, we like to refer to [2, 8] on the stability of mild solutions for infinite dimensional stochastic functional differential equations.

In infinite dimensional setting, for the purpose of deriving stability results, a suitable construction of Lyapunov functionals rather than functions is a natural generalization of the Lyapunov direct method in finite dimensional spaces. We present below a Lyapunov functional type of argument of stability to show that the situation in treating (1) by this approach could become very complicated. Suppose $u(t; \xi)$ is the solution of (1) through $(0, \xi)$ and $u_t(\xi) = \{u(t+\theta; \xi) : -r \leq \theta \leq 0\}$. Let us study a typical stability result which is a direct stochastic version of Theorem 2.1 in Chapter 5 in [9]. The reader is referred to [9] for more details.

PROPOSITION 1 *Let $p \geq 2$ and the standard hypothesis (H1) imposed in Section 2 hold. Suppose $v(\cdot)$, $l(\cdot) : \mathbf{R}^+ \to \mathbf{R}^+$ are two continuous nonde-*

creasing functions, $v(s)$ and $l(s)$ are positive for $s > 0$, $v(s)$ is convex and $l(s)$ is concave with $v(0) = l(0) = 0$. If there is a continuous functional $V : \mathbf{R} \times C([-r, 0]; H) \to \mathbf{R}$ such that

$$\begin{aligned} v\left(\|\varphi(0)\|_H^p\right) &\leq V(t, \varphi) \leq l\left(\|\varphi\|_C^p\right), \quad \forall \varphi \in C([-r, 0]; H), \\ EV(t, u_t(\xi)) &\leq EV(s, u_s(\xi)), \quad \forall t \geq s \geq 0. \end{aligned}$$

Then the (mild) solution of (1) is p-th moment stable.

In spite of the formal simplicity of the above result, it is hard to apply this proposition directly to practical problems even though H is finite dimensional, e.g., $H = \mathbf{R}^n$. The reason is twofold. On the one hand, instead of the usual Lyapunov functions in finite dimensional spaces, a Lyapunov functional as above must be constructed properly, a case which is usually not easy to handle. On the other hand, the conditions as above are difficult to justify because of the inclusion of the solution itself which is not known explicitly in most situations. This proposition certainly loses the advantage of the Lyapunov direct method in the sense that it is unnecessary to solve the equations explicitly in order to determine the stability of solutions.

One of the most effective ways to deal with these problems is a method originated by Razumikhin [10, 11]. The argument of Razumikhin-type has been well developed and showed significant advantage for the stability of (stochastic) functional differential equations in finite dimensions (see [9, 12, 13]). The Lyapunov functions of Razumikhin-type have been shown to be rather powerful to treat functional differential equations, and as a consequence, that they really bring forth the advantage of Lyapunov direct method. In [13] Mao has shown a smart argument of Razumikhin-type to exponential stability of finite dimensional stochastic functional differential equations. However by virtue of Mao's argument, it is not trivial to treat the case in infinite dimensions, because there is not Itô's formula applicable to mild solutions of stochastic functional differential equations in infinite dimensions. So far there have been almost no results of Razumikhin-type on the stability of mild solutions of infinite dimensional stochastic functional differential equations. To the best of our knowledge this is the first time the possibility of using Lyapunov functions of Razumikhin-type to determine sufficient conditions of stability for stochastic functional differential equations in infinite dimensions has been explored. By virtue of the new criteria derived later on, we can show the exponential stability of stochastic delay differential equations and stochastic delay partial differential equations. In particular, by using the results derived in this paper we may essentially improve some stability results in [1, 2].

2. Preliminary results

Let $(\Omega, \mathcal{F}, \{\mathcal{F}_t\}_{t\geq 0}, P)$ be a complete probability space with a filtration $\{\mathcal{F}_t\}_{t\geq 0}$ satisfying the usual conditions (i.e., it is right continuous and $\mathcal{F}_0$ contains all P-null sets). Let K be a real separable Hilbert space. With the symbol $\{W(t), t \geq 0\}$ we denote a K-valued $\{\mathcal{F}_t\}_{t\geq 0}$-Wiener process defined on the probability space $(\Omega, \mathcal{F}, \{\mathcal{F}_t\}_{t\geq 0}, P)$ with covariance operator Q, i.e.,

$$E(W(t), x)_K (W(s), y)_K = (t \wedge s)(Qx, y)_K \, \forall x, y \in K,$$

where Q is a nonnegative finite trace class operator from K into itself. In particular, we call $\{W(t)\}_{t\geq 0}$ a K-valued Q-Wiener process with respect to $\{\mathcal{F}_t\}_{t\geq 0}$.

Let H be a real Hilbert space and we denote by $\langle \cdot, \cdot \rangle$ its inner product and $\|\cdot\|_H$ its norm, respectively. Assume r is a given positive constant. In the present paper, we shall consider the following infinite dimensional stochastic functional differential equation on $I = [-r, T]$, (here $T \geq 0$ and $t \in [0, T]$)

$$\begin{aligned} du(t) &= Au(t)\,dt + F(t, u(t), u_t)\,dt + G(t, u(t), u_t)\,dW(t), \\ u_0 &= \xi \in C^b_{\mathcal{F}_0}([-r, 0]; H). \end{aligned} \tag{2}$$

Throughout this paper, we shall impose the following assumptions:

(H1) A is the infinitesimal generator of a C_0-semigroup $\{T(t), t \geq 0\}$ of bounded linear operators on H satisfying $\|T(t)\| \leq M \cdot e^{\lambda t}$ for some $M \geq 1$, $\lambda \in \mathbf{R}^1$. The coefficients $F : \mathbf{R}^+ \times H \times C([-r, 0]; H) \to H$ and $G : \mathbf{R}^+ \times H \times C([-r, 0]; H) \to \mathcal{L}(K, H)$ are two measurable nonlinear mappings satisfying the following Lipschitz condition

$$\begin{aligned} \|F(t, x, y) - F(t, x', y')\|_H + \|G(t, x, y) - G(t, x', y')\|_2 \\ \leq k(\|x - x'\|_H + \|y - y'\|_H), \end{aligned} \tag{3}$$

for some constant $k > 0$ and arbitrary $x, x', \in H$, $y, y' \in C([-r, 0]; H)$ and $t \in \mathbf{R}^+$. Here $\|\cdot\|_2$ denotes the Hilbert-Schmidt norm of a nuclear operator, i.e., $\|G(t, x, y)\|_2^2 = tr(G(t, x, y)QG(t, x, y)^*)$, $x \in H$, $y \in C([-r, 0]; H)$. Denote by $C^b_{\mathcal{F}_0}([-r, 0]; H)$ the family of all bounded, $\mathcal{F}_0$-measurable, $C([-r, 0]; H)$-valued random variables. For $p \geq 2$ and $t \geq 0$, denote by $L^p_{\mathcal{F}_t}([-r, 0]; H)$ the family of all $\mathcal{F}_t$-measurable $C([-r, 0]; H)$-valued random variables $\phi = \{\phi(\theta) : -r \leq \theta \leq 0\}$ such that

$$\sup_{-r \leq \theta \leq 0} E\|\phi(\theta)\|_H^p < \infty.$$

We introduce two kinds of solutions of (2) as follows similarly to [14]:

DEFINITION 2 *A stochastic process $u(t)$, $t \in I$, is called a strong solution of (2) if (i) $u(t)$ is adapted to $\mathcal{F}_t$;*

(ii) $u(t)$ is continuous in t almost surely;

(iii) $u(t) \in \mathcal{D}(A)$ on $I \times \Omega$ with $\int_0^T \|Au(t)\|_H dt < \infty$ almost surely and

$$\begin{aligned} u(t) &= \xi(0) + \int_0^t (Au(s) + F(s, u(s), u_s)) ds + \\ & \int_0^t G(s, u(s), u_s) dW(s), \\ u(t) &= \xi(t), \quad t \in [-r, 0], \end{aligned}$$

for all $t \in I$ with probability one.

In general, this concept is rather strong and a weaker one described below is more appropriate for practical purposes.

DEFINITION 3 *A stochastic process $u(t)$, $t \in I$, is called a mild solution of (2) if (i) $u(t)$ is adapted to $\mathcal{F}_t$;*

(ii) $u(t)$ is measurable with $\int_0^T \|u(t)\|_H^2 dt < \infty$ almost surely and

$$\begin{aligned} u(t) &= T(t)\xi(0) + \int_0^t T(t-s) F(s, u(s), u_s) ds \\ &+ \int_0^t T(t-s) G(s, u(s), u_s) dW(s), \\ u(t) &= \xi(t), \quad t \in [-r, 0], \end{aligned}$$

for all $t \in I$ with probability one.

Note that if $\{u(t), t \in I\}$ is a strong solution of (2), then it is also a mild solution. The following existence and uniqueness theorem can be obtained similarly by an adapted argument from [6] or [15]. The reader is referred to them for further details on this aspect.

THEOREM 4 *Let $\{\xi(t), t \in [-r, 0]\}$ be a given $\mathcal{F}_0$-measurable initial datum with $\sup_{-r \le t \le 0} E\|\xi(t)\|_H^2 < \infty$. Suppose the hypothesis (H1) holds, then (2) has a unique mild solution $u(t; \xi)$, or simply $u(t)$, in $C(0, T; L^2(\Omega, \mathcal{F}, P; H))$.*

For our purposes, we can introduce Itô's formula which will play an important role in our stability analysis as follows.

Let $C^2(R \times H)$ denote the space of all real-valued functions V on $\mathbf{R} \times H$ with properties:

(i) $V(t, x) \in C^2(\mathbf{R} \times H)$ is twice (Fréchet) differentiable in x and once differentiable in t;

(ii) $V_x(t,x)$ and $V_{xx}(t,x)$ are both continuous in H and $\mathcal{L}(H) = \mathcal{L}(H,H)$, respectively.

THEOREM 5 *(Itô's formula) Suppose* $V \in C^2(\mathbf{R} \times H)$ *and* $\{u(t), t \geq 0\}$ *is a strong solution of (2), then*

$$\begin{aligned} V(t,u(t)) &= V(0,\xi(0)) + \int_0^t \mathcal{L}V(s,u(s),u_s)\,ds \\ &\quad + \int_0^t \langle V_x(s,u(s)), G(s,u(s),u_s)\,dW(s)\rangle, \end{aligned}$$

where $\forall x \in \mathcal{D}(A),\quad y \in C([-r,0];H),\quad t \geq 0,$

$$\begin{aligned} \mathcal{L}V(t,x,y) &= V_t(t,x) + \langle V_x(t,x), Ax + F(t,x,y)\rangle \\ &\quad + \frac{1}{2} tr\left(V_{xx}(t,x)\, G(t,x,y)\, QG^*(t,x,y)\right). \end{aligned} \tag{4}$$

Since Itô's formula is only applicable to strong solutions, we introduce the following approximating systems of (2), for $t \geq 0$,

$$\begin{aligned} du(t) &= Au(t)\,dt + R(n)\,F(t,u(t),u_t)\,dt \\ &\quad + R(n)\,G(t,u(t),u_t)\,dW(t), \\ u(t) &= R(n)\,\xi(t) \in \mathcal{D}(A), \quad t \in [-r,0] \end{aligned} \tag{5}$$

where $n_0 \leq n \in \rho(A)$, the resolvent set of A and $R(n) = nR(n,A)$, $R(n,A)$ is the resolvent of A. A similar operator $\mathcal{L}_n$ to (4) in correspondence with this equation is

$$\begin{aligned} \mathcal{L}_n V(t,x,y) &= V_t(t,x) + \langle V_x(t,x), Ax + R(n)\,F(t,x,y)\rangle \\ &\quad + \frac{1}{2} tr\left(V_{xx}(t,x)\, R(n)\, G(t,x,y)\, Q\, (R(n)\, G(t,x,y))^*\right), \\ \forall x &\in \mathcal{D}(A), \quad y \in C([-r,0];H), \quad t \geq 0. \end{aligned}$$

THEOREM 6 *([15]) Under the hypotheses of Theorem 4, (5) has a unique strong solution* $u^n(t)$ *in* $C(0,T;L^2(\Omega,\mathcal{F},P;H))$ *for all* $T \geq 0$*. Moreover,* $u^n(t)$ *converges to the mild solution* $u(t)$ *of (2) in* $C(0,T;L^2(\Omega,\mathcal{F},P;H))$ *as* $n \to \infty$*, i.e.,*

$$\lim_{n\to\infty} E\left(\sup_{t\in[0,T]} \|u^n(t) - u(t)\|_H^2\right) = 0.$$

3. The main results

In this section, we shall carry out an argument of Razumikhin type to study the stability of the mild solutions of (2) in the sense of p-th moment and pathwise with probability one. Based on the ideas of constructing Lyapunov functions rather than functionals, we study the stability of the equation (2) in the spirit of Razumikhin in finite dimensions.

Let $C^{1,2}([-r,\infty)\times H;\mathbf{R}^+)$ denote the family of all nonnegative functions $V(t,x)$ on $[-r,\infty)\times H$ which are continuously twice Fréchet differentiable in x and once differentiable in t. Then for $V\in C^{1,2}([-r,\infty)\times H;\mathbf{R}^+)$, we can introduce, similarly to (4), an operator $\mathcal{L}$ on $C^{1,2}([-r,\infty)\times H;\mathbf{R}^+)$ by

$$\begin{aligned}&\mathcal{L}V(t,\varphi(0),\varphi)\doteq V_t(t,\varphi(0))+\langle V_x(t,\varphi(0)),A\varphi(0)\\&+F(t,\varphi(0),\varphi)\rangle_H+,\\&+\frac{1}{2}tr\left[V_{xx}(t,\varphi(0))G(t,\varphi(0),\varphi)QG^*(t,\varphi(0),\varphi)\right],\varphi(0)\in\mathcal{D}(A).\end{aligned}$$

Here $t>0$. We furthermore assume that $F(t,0,0)\equiv 0$ and $G(t,0,0)\equiv 0$, we can assert the following the exponential stability for (2).

THEOREM 7 *Let the standard hypothesis (H1) hold and $p\geq 2$, λ, c, c_1 all be positive numbers and $q>1$. If there exists a continuous function $V\in C^{1,2}([-r,\infty)\times H;\mathbf{R}^+)$ satisfying*

$$\begin{aligned}c\|x\|_H^p&\geq V(t,x)+\|x\|_H\|V_x(t,x)\|_H+\|x\|_H^2\|V_{xx}(t,x)\|_H\\c_1\|x\|_H^p&\leq V(t,x),\quad\forall(t,x)\in[-r,\infty)\times H,\end{aligned}\tag{6}$$

and for arbitrary $t\geq 0$

$$E\mathcal{L}V(t,\phi(t),\phi)\leq-\lambda EV(t,\phi(0)),\tag{7}$$

whenever $\phi=\{\phi(\theta):-r\leq\theta\leq 0\}\in L^p_{\mathcal{F}_t}([-r,0];\mathcal{D}(A))$ satisfies

$$EV(t+\theta,\phi(\theta))<qEV(t,\phi(0)),\ \text{for all}\ -r\leq\theta\leq 0.$$

Here for $\phi\in L^p_{\mathcal{F}_t}([-r,0];\mathcal{D}(A))$, $\quad\phi(0)\in\mathcal{D}(A)$, $\quad t\geq 0$

$$\begin{aligned}\mathcal{L}V(t,\phi(t),\phi)\doteq V_t(t,\phi(0))+\langle V_x(t,\phi(0)),A\phi(0)+F(t,\phi(0),\phi)\rangle\\+\frac{1}{2}tr[V_{xx}(t,\phi(0))G(t,\phi(0),\phi)QG^*(t,\phi(0),\phi)].\end{aligned}$$

Then the mild solution $u(t)$ of (2) is p-th moment exponentially stable, i.e., for all $\xi\in C^b_{\mathcal{F}_0}([-r,0];H)$

$$E\|u(t;\xi)\|_H^p\leq\frac{c}{c_1}E\|\xi\|_C^p e^{-\gamma t},\quad\forall t\geq 0,$$

where $\gamma=\min\left\{\lambda,\frac{\log(q)}{r}\right\}$.

Proof. Fix the initial data $\xi \in C^b_{\mathcal{F}_0}([-r,0];H)$ arbitrarily and write $u(t;\xi) = u(t)$ simply. Let $\varepsilon \in (0,\gamma)$ be arbitrary and set $\bar{\gamma} = \gamma - \varepsilon$. Define

$$U(t) = \max_{-r \le \theta \le 0} \left[e^{\bar{\gamma}(t+\theta)} EV(t+\theta, u(t+\theta)) \right], \quad \text{for } t \ge 0.$$

Obviously $U(t)$ is well defined and continuous. We can claim that

$$D^+U(t) \doteq \limsup_{h \to 0+} \frac{U(t+h) - U(t)}{h} \le 0, \quad \forall t \ge 0. \tag{8}$$

To show this, for each fixed $t_0 \ge 0$, define

$$\bar{\theta} = \max \left\{ \theta \in [-r,0] : U(t_0) = e^{\bar{\gamma}(t_0+\theta)} EV(t_0+\theta, u(t_0+\theta)) \right\}.$$

Obviously, $\bar{\theta}$ is well defined, $\bar{\theta} \in [-r,0]$ and

$$U(t_0) = e^{\bar{\gamma}(t_0+\bar{\theta})} EV(t_0+\bar{\theta}, u(t_0+\bar{\theta})), \text{a.s.}.$$

If $\bar{\theta} < 0$, for all $\bar{\theta} < \theta \le 0$ one has

$$e^{\bar{\gamma}(t_0+\theta)} EV(t_0+\theta, u(t_0+\theta)) < e^{\bar{\gamma}(t_0+\bar{\theta})} EV(t_0+\bar{\theta}, u(t_0+\bar{\theta})).$$

It is therefore easy to observe that for any $h > 0$ small enough

$$e^{\bar{\gamma}(t_0+h)} EV(t_0+h, u(t_0+h)) \le e^{\bar{\gamma}(t_0+\bar{\theta})} EV(t_0+\bar{\theta}, u(t_0+\bar{\theta})).$$

Hence

$$U(t_0+h) \le U(t_0) \text{ and } D^+U(t_0) \le 0.$$

If $\bar{\theta} = 0$, then

$$e^{\bar{\gamma}(t_0+\theta)} EV(t_0+\theta, u(t_0+\theta)) \le e^{\bar{\gamma}t_0} EV(t_0, u(t_0)), \quad \forall \theta \in [-r,0].$$

So

$$\begin{aligned} EV(t_0+\theta, u(t_0+\theta)) &\le e^{-\bar{\gamma}\theta} EV(t_0, u(t_0)) \\ &\le e^{\bar{\gamma}r} EV(t_0, u(t_0)), \quad \forall \theta \in [-r,0]. \end{aligned} \tag{9}$$

Note that either $EV(t_0, u(t_0)) = 0$ or $EV(t_0, u(t_0)) > 0$. In the case of $EV(t_0, u(t_0)) = 0$, (6) and (9) imply that $u(t_0+\theta) = 0$ a.s. for all $\theta \in [-r,0]$. Recalling the fact that $F(t_0,0,0) \equiv 0$ and $G(t_0,0,0) \equiv 0$, it follows that $u(t_0+h) = 0$ a.s. for all $h > 0$, hence $U(t_0+h) = 0$ and $D^+U(t_0) = 0$. On the other hand, in the case of $EV(t_0, u(t_0)) > 0$, (9) implies

$$\begin{aligned} EV(t_0+\theta, u(t_0+\theta)) &\le e^{\bar{\gamma}r} EV(t_0, u(t_0)) \\ &< qEV(t_0, u(t_0)), \quad \forall \theta \in [-r,0], \end{aligned}$$

since $e^{\bar{\gamma}r} < q$. Let $\beta = q - e^{\bar{\gamma}r} > 0$. It follows from the continuity of $EV(t_0, u(t_0))$ and (6) that for some $h > 0$ sufficiently small

$$EV(t_0+\theta, u(t_0+\theta)) \leq \left(e^{\bar{\gamma}r} + \frac{\beta}{2}\right) EV(t_0, u(t_0)), \quad \forall \theta \in [0, h].$$

Now we need to introduce the strong solutions $u^n(t)$ of (5). By virtue of (H1), (6) and Theorem 6, there exists a sub-sequence of $\{n\}$ in $\rho(A)$ (still denote by $\{n\}$) such that $u^n(t) \to u(t)$ almost surely as $n \to \infty$ in $C(0, T; H)$ uniformly with respect to $t \in [0, T]$. Consequently, for some positive constant $\delta \in \left(0, \frac{\beta}{4+2\beta} EV(t_0, u(t_0))\right)$, there are a sufficiently small constant $h > 0$ and a large number $N > 0$ such that for $n \geq N$, one has that for any $s \in [t_0, t_0+h]$

$$\begin{aligned}
EV(s, u(s)) &> EV(t_0, u(t_0)) - \delta > 0, \\
EV(s+\theta, u(s+\theta)) &< EV(t_0+\theta, u(t_0+\theta)) + \delta, \quad \forall \theta \in [-r, 0], \\
e^{\bar{\gamma}r} EV(t_0, u(t_0)) &< e^{\bar{\gamma}r} EV(s, u(s)) + \delta, \\
EV(s, u^n(s)) &> EV(s, u(s)) - \delta > 0, \\
e^{\bar{\gamma}r} EV(s, u(s)) &< e^{\bar{\gamma}r} EV(s, u^n(s)) + \delta, \\
EV(s+\theta, u^n(s+\theta)) &< EV(s+\theta, u(s+\theta)) + \delta, \quad \forall \theta \in [-r, 0],
\end{aligned}$$

which immediately imply

$$EV(s+\theta, u^n(s+\theta)) < qEV(s, u^n(s)), \quad \forall \theta \in [-r, 0]. \tag{10}$$

By the condition (7), (10) implies that

$$E\mathcal{L}V(s, u^n(s), u_s^n) \leq -\lambda EV(s, u^n(s)), \quad \forall s \in [t_0, t_0+h]. \tag{11}$$

Applying Itô's formula to the function $e^{\bar{\gamma}t} V(t, u)$ along the strong solutions $u^n(t)$ of (5), one can derive that for any $\bar{h} \in [0, h]$,

$$\begin{aligned}
&e^{\bar{\gamma}(t_0+\bar{h})} EV(t_0+\bar{h}, u^n(t_0+\bar{h})) - e^{\bar{\gamma}t_0} EV(t_0, u^n(t_0)) \\
&= \int_{t_0}^{t_0+\bar{h}} e^{\bar{\gamma}s} [\bar{\gamma} EV(s, u^n(s)) + E\mathcal{L}_n V(s, u^n(s), u_s^n)] \, ds \\
&= \int_{t_0}^{t_0+\bar{h}} e^{\bar{\gamma}s} [\bar{\gamma} EV(s, u^n(s)) + E\mathcal{L}V(s, u^n(s), u_s^n)] \, ds \\
&+ \int_{t_0}^{t_0+\bar{h}} e^{\bar{\gamma}s} E \langle V_x(s, u^n(s)), (R(n) - I) F(s, u^n(s), u_s^n) \rangle \, ds \\
&+ \frac{1}{2} \int_{t_0}^{t_0+\bar{h}} e^{\bar{\gamma}s} Etr[V_{xx}(s, u^n(s)) R(n) G(s, u^n(s), u_s^n) \\
&Q(R(n) G(s, u^n(s), u_s^n))^*] ds \\
&- \frac{1}{2} \int_{t_0}^{t_0+\bar{h}} e^{\bar{\gamma}s} Etr[V_{xx}(s, u^n(s)) G(s, u^n(s), u_s^n) Q G^*(s, u^n(s), u_s^n)] \, ds.
\end{aligned}$$

By virtue of (11), one can deduce

$$
\begin{aligned}
& e^{\bar{\gamma}(t_0+\bar{h})} EV\left(t_0+\bar{h}, u^n\left(t_0+\bar{h}\right)\right) \\
& \leq e^{\bar{\gamma} t_0} EV\left(t_0, u^n\left(t_0\right)\right)+(\bar{\gamma}-\lambda) \int_{t_0}^{t_0+\bar{h}} e^{\bar{\gamma} s} EV\left(s, u^n(s)\right) ds \\
& +\int_{t_0}^{t_0+\bar{h}} e^{\bar{\gamma} s} E\left\langle V_x\left(s, u^n(s)\right),(R(n)-I) F\left(s, u^n(s), u_s^n\right)\right\rangle ds \\
& +\frac{1}{2} \int_{t_0}^{t_0+\bar{h}} e^{\bar{\gamma} s} Etr[V_{xx}\left(s, u^n(s)\right) R(n) G\left(s, u^n(s), u_s^n\right) \\
& Q(R(n) G(s, u^n(s), u_s^n))^*] ds \\
& -\frac{1}{2} \int_{t_0}^{t_0+\bar{h}} e^{\bar{\gamma} s} Etr\left[V_{xx}\left(s, u^n(s)\right) G\left(s, u^n(s), u_s^n\right) Q G^*\left(s, u^n(s), u_s^n\right)\right] ds,
\end{aligned}
$$

which, letting $n \rightarrow \infty$, immediately yields

$$
\begin{aligned}
e^{\bar{\gamma}(t_0+\bar{h})} EV\left(t_0+\bar{h}, u\left(t_0+\bar{h}\right)\right) & \leq e^{\bar{\gamma} t_0} EV\left(t_0, u\left(t_0\right)\right) \\
& +\int_{t_0}^{t_0+\bar{h}}(\bar{\gamma}-\lambda) e^{\bar{\gamma} s} EV(s, u(s)) ds \\
& \leq e^{\bar{\gamma} t_0} EV\left(t_0, u\left(t_0\right)\right).
\end{aligned}
$$

Then it follows that

$$
e^{\bar{\gamma} s} EV(s, u(s)) \leq e^{\bar{\gamma} t_0} EV\left(t_0, u\left(t_0\right)\right), \quad \forall s \in\left[t_0, t_0+h\right].
$$

So it must hold that $EU\left(t_0+h\right)=EU\left(t_0\right)$ for any $h>0$ sufficiently small, and hence $D^{+} U\left(t_0\right)=0$. Since t_0 is arbitrary, the inequality (8) is shown to hold for any $t \geq 0$. It now follows immediately from (8) that $U(t) \leq U(0)$, for any $t \geq 0$. Also, (6) implies that

$$
e^{\bar{\gamma} t} EV(t, u(t)) \leq U(t) \leq U(0) \leq c E\|\xi\|_C^p, \quad \forall t \geq 0.
$$

Note that ε is arbitrary, it thus follows that

$$
EV(t, u(t)) \leq c E\|\xi\|_C^p e^{-\gamma t}, \quad \forall t \geq 0,
$$

which, by virtue of (6), immediately yields that

$$
E\|u(t)\|_H^p \leq \frac{c}{c_1} E\|\xi\|_C^p e^{-\gamma t}, \quad \forall t \geq 0.
$$

Therefore the desired result is obtained. The proof is complete.

By virtue of the above theorem we can give the almost sure exponential stability for stochastic functional differential equations, similarly to Theorem 2.2 in [13]. We thus omit it at the moment.

References

[1] T. Taniguchi. Almost sure exponential stability for stochastic partial functional differential equations. *Stoch. Anal. Appl.* 16(5):965-975, 1998.

[2] K. Liu, A. Truman. Moment and almost sure Lyapunov exponents of mild solutions of stochastic evolution equations with variable delays via approximation approaches. *J. Math. Kyoto Univ.* 41:749-768, 2002.

[3] G. Da Prato, J. Zabczyk. *Stochastic Equations in Infinite Dimensions.* Encyclopedia of Mathematics and its Applications, Cambridge University Press, 1992.

[4] E. Pardoux. Equations aux Dérivées Partielles Stochastiques Nonlinéaires Monotones. Thesis, Université Paris Sud, 1975.

[5] J. Real. Stochastic partial differential equations with delays. *Stochastics.* 8(2):81-102, 1982-1983.

[6] T. Caraballo, K. Liu. Exponential stability of mild solutions of stochastic partial differential equations with delays. *Stoch. Anal. Appl.* 17:743-764, 1999.

[7] T. Caraballo, K. Liu, A. Truman. Stochastic functional partial differential equations: existence, uniqueness and asymptotic decay property. *Proc. R. Soc. Lond.* A 456:1775-1802, 2000.

[8] T. Taniguchi, K. Liu, A. Truman. Existence, uniqueness and asymptotic behavior of mild solutions to stochastic functional differential equations in Hilbert spaces. *J. Diff. Eq.* 181:72-91, 2002.

[9] J. K. Hale, S. M. Verduyn Lunel. *Introduction to Functional Differential Equations.* Springer-Verlag, New York, 1993.

[10] B.S. Razumikhin. On the stability of systems with a delay. *Prikl. Mat. Meh.* 20:500-512, 1956. (translated into English in J. Appl. Math. Mech.)

[11] B.S. Razumikhin. Applications of Lyapunov's method to problems in the stability of systems with a delay. *Automat. i. Telemeh.* 21:740-749, 1960. (translated into English in Automat. Remote Control 21:515-520, 1960.)

[12] T. Taniguchi. Moment asymptotic behavior and almost sure Lyapunov exponent of stochastic functional differential equations with finite delays via Lyapunov-Razumikhin method. *Stochastics.* 58:191-208, 1996.

[13] X. R. Mao. Razumikhin-type theorems on exponential stability of stochastic functional differential equations. *Stoch. Proc. Appl.* 65:233-250, 1996.

[14] A. Ichikawa. Stability of semilinear stochastic evolution equations. *J. Math. Anal. Appl.* 90:12-44, 1982.

[15] K. Liu. Lyapunov functionals and asymptotic stability of stochastic delay evolution equations. *Stochastics.* 63:1-26, 1998.

AN OPTIMAL CONTROL PROBLEM IN MEDICAL IMAGE PROCESSING

K. Bredies,[1] D. A. Lorenz,[1] and P. Maass[1]
[1] *University of Bremen, Bibliothekstrasse 2, 28359 Bremen, Germany,* {*kbredies,dlorenz,pmaass*} *@math.uni-bremen.de*

Abstract As a starting point of this paper we present a problem from mammographic image processing. We show how it can be formulated as an optimal control problem for PDEs and illustrate that it leads to penalty terms which are non-standard in the theory of optimal control of PDEs.

To solve this control problem we use a generalization of the conditional gradient method which is especially suitable for non-convex problems. We apply this method to our control problem and illustrate that this method also covers the recently proposed method of surrogate functionals from the theory of inverse problems.

Keywords: generalized conditional gradient method, surrogate functionals, image processing, optimal control of PDEs

1. Motivation from medical imaging

For many years medical imaging has aimed at developing fully automatic, software based diagnostic systems. However, the success of those automatic systems is rather limited and the human expert is as much responsible for the final diagnosis as in previous years. Hence, growing effort has been devoted to enhancing the techniques for presenting the medical images as well as additional information.

In Germany a particular effort is made in mammography, i. e. X-ray scans of the female breast for early detection of breast cancer. The process of examination by the medical experts is divided into a very short recognition phase (< 1 sec.) and a second verification phase (≈ 1 min.).

During the recognition phase, the expert first recognizes the coarse features, then more and more fine features. Tests have shown, that the experts usually form their decisions during this very short recognition phase. Nevertheless, the verification phase is the more critical one. The critical and difficult cases, where the recognition phase does not end with a preliminary diagnosis, most often applies to women in the early stages of cancer. During the verification phase

Please use the following format when citing this chapter:

Bredies, K., Lorenz, D.A., and Maass, P., 2006, in IFIP International Federation for Information Processing, Volume 202, Systems, Control, Modeling and Optimization, eds. Ceragioli, F., Dontchev, A., Furuta, H., Marti, K., Pandolfi, L., (Boston: Springer), pp. 249-259.

the expert shifts forwards and backwards, thereby alternating in examining small details and in catching an overall impression of the location of critical patterns within the organ.

This process can be supported by presenting the expert different versions of the original image during close up and normal subphases. More precisely, the expert sees a version with contrast enhanced small details in a close up phase ('fine scale'), while he sees an image which preserves all major edges but smoothes within regions ('coarse scale') during the normal phase. For enhancing fine details in mammography images a variety of algorithm have been proposed. Many of them are based on the wavelet transform due to its property of dividing an image into different scale representations, see for example [8] and references therein.

In this work we deal with the development of an optimized presentation for one cycle of the verification phase. To put the problem in mathematical terms, we start with a given image y_0 assumed to be a function defined on $\Omega := [0, 1]^2$. The fine scale and the coarse scale image are denoted y_f and y_c respectively. Under the natural assumption of finite energy images we model them as functions in $L^2(\Omega)$. The goal is, to produce a movie (i. e. a time dependent function) $y : [0, 1] \to L^2(\Omega)$, from the given images y_0, y_f and y_c such that

- the movie starts in y_0, i. e. $y(0) = y_0$,
- the movie sweeps to the fine scale image and to the coarse scale image, e. g. $y(t) \approx y_f$ for $t \in [.2, .4]$ and $y(t) \approx y_c$ for $t \in [.6, .8]$,
- the movie sweeps in a "natural" way.

An example for a mammography image, a fine scale, and coarse scale image is shown in Figure 1. As a first guess one could try to make a linear interpolation between the fine scale and the coarse scale representation. This method has one serious drawback: It does not take the scale sweep into account, i. e. all fine details are just faded in rather than developing one after another.

Hence, more advanced methods have to be employed. In this article we show a solution of this presentation problem based on optimal control of PDEs. To simplify the presentation we only model a sweep between two images in the following.

2. Modeling as an optimal control problem

2.1 PDEs and control problems in image processing

Parabolic partial differential equations are a widely used tool in image processing. Diffusion equations like the heat equation [15], the Perona-Malik

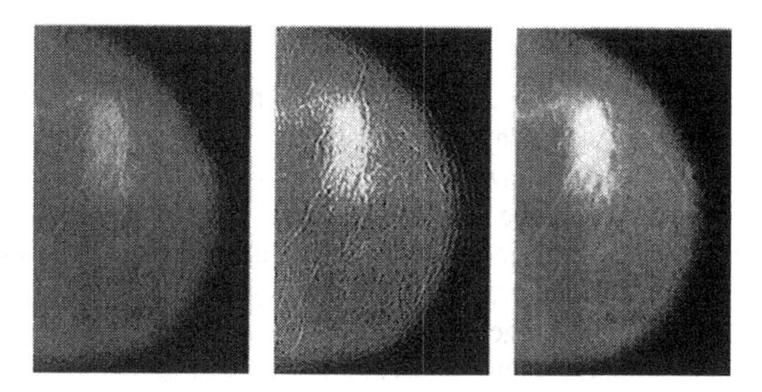

Figure 1. A mammography image. Left: original image y_0, middle: fine scale image y_f, right: coarse scale image y_c

equation [11] or anisotropic equations [14] are used for smoothing, denoising and edge enhancing.

The smoothing of a given image $y_0 \in L^2(\Omega)$ with the heat equation is done by the solution of the equation

$$\begin{aligned} y_t - \Delta y &= 0 \text{ in } [0,1] \times \Omega \\ y_\nu &= 0 \text{ on } [0,1] \times \partial\Omega \\ y(0) &= y_0 \end{aligned}$$

where y_ν stands for the normal derivative, i. e. we impose homogeneous Neumann boundary conditions.

The solution $y : [0,1] \to L^2(\Omega)$ gives a movie which starts at the image y_0 and becomes smoother with time t. This evolution is also called scale space and is analyzed by the image processing community in detail since the 1980s. Especially the heat equation does not create new features with increasing time, see e. g. [6] and the references therein.

Thus, the heat equation is well suited to model a sweep from a fine scale image y_f to a coarse scale image y_c. Hence, we take the image y_f as initial value. To make the movie y end at a certain coarse scale image y_c instead of its given endpoint $y(1)$ we propose the following optimal control problem:

$$\begin{aligned} \text{Minimize} \quad & J(y,u) = \frac{1}{2}\int_\Omega |y(1) - y_c|^2 dx + \frac{\alpha}{2}\int_0^1 \int_\Omega |u|^2 \, dx dt \\ \text{subject to} \quad & y_t - \Delta y = u \text{ in } [0,1] \times \Omega \\ & y_\nu = 0 \text{ on } [0,1] \times \partial\Omega \\ & y(0) = y_f. \end{aligned}$$

In other words, the diffusion process is forced to end in y_c with the help of a heat source u.

2.2 Adaption to image processing

The above described problem is classical in the theory of optimal control of PDEs, though not well adapted to image processing. The solution of this problem may have several drawbacks: The control u will be smooth due to the regularization and have a large support. This will result in very smooth changes in the image sequence y and, more worse, in global changes in the whole image. To overcome these difficulties, different norms can be used for regularization. A widely used choice in image processing is to use Besov norms because they are appropriate to model images. Besov norms can be defined in different ways, e. g. in terms of moduli of smoothness [13] or in terms of Littlewood-Paley decompositions [7]. Here we take another viewpoint and define the Besov spaces via norms of wavelet expansions [3, 10]. For a sufficient smooth wavelet ψ the Besov semi norm of a function f on a set $M \subset \mathbf{R}^d$ is defined as

$$|f|^q_{B^s_{p,q}(M)} = \sum_j \left(2^{sjp} 2^{j(p-2)d/2} \sum_{i,k} |\langle f, \psi_{i,j,k} \rangle|^p \right)^{q/p}$$

where j is the scale index, k indicates translation and i stands for the directions. The Besov space $B^s_{p,q}(M)$ is defined as the functions $f \in L^p(M)$ that has a finite Besov semi norm. See [7, 10] for a more detailed introduction to wavelets and Besov spaces.

One particular Besov space plays a special role in image processing: the space $B^{d/2}_{1,1}(M)$. This is because it is close to the space of functions of bounded variation $BV(M)$ which used to model images [12]. Especially one has $B^{d/2}_{1,1}(M) \subset BV(M)$. In the following we use the scale $B^s_{p,p}([0,1] \times \Omega)$ of Besov spaces and the wavelet norms

$$|f|^p_{B^s_{p,p}([0,1]\times\Omega)} = \sum_{i,j,k} w_j |\langle f | \psi_{i,j,k} \rangle|^p$$

with the weighting sequence $w_j = 2^{sjp} 2^{j(p-2)3/2}$.

2.3 The solution of the PDE and the control-to-state mapping

The solution of the heat equation is a classical task. If we assume that the initial value y_f is in $L^2(\Omega)$ and the control u is in $L^2([0,1] \times \Omega)$ the solution y is in $L^2(0, 1, H^1(\Omega)) \cap C([0, 1], L^2(\Omega))$. Especially y is continuous with respect to time and the point evaluation $y(1)$ makes sense, see e. g. [9].

In our case the solution operator $u \mapsto y$ is affine linear, due to the non-zero initial value. We make the following modifications to come back to a linear problem: We split the solution into two parts. The non-controlled part y^n is the solution of

$$\begin{aligned} y_t^n - \Delta y^n &= 0 \\ y^n(0) &= y_f \end{aligned}$$

and the homogeneous part y^h is the solution of

$$\begin{aligned} y_t^h - \Delta y^h &= u \\ y^h(0) &= 0 \end{aligned} \tag{1}$$

(both with homogeneous Neumann boundary conditions). Then the solution operator $G : u \mapsto y^h$ of equation (1) is linear and continuous from $L^2([0,1], L^2(\Omega))$ to $L^2(0,1,H^1(\Omega)) \cap C([0,1], L^2(\Omega))$. With the help of the point evaluation operator we have the control-to-state mapping $K : u \mapsto y^h(1)$ linear and continuous from $L^2([0,1], L^2(\Omega))$ to $L^2(\Omega)$. Then the solution is $y = y^n + y^h$ and we can focus on the control problem for y^h.

Together with the thoughts of the previous subsection we end up with the following minimization problem:

$$\text{minimize } J(u) = \frac{1}{2}||Ku - y_c + y^n(1)||^2_{L^2(\Omega)} + \alpha|u|^p_{B^s_{p,p}([0,1]\times\Omega)}. \tag{2}$$

3. Solution of the optimal control problem

The minimization of the functional (2) is not straightforward. The non-quadratic constraint leads to a nonlinear normal equation which can not be solved explicitly. Here we use a generalization of the conditional gradient method for the minimization.

3.1 The generalized conditional gradient method

The classical conditional gradient method deals with minimization problems of the form

$$\min_{u\in C} F(u), \tag{3}$$

where C is a bounded convex set and F is a possible non-linear function. One notices that this constrained problem can actually be written as an "unconstrained" one with the help of the indicator functional

$$I_C(u) = \begin{cases} 0 & u \in C \\ \infty & u \notin C \end{cases}.$$

With $\Phi = I_C$, problem (3) thus can be reformulated as

$$\min_{u\in H} F(u) + \Phi(u). \tag{4}$$

To illustrate the proposed generalization, we summarize the key properties of F and Φ: F is smooth while Φ may contain non-differentiable parts. The minimization problem with Φ alone is considered be solved easily while the minimization of F is comparatively hard. The influence of Φ is rather small in comparison to F.

With these assumptions in mind, the conditional gradient method can also be motivated as follows. Let $u \in H$ be given such that $\Phi(u) < \infty$. We like to find an update direction by a linearized problem. Since Φ is not differentiable, we only linearize F:

$$\min_{v \in H} \langle F'(u) | v \rangle + \Phi(v). \tag{5}$$

The minimizer of this problem serves as an update direction.

So this "generalized conditional gradient method" in the $(n+1)$–st step reads as follows: Let $u_n \in H$ be given such that $\Phi(u_n) < \infty$.

1 Determine the solution of (5) and denote it v_n.

2 Set s_n as a solution of

$$\min_{s \in [0,1]} F(u_n + s(v_n - u_n)) + \Phi(u_n + s(v_n - u_n)).$$

3 Let $u_{n+1} = u_n + s_n(v_n - u_n)$.

To ensure existence of a solution in Step 1 we state the following condition:

ASSUMPTION 1 *Let the functional $\Phi : H \to]-\infty, \infty]$ be proper, convex, lower semi-continuous and coercive with respect to the norm.*

Standard arguments from convex analysis yield the existence of a minimizer in Step 1 of the algorithm [5]. So if F is Gâteaux-differentiable in H, the algorithm is well-defined. The convergence of the generalized conditional gradient method is analyzed in detail by the authors in [2]. The main result there is the following theorem.

THEOREM 2 *Let Φ satisfy Assumption 1 and let us assume that every set $E_t = \{u \in H \mid \Phi(u) \leq t\}$ is compact. Let F be continuously Fréchet differentiable, let $F + \Phi$ be coercive and u_0 be given such that $\Phi(u_0) < \infty$. Denote (u_n) the sequence generated by the generalized conditional gradient method.*

Then every convergent subsequence of (u_n) converges to a stationary point of $F + \Phi$. At least one such subsequence exists.

Two remarks are in order: First, we notice that the theorem is also valid if the functional F is not convex. Second, the theorem only gives convergence to a stationary point which may seem unsatisfactory, specially if one wants to minimize non-convex functions. But this does not have to be a drawback, as we will see in the next section:

3.2 Application to the control problem

To apply the generalized conditional gradient method to the optimal control problem (2) we split the functional J into two parts as follows:

$$\begin{aligned} F(u) &= \frac{1}{2}\|Ku - y^*\|^2_{L^2(\Omega)} - \frac{\lambda}{2}\|u\|^2_{L^2([0,1],L^2(\Omega))} \\ \Phi(u) &= \frac{\lambda}{2}\|u\|^2_{L^2([0,1],L^2(\Omega))} + \alpha|u(t)|^p_{B^s_{p,p}([0,1]\times\Omega)} \end{aligned}$$

where $y^* = y_c - y^n(1)$ and $\lambda > 0$. The non-differentiable part Φ clearly fulfills Condition 1. Moreover, the level sets E_t are compact, if $s > 3(2-p)/(2p)$ because then the embedding $B^s_{p,p}([0,1]\times\Omega) \subset L^2([0,1]\times\Omega)$ is compact.

The derivative of F is given by

$$F'(u) = K^*(Ku - y^*) - \lambda u,$$

and hence the minimization problem in step one of the generalized gradient method reads as

$$\min_v \langle K^*(Ku - y^*) - \lambda u | v\rangle + \frac{\lambda}{2}\|v\|^2 + \alpha|v(t)|^p_{B^s_{p,p}([0,1]\times\Omega)}.$$

which can be written as in terms of wavelet expansions

$$\min_v \sum_{i,j,k}\Bigg(\langle K^*(K(u-y^*) - \lambda u|\psi_{i,j,k}\rangle\langle v|\psi_{i,j,k}\rangle \\ + \frac{\lambda}{2}\langle v|\psi_{i,j,k}\rangle^2 + \alpha w_j|\langle v|\psi_{i,j,k}\rangle|^p\Bigg).$$

This minimization problem is equivalent to

$$\min_v \sum_{i,j,k}\left(\frac{1}{2}|\langle K^*(K(u-y^*) - \lambda u - \lambda v|\psi_{i,j,k}\rangle|^2 + \alpha w_j|\langle v|\psi_{i,j,k}\rangle|^p\right).$$

Now the minimization problem reduces to pointwise minimization in the basis coefficients of v. This can be done analytically by the expression

$$v = \mathbf{S}_{\alpha w_j/\lambda}(u) = \sum_{i,j,k} S_{\alpha w_j/\lambda}(\langle u - \lambda^{-1}K^*(Ku - y^*)|\psi_{i,j,k}\rangle)\psi_{i,j,k}$$

where S_α is the a shrinkage function given by

$$S_\alpha(x) = \begin{cases} G^{-1}_{\alpha,p}(x) & p > 1 \\ \operatorname{sign}(x)\max(|x| - \alpha, 0) & p = 1 \end{cases}$$

with $G_{\alpha,p}(x) = x + \alpha p \operatorname{sign}(x)|x|^{p-1}$.

To sum up, the generalized conditional gradient method for the control problem reads as:

- Initialize with u_0 and choose parameter $\lambda > 0$.
- Calculate direction $v_n = \mathbf{S}_{\alpha w_j/\lambda}(u_n)$
- Calculate step size s_n such that $J(u_n + s_n(v_n + u_n))$ is minimal.
- Update $u_{n+1} = u_n + s_n(v_n - u_n)$ and set $n \leftarrow n + 1$.

REMARK 3 *The choice of the step size can be done analytically for $p = 2$ and $p = 1$ easily with just one evaluation of the operator K (again we refer to [2] for details). Furthermore, one can choose an approximate step size which decreases the functional J nevertheless. This is always the case is the step size is small enough.*

3.3 Equivalence with the method of surrogate functionals

In this subsection we will point out, that the above proposed algorithm for minimization problems of the form

$$J(u) = \frac{1}{2}\|Ku - f\|^2 + \alpha|u|^p_{B^s_{p,p}}$$

is an extension of the method of surrogate functionals as proposed recently by Daubechies, Defries, De Mol [4]. This method is based on the so called surrogate functional

$$\tilde{J}(u, a) = J(u) - \frac{1}{2}\|Ku - Ka\|^2 + \frac{1}{2}\|u - a\|^2.$$

One notices that $\tilde{J}(u, u) = J(u)$. The minimization of J is done by alternate minimization of $\tilde{J}$ with respect to the variables a and u. Since minimization with respect to a just gives $u = a$ one can rewrite this iteration as

$$u_{n+1} = \operatorname{argmin} \tilde{J}(u, u_n).$$

In [4] it is proved, that this minimization leads to the iteration

$$u_{n+1} = \mathbf{S}_{\alpha w_j}(u_n).$$

Hence, the method of surrogate functionals produces the same sequence as the generalized conditional gradient method with $\lambda = 1$ and step size equals to one. Furthermore, in [4] is it proved, that the produced sequence converges strongly to a minimized of J if the sequence of weights w_j is strictly bounded away from zero.

Our result shows, that the method of surrogate functionals can be interpreted as a gradient descend method. Again, we refer to [2] for a more detailed description.

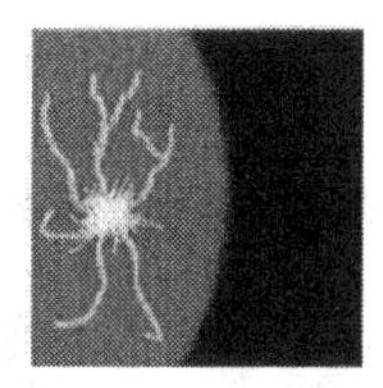
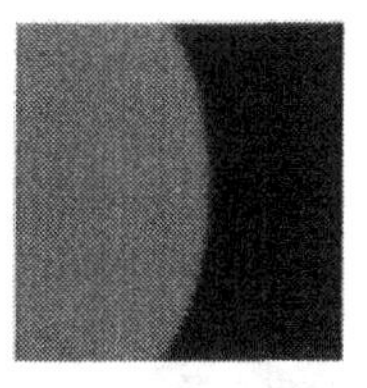

Figure 2. Images used for illustration. Left: fine scale image, right: coarse scale image.

4. Application

Here we show the application of the above described methodology. Since the effects can be seen more clearly in artificial images, we will not use original images. The artificial images we used are shown in Figure 2.

For illustration we use the values $p = 1$ and $s = 3/2 + \varepsilon > 3/2$ in the minimization problem (2), since this is close to the BV-norm and we have $B^{3/2+\varepsilon}_{1,1}([0,1] \times \Omega) \subset L^2([0,1] \times \Omega)$ compactly.

The results are presented in Figure 3. The figure shows a comparison of the linear interpolation, the pure result of the application of the heat equation and the result of the optimal control problem. One sees that the linear interpolation is only fading out the details. In the uncontrolled result (middle column) the details are vanishing one after another but the process does not end in the desired endpoint. The result of the optimal control problem (right column) exhibits both a nice vanishing of the details and end in the given endpoint.

5. Conclusion

We have seen that the application of the theory of optimal control of PDEs to image processing problems is a fruitful field of research. Besides promising result, even for easy models like the linear heat equation, new interesting mathematical problems arise, like the treatment of non-quadratic penalty terms. For future research, better adapted PDEs (like the anisotropic diffusion equations) could be investigated.

References

[1]

[2] K. Bredies, D. A. Lorenz, and P. Maass. Equivalence of a generalized conditional gradient method and the method of surrogate functionals. Preprint of the DFG Priority Program 1114 No. 135, University of Bremen, 2005.

[3] A. Cohen. *Numerical Analysis of Wavelet Methods*. Elsevier Science B. V., 2003.

[4] I. Daubechies, M. Defrise, and C. De Mol. An iterative thresholding algorithm for linear inverse problems with a sparsity constraint. *Communications in Pure and Applied Mathematics*, 57(11):1413–1457, 2004.

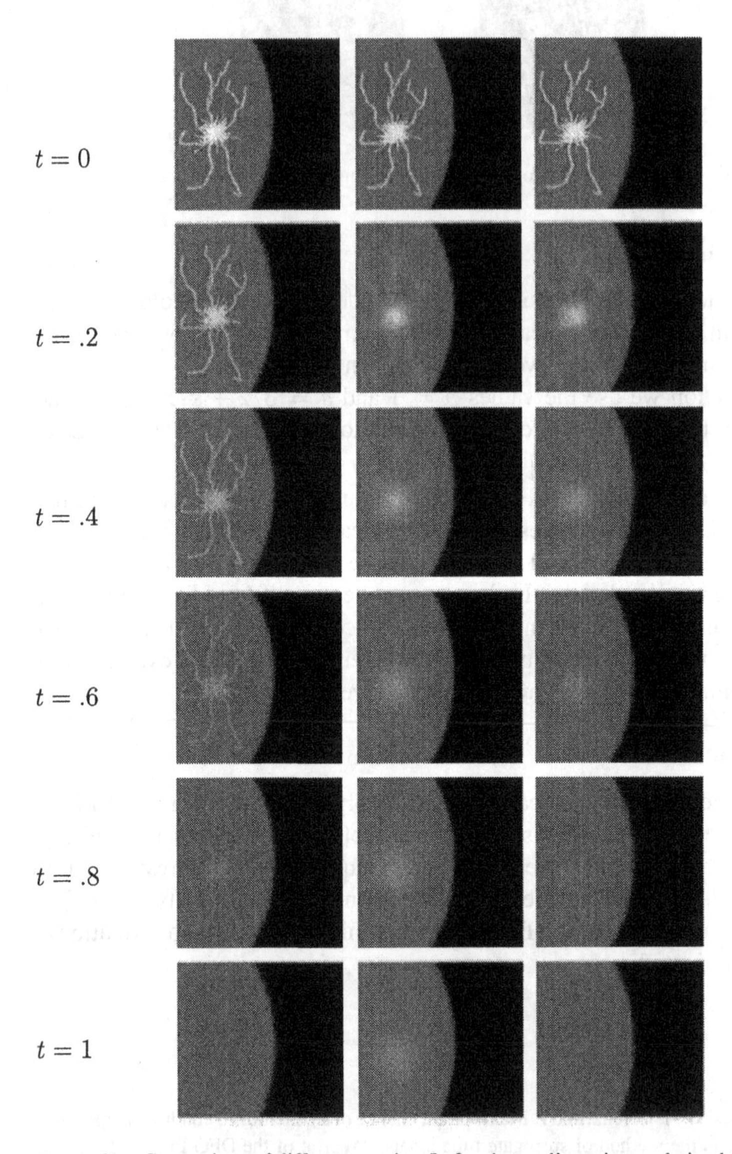

Figure 3. Comparison of different movies. Left column: linear interpolation between y_f and y_c from Figure 2. Middle column: Solution of the heat equation with initial value y_f. Right column: Solution of the optimal control problem (2).

[5] I. Ekeland and R. Temam. *Convex analysis and variational problems*. North-Holland, Amsterdam, 1976.

[6] L. Florack and A. Kuijper. The topological structure of Scale-Space images. *Journal of Mathematical Imaging and Vision*, 12:65–79, 2000.

[7] M. Frazier, B. Jawerth, and G. Weiss. *Littlewood-Paley theory and the study of function spaces*. Number 79 in Regional Conference Series in Mathematics. American Mathematical Society, 1991.

[8] P. Heinlein, J. Drexl, and W. Schneider. Integrated wavelets for enhancement of microcalcifications in digital mammography. *IEEE Transactions on Medical Imaging*, 22(3):402–413, March 2003.

[9] J.-L. Lions. *Optimal Control of Systems Governed by Partial Differential Equations*. Springer, 1971.

[10] Y. Meyer. *Wavelets and Operators*, volume 37 of *Cambridge Studies in Advanced Mathematics*. Cambridge University Press, 1992.

[11] P. Perona and J. Malik. Scale-space and edge detection using anisotropic diffusion. *IEEE Transactions on Pattern Analysis and Machine Intelligence*, 12(7):629–639, 1990.

[12] L. I. Rudin, S. J. Osher, and E. Fatemi. Nonlinear total variation based noise removal algorithms. *Physica D*, 60:259–268, 1992.

[13] H. Triebel. *Theory of Function Spaces II*. Monographs in Mathematics. Birkhäuser, 1992.

[14] J. Weickert. *Anisotropic diffusion in image processing*. Teubner, Stuttgart, 1998.

[15] A. P. Witkin. Scale-space filtering. In *Proceedings of the International Joint Conference on Artificial Intelligence*, pages 1019–1021, 1983.

DYNAMICAL RECONSTRUCTION AND FEEDBACK ROBUST CONTROL OF PARABOLIC INCLUSIONS

V. Maksimov[1]

[1] *Institute of Mathematics and Mechanics, Ural Branch, Russian Academy of Sciences, Ekaterinburg, Russia, maksimov@imm.uran.ru**

Abstract Two types of problems for parabolic inclusions, namely, problems of robust control under the action of uncontrolled disturbances and problems of dynamical identification of inputs, are discussed. Algorithms for solving such problems stable with respect to informational noises and computational errors are presented. The algorithms oriented to computer realization allow one to simulate a solving process in the "real time" mode. They adaptively take into account inaccurate measurements of phase trajectories and are regularizing in the following sense: the more precise is incoming information, the better is algorithm's output. The algorithms are based on the method of auxiliary positionally-controlled models [1, 2, 4–7]. The basic elements of the algorithms are represented by stabilization procedures (functioning by the feedback principle) for appropriate Lyapunov functionals.

keywords: parabolic inclusions, feedback control, reconstruction

1. Introduction

Let a dynamical system be described by the parabolic inclusion

$$\dot{x}(t) + \partial\varphi(x(t)) \ni Bu(t) - Cv(t) + f(t), \quad t \in T = [t_0, \vartheta]. \tag{1}$$

Here $H = H^*$ is a real Hilbert space with the norm $|\cdot|_H$ and scalar product $(\cdot;\cdot)_H$, $f(\cdot) \in L_2(T; H)$ is a given function, $\varphi : H \to \overline{R} = \{r \in R : -\infty < r \leq +\infty\}$ is a lower semicontinuous convex function, $\partial\varphi$ is the subdifferential of φ. Let $x(t_0) = x_0 \in D(\varphi) = \{x \in H : \varphi(x) < +\infty\}$ be an initial state. Let $(U, |\cdot|_U)$ and $(V, |\cdot|_V)$ be uniformly convex Banach spaces; $B \in \mathcal{L}(U; H)$, $C \in \mathcal{L}(V; H)$ be linear continuous operators. It is known that there exists (for any $\{u(\cdot), v(\cdot)\} \in L_2(T; U) \times L_2(T; V)$) a unique solution

*This work was supported in part by the Russian Foundation for Basic Research (grant #04-01-00059), Program on Basic Research of the Presidium of the Russian Acad. Sci. #19, Program of supporting leading scientific schools of Russia, and the Ural-Siberian Interdisciplinary Project.

Please use the following format when citing this chapter:

Maksimov, V., 2006, in IFIP International Federation for Information Processing, Volume 202, Systems, Control, Modeling and Optimization, eds. Ceragioli, F., Dontchev, A., Furuta, H., Marti, K., Pandolfi, L., (Boston: Springer), pp. 261-267.

$x(\cdot) = x(\cdot; t_0, x_0, u(\cdot), v(\cdot))$ of inclusion (1) with the following properties [3, 8]:

$$x(\cdot) \in W(T), \quad x(t) \in D(\varphi) \quad \forall t \in T, \quad t \to \varphi(x(t)) \in AC(T).$$

Here $AC(T)$ is the set of absolutely continuous functions $z(\cdot) : T \to R$, $W(T) = \{z(\cdot) \in L_2(T; H) : z_t(\cdot) \in L_2(T; H)\}$.

The paper is devoted to two problems: the problem of robust control and the problem of dynamical reconstruction of an input. Now we formulate the problems and we outline the ideas used in the solution. Let a uniform net $\Delta = \{\tau_i\}_{i=0}^m$, $\tau_i = \tau_{i-1} + \delta$, $\tau_0 = t_0$, $\tau_m = \vartheta$ with diameter $\delta = \delta(\Delta) = \tau_i - \tau_{i-1}$ be fixed on a given time interval T. Let a solution of inclusion (1) $x(\cdot) = x(\cdot; t_0, x_0, u(\cdot), v(\cdot))$ depend on a time-varying control $u(\cdot) \in L_2(T; U)$ and an unknown disturbance $v(\cdot) \in L_2(T; V)$. The function $x(\cdot)$ is unknown. At moments $\tau_i \in \Delta$ the phase state $x(\tau_i)$ is inaccurately measured. Results of measurements $\xi_i^h \in H$, $i \in [0 : m-1]$ satisfy the inequalities

$$|\xi_i^h - x(\tau_i)|_H \le h. \tag{2}$$

Here, $h \in (0, 1)$ is a level of informational noise.

The problem of robust control consists in the following. A nonempty set $N \subset H$ and a number $\varepsilon > 0$ are given. It is required to construct an algorithm of feedback control $u = u(t) \in P$, $t \in T$, of inclusion (1) providing fulfillment of the following condition. Whatever an unknown disturbance $v(\cdot)$, $v = v(t) \in Q$, $t \in T$, may be, the distance between the phase state $x(t) = x(t; t_0, x_0, u(\cdot), v(\cdot))$ at the moment $t = \vartheta$ and the set N should not exceed the value of ε. Here P and Q are given bounded and closed sets from spaces of controls U and V, respectively.

The problem of dynamical reconstruction of an input is as follows. Let in inclusion (1) control $u = u(t) = 0$, $t \in T$. It is required to design a dynamical algorithm of reconstruction of an unknown input $v = v(\cdot)$ in the "real time" mode.

The scheme of algorithm for solving the problem of robust control is given in the figure below [1, 4, 5].

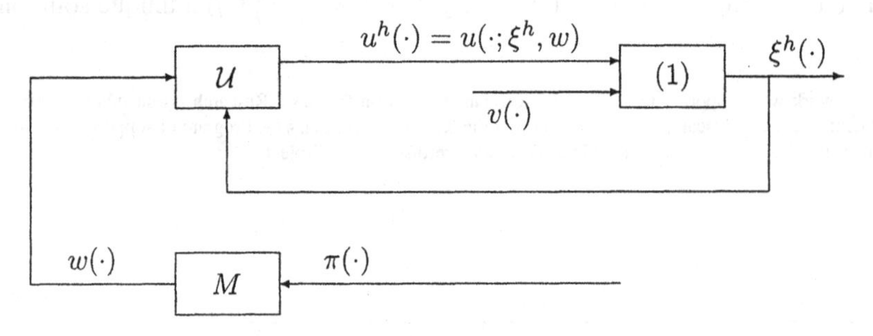

In the beginning, an auxiliary system M (called a model) is introduced. The model has an input $\pi(\cdot)$ and an output $w(\cdot)$. The process of synchronous feedback control of inclusion (1) and M is organized on the interval T. This process is decomposed into $(m-1)$ identical steps. At the i-th step carried out during the time interval $\delta_i = [\tau_i, \tau_{i+1})$, the following actions are fulfilled. First, at the time moment τ_i, according to the chosen rule $\mathcal{U}$, the element $u_i = \mathcal{U}(\tau_i, \xi_i^h, w_i)$ is calculated. Then (till the moment τ_{i+1}) the control $u(t) = u_i$, $\tau_i \leq t < \tau_{i+1}$, is fed onto the input of inclusion (1). The values ξ_{i+1}^h and $w_{i+1} = w(\tau_{i+1})$ are treated as algorithm's output at the i-th step.

An analogous scheme is applicable to solving reconstruction problem. In this case, an auxiliary system M (a model) is also introduced. The problem of reconstruction is substituted by the problem of designing an algorithm of feedback control of the model. This algorithm is identified with a function $\mathcal{U}$, which is chosen in such a way that the control $u = u^h(\cdot)$ approximates the unknown disturbance $v(\cdot)$: $u^h(t) = u_i = \mathcal{U}(\tau_i, \xi_i^h, w(t)), \quad t \in \delta_i$.

2. Statement of the problems

Before we pass to the mathematical formulation of the problems, let us give some definitions. Denote by $u_{a,b}(\cdot)$ a function $u(t)$, $t \in [a, b]$. The symbol $P_{a,b}(\cdot)$ stands for the restriction of a set $P_T(\cdot)$ onto the segment $[a, b] \subset T$. Elements of the product $T \times H \times H$ are called positions and denote by p. Any possible function (multifunction) $\mathcal{U} : T \times H \times H \to P$ is said to be a feedback strategy. Feedback strategies correct controls at corresponding time moments. Any strongly measurable (by Lebesgue) functions $u(\cdot) : T \to P$ and $v(\cdot) : T \to Q$ are called an open-loop control and a disturbance, and are denoted by $u_T(\cdot)$ and $v_T(\cdot)$, respectively. Let a model M be described by the inclusion

$$\dot{w}(t) + \partial\varphi(w(t)) \ni \pi(t) + f(t), \quad t \in T, \quad w(t_0) = x_0, \tag{3}$$

where $\pi(\cdot) \in L_2(T; H)$ is an open-loop control. Hereinafter, denote a solution of inclusion (3) by the symbol $w(\cdot) = w(\cdot; t_0, x_0, \pi(\cdot))$. A solution $x(\cdot)$ of inclusion (1) starting from an initial state (t_*, x_*) and corresponding to a piecewise constant control $u(\cdot)$,

$$u(t) = u_i \in \mathcal{U}(p_i) \in U, \quad t \in [\tau_i, \tau_{i+1}), \quad i \in [0 : m-1],$$

$$p_i = (\tau_i, \xi_i^h, w_i), \quad w_i = w(\tau_i; t_0, x_0, \pi(\cdot)), \quad |\xi_i^h - x(\tau_i)|_H \leq h,$$

and to a disturbance $v(\cdot) : [t_*, \vartheta] \to Q$ is called an (h, Δ, w)-motion $x_{\Delta,w}^h(\cdot; t_*, x_*, \mathcal{U}, v_{t_*,\vartheta}(\cdot))$ generated by a positional strategy $\mathcal{U}$ on a partition Δ. The set of all (h, Δ, w)-motions is denoted by $X_{[t_*,\vartheta]}(t_*, x_*, \mathcal{U}, h, \Delta, w)$.

The problem of robust control (**Problem 1**) consists in constructing an open-loop control $\pi(\cdot) \in P_T(\cdot)$ in the model and a feedback strategy $\mathcal{U} : T \times H \times$

$H \to P$ with the following properties: whatever a value $\varepsilon > 0$ may be, one can indicate (explicitly) numbers $h_* > 0$ and $\delta_* > 0$ such that the inequalities

$$\mathrm{dist}(x^h_{\Delta,w}(\vartheta), N) \le \varepsilon \qquad \forall x^h_{\Delta,w}(\cdot) \in X_T(t_0, x_0, \mathcal{U}, h, \Delta, w) \tag{4}$$

are fulfilled uniformly with respect to all measurements ξ_i^h with properties (2) if $h \le h_*$ and $\delta = \delta(\Delta) \le \delta_*$.

Here the symbol $\mathrm{dist}(x, N)$ denotes the distance from x to N, i.e., $\mathrm{dist}(x, N) = \inf_{y \in N}\{|x - y|_H\}$.

Let us turn to the problem of reconstruction. In this case, a disturbance $v(\cdot)$ to be reconstructed is an element of the space $L_2(T; V)$. The solution of (1) generated by a disturbance $v(\cdot) \in L_2(T; V)$ is denoted by the symbol $x(\cdot) = x(\cdot; t_0, x_0, v(\cdot))$ (we have $u(\cdot) = 0$ in (1)). A model M is described by the relation

$$\dot{w}(t) + \partial\varphi(w(t)) \ni -Cu^h(t) + f(t), \quad t \in T, \quad w(t_0) = x_0. \tag{5}$$

Let us a family of partitions

$$\Delta_h = \{\tau_{i,h}\}_{i,h=0}^{m_h}, \; \tau_{i,h} = \tau_{i-1,h} + \delta, \; \tau_{0,h} = t_0, \; \tau_{m_h,h} = \vartheta \tag{6}$$

with fixed diameters $\delta = \delta(h)$ be fixed. The control $u^h(\cdot)$ is defined as follows:

$$u^h(t) = \mathcal{U}(p_i) \in V, \quad t \in \delta_{i,h} = [\tau_{i,h}, \tau_{i+1,h}), \tag{7}$$

where the position $p_i = (\tau_{i,h}, \xi_i^h, w(t))$ for $t \in \delta_{i,h}$. Let $v_*(\cdot; x(\cdot))$ be a minimal $L_2(T; V)$-norm element of the set $V_*(x(\cdot))$ of all functions $v(\cdot) \in L_2(T; V)$ generating the solution $x(\cdot)$: $V_*(x(\cdot)) = \{\tilde{v}(\cdot) \in L_2(T; V) : x(\cdot) = x(\cdot; t_0, x_0, \tilde{v}(\cdot))\}$.

The problem of dynamical reconstruction (**Problem 2**) consists in constructing a feedback strategy $\mathcal{U} : T \times H \times H \to V$ such that the control $u = u^h(\cdot)$ defined by (7) possesses the property

$$u^h(\cdot) \to v_*(\cdot; x(\cdot)) \text{ in } L_2(T; V) \quad \text{as} \quad h \to 0.$$

3. Algorithm for solving Problem 1

In this section, an algorithm for solving Problem 1 is designed. This algorithm is based on constructions of the theory of guaranteed control [1, 4, 5]. Let the following condition be fulfilled.

Condition 1 ("Entire domination") *There exists a closed set $D \subset H$ such that $BP = CQ + D$.*

Here we use the following notation: $BP = \{Bu : u \in P\}$, $CQ = \{Cv : v \in Q\}$, $CQ + D = \{u : u = u_1 + u_2, \; u_1 \in CQ, \; u_2 \in D\}$.

Let us describe the procedure used in the construction of the (h, Δ, w)-motion $x^h_{\Delta,w}(t; t_0, x_0, \mathcal{U}, v_{t_0,t}(\cdot))$ corresponding to a fixed partition Δ and a strategy $\mathcal{U}$ of the form:

$$\mathcal{U}(t, x, w) = \arg\min\{(x - w(t), Bu)_H : u \in P\}. \tag{8}$$

We assume $u^h_0 \in P$ in the interval $[t_0, \tau_1)$. Under the action of this control as well as of an unknown open-loop disturbance $v_{t_0,\tau_1}(\cdot)$, some (h, Δ, w)-motion $\{x^h_{\Delta,w}(\cdot; t_0, x_0, u^h_0, v_{t_0,\tau_1}(\cdot))\}_{t_0,\tau_1}$ is realized. At the moment $t = \tau_1$ we find u_1 from the condition

$$u^h_1 \in \mathcal{U}(p_1), \quad p_1 = (\tau_1, \xi^h_1, w(\tau_1)), \quad |\xi^h_1 - x^h_{\Delta,w}(\tau_1)|_H \leq h.$$

Then we calculate the realization of the (h, Δ, w)-motion $\{x^h_{\Delta,w}(\cdot; \tau_1, x^h_{\Delta,w}(\tau_1), u^h_1, v_{\tau_1,\tau_2}(\cdot))\}_{\tau_1,\tau_2}$. Let the (h, Δ, w)-motion $x^h_{\Delta,w}(\cdot)$ be defined in the interval $[t_0, \tau_i]$, $(\tau_i = \tau_{i,h})$. At the moment $t = \tau_i$ we assume

$$u^h_i \in \mathcal{U}(p_i), \quad p_i = (\tau_i, \xi^h_i, w(\tau_i)), \quad |\xi^h_i - x^h_{\Delta,w}(\tau_i)|_H \leq h.$$

As the result of the action of this control and of an unknown open-loop disturbance $v_{\tau_i,\tau_{i+1}}(\cdot)$ the (h, Δ, w)-motion of system (1) $\{x^h_{\Delta,w}(\cdot; \tau_i, x^h_{\Delta,w}(\tau_i), u^h_i, v_{\tau_i,\tau_{i+1}}(\cdot))\}_{\tau_i,\tau_{i+1}}$ is realized in the interval $[\tau_i, \tau_{i+1}]$. The procedure of forming the (h, Δ, w)-motion stops at the moment ϑ.

Let $W(t; t_0, x_0)$ be the set of attainability for inclusion (3), i.e.,

$$W(t; t_0, x_0) = \{w(t) : w(t) = w(t; t_0, x_0, \pi_{t_0,t}(\cdot)), \pi_{t_0,t}(\cdot) \in D_{t_0,t}(\cdot)\},$$

$$D_{t_0,t}(\cdot) = \{u(\cdot) \in L_2([t_0, t]; H) : u(\tau) \in D \text{ for a. a. } \tau \in [t_0, t]\}.$$

THEOREM 1 *If $W_D(\vartheta; t_0, x_0) \cap N \neq \emptyset$ then Problem 1 has a solution and the positional strategy $\mathcal{U} : T \times H \times H \to P$ defined according to* (8) *solves this Problem. Otherwise, Problem 1 does not have a solution.*

Proof (outline). The proof of the first part of this theorem is performed according to the known scheme (see, for example, [4]). The basic elements of the algorithms are represented by stabilization procedures (functioning by the feedback principle) for appropriate Lyapunov functionals. Let $\pi^*(\cdot) \in D_T(\cdot)$ be an open-loop control with the following property: $w(\vartheta; t_0, x_0, \pi^*(\cdot)) \in N$. Here $w(\cdot) = w(\cdot; t_0, x_0, \pi^*(\cdot))$ denotes the solution of (3) for $\pi(\cdot) \equiv \pi^*(\cdot)$. The cornerstone is the proof of "smallness" of variation of the Lyapunov functional $\varepsilon(t) = |x^h_{\Delta,w}(t) - w(t)|^2_H$ in the interval T. Namely, the strategy $\mathcal{U}$ (8) provides fulfillment of the inequalities

$$\varepsilon(t) \leq \varepsilon(\tau_i) + C(t - \tau_i)(h + (t - \tau_i)^{1/2})$$

for $t \in [\tau_i, \tau_{i+1}]$, $i \in [0 : m-1]$. The inequalities, in turn, imply the estimate $\varepsilon(t) \le C(h+\delta)$, where a constant C does not depend on $x^h_{\Delta,w}(\cdot)$ and $w(\cdot)$. Hence, in virtue of condition 1, it follows that relation (4) is true.

Let us prove the second part of the theorem. For this purpose, it is sufficient to indicate a rule $\mathcal{V} : T \times H \times H \to Q$ forming a disturbance $v(\cdot)$ such that, for any feedback control $u(\cdot)$, the solution of inclusion (1) avoids the set N at the moment ϑ. Along with inclusion (1), we consider the inclusion

$$\dot{y}(t) + \partial\varphi(y(t)) \ni Bu^e(t) - Cv^e(t) + f(t). \tag{9}$$

We construct the procedure of synchronous feedback control of (1) and (9). Let us fix a partition Δ of the interval T. In the interval $[t_0, \tau_1]$, we assume $u^e(t) = u^{(0)} \in P$, $v^e(t) = v^{(0)} \in Q$, $v(t) = v_0 \in Q$, where $u^{(0)}$, v_0 are arbitrary elements from the sets P and Q, $v^{(0)}$ is an element such that $Bu^{(0)} - Cv^{(0)} \in D$. Under the action of these constant controls and of an unknown control $u_{t_0,\tau_1}(\cdot)$, some solution $\{x(\cdot; t_0, x_0, u_{t_0,\tau_1}(\cdot), v_0)\}_{t_0,\tau_1}$ of inclusion (1) and some solution $y^h_\Delta(t) = y(t; t_0, x_0, u^{(0)}, v^{(0)})$, $t_0 \le t \le \tau_1$, of inclusion (9) are realized. By analogy with (h, Δ, w)-motion $x^h_{\Delta,w}$, the former is denoted by $x^h_{\Delta,y}$ and is called an (h, Δ, y)-motion: $x^h_{\Delta,y}(\cdot) = x(\cdot; t_0, x_0, u_{t_0,\vartheta}(\cdot), \mathcal{V})$ and the latter is called an (h, Δ)-motion of the model. The set of all $x^h_{\Delta,y}(\cdot)$ is denoted by $X_T(t_0, x_0, \mathcal{V}, \Delta, y)$. Let the (h, Δ)-motion $y^h_\Delta(\cdot)$ and (h, Δ, y)-motion $x^h_{\Delta,y}(\cdot)$ be defined in the interval $[t_0, \tau_i]$. To form $\{y^h_\Delta(\cdot)\}_{\tau_i,\tau_{i+1}}$, $\{x^h_{\Delta,y}(\cdot)\}_{\tau_i,\tau_{i+1}}$, we proceed in the following way. At the moment $t = \tau_i$ we find $u^{(i)}$, $v^{(i)}$, and v_i from the rule

$$u^{(i)} = \arg\min\{(y^h_\Delta(\tau_i) - \xi^h_i, Bu)_H : u \in P\}, \quad |\xi^h_i - x^h_{\Delta,y}(\tau_i)|_H \le h,$$

$v^{(i)}$ is an arbitrary element from the set Q such that $Bu^{(i)} - Cv^{(i)} \in D$,

$$v_i \in \mathcal{V}(t, \xi^h_i, y^h_\Delta(\tau_i)),$$

$$\mathcal{V}(\tau_i, x, y^h_\Delta(\tau_i)) = \arg\min\{(y^h_\Delta(\tau_i) - x, Cv)_H : v \in Q\}.$$

After that, we assume that $u^e(t) = u^{(i)}$, $v^{(e)}(t) = v^{(i)}$ for $t \in [\tau_i, \tau_{i+1})$ in (9), then we calculate the (h, Δ)-motion in the interval $[\tau_i, \tau_{i+1}]$: $y^h_\Delta(t) = y(t; \tau_i, y^h_\Delta(\tau_i), u^{(i)}, v^{(i)})$, $\tau_i \le t \le \tau_{i+1}$. Simultaneously, the control $v(t) = v_i$ for $t \in [\tau_i, \tau_{i+1})$ is fed onto the input of (1). As a result of the action of this control and of an unknown control $u_{\tau_i,\tau_{i+1}}(\cdot)$, some (h, Δ, y)-motion $\{x^h_{\Delta,y}(\cdot; \tau_i, x^h_{\Delta,y}(\tau_i), u_{\tau_i,\tau_{i+1}}(\cdot), v_i)\}_{\tau_i,\tau_{i+1}}$ is realized in the interval $[\tau_i, \tau_{i+1}]$. This procedure stops at the moment ϑ. Analogously to [4], we establish that for every $\varepsilon \in (0, \varepsilon_*)$ one can indicate (explicitly) numbers $h_1 = h_1(\varepsilon) > 0$ and $\delta_1 = \delta_1(\varepsilon) > 0$ such that the inequality

$$\text{dist}(x^h_{\Delta,y}(\vartheta), N) \ge \varepsilon \quad \forall x^h_{\Delta,y}(\cdot) \in X_T(t_0, x_0, \mathcal{V}, \Delta, y)$$

is fulfilled if $h \leq h_1$, $\delta = \delta(\Delta) \leq \delta_1$. Toward this aim, the Lyapunov functional $\varepsilon^{(1)}(t) = |x^h_{\Delta,y}(t) - y^h_\Delta(t)|^2_H$ is estimated in the interval T, and its "small" variation is established. The theorem is proved.

4. Algorithms for solving Problem 2

Let us describe algorithms for solving Problem 2. Recall that we consider inclusion (1) with unknown $v(\cdot)$ and $u(\cdot) = 0$. In this section, we assume that U is a Hilbert space with the scalar product $(\cdot,\cdot)_U$. Constructions described below are based on the approach developed in [2, 6, 7].

Let a partition Δ_h of the form (6) and a function $\alpha(h) : (0,1) \to R^+$ be fixed. Let the following condition be fulfilled:

$$\alpha(h) \to 0, \quad \delta(h) \to 0, \quad \delta(h)\alpha^{-2}(h) \to 0, \tag{10}$$

$$h^2\delta^{-1}(h)\alpha^{-1}(h) \to 0 \text{ as } h \to 0.$$

A positional strategy $\mathcal{U} : T \times H \times H \to V$ is defined by the rule

$$\mathcal{U}(p_i) = \alpha^{-1}(h)C^*(\xi^h_i - w(t)), \tag{11}$$

where $p_i = (\tau_i, \xi^h_i, w(t))$ is the position for $t \in \delta_i = [\tau_i, \tau_{i+1})$, $\tau_i = \tau_{i,h}$, $w(\cdot)$ is the solution of inclusion (5) with $u(\cdot)$ defined by (7), (11).

THEOREM 2 *Let condition (10) be fulfilled. Then the positional strategy $\mathcal{U}$ of the form (7), (11) solves Problem 2.*

The proof of this theorem is performed according to the scheme from [2].

References

[1] Yu.S. Osipov. On a positional control in parabolic systems. *Prikl. Math. Mechan.*, 41(2):23-27, 1977 (in Russian).

[2] V.I. Maksimov, L. Pandolfi. On reconstruction of unbounded controls in nonlinear dynamical systems. *Prikl. Math. Mechan.*, 65(3):385-390, 2001 (in Russian).

[3] H. Brézis. Propriétés régularisantes de certains semi-groupes non linéaires, *Israel J. Math.*, 9(4):513-534, 1971.

[4] N.N. Krasovskii, A. I. Subbotin. *Game-theoretical control problems.* Springer, Berlin. 1988

[5] N.N. Krasovskii. *Controlling of a dynamical system.* Nauka, Moscow, 1985 (in Russian).

[6] Yu.S. Osipov, A. V. Kryazhimskii. *Inverse problems for ordinary differential equations: dynamical solutions.* Gordon and Breach, London, 1995.

[7] V.I. Maksimov. *Dynamical Inverse Problems of Distributed Systems.* VSP, Boston, 2002.

[8] V. Barbu. *Optimal control of variational inequalities.* Research Notes in Mathematics, Pitman Advanced Publishing Program, London, 1984.

is fulfilled [illegible] $\leq \delta$ [illegible]. Toward [illegible] the [illegible] functional [illegible] is estimated in the [illegible] and [illegible] satisfied. The [illegible].

[illegible] Algorithms for solving Problem 2

[illegible] algorithms for solving Problem 2. Recall that we consider [illegible]. In this section, we assume [illegible] product [illegible].

[illegible]

[illegible] the [illegible] and [illegible]

[illegible] possible [illegible]

[illegible] is per[illegible] from [illegible]

[illegible]

[1] [illegible]

[2] V. [illegible]

[3] [illegible]

[4] M. [illegible], Springer, Berlin, 1988

[5] M. N. [illegible] (in Russian)

[6] Yu. S. [illegible] differential equations [illegible]

[7] V. I. M[illegible] 2007

[8] V. Z[illegible] Noises [illegible] Publishers [illegible]

NONSEARCH PARADIGM FOR LARGE-SCALE PARAMETER-IDENTIFICATION PROBLEMS IN DYNAMICAL SYSTEMS RELATED TO ONCOGENIC HYPERPLASIA

E. Mamontov [1] and A. Koptioug[2]

[1]*Department of Physics, Gothenburg University, Kemivägen 9, SE-412 96 Gothenburg, Sweden, yem@physics.gu.se,* [2]*Department of Electronics, Institute of Information Technology and Media, Mid Sweden University, Akademigatan 1, SE-831 25 Östersund, Sweden, Andrei.Koptioug@miun.se*

Abstract In many engineering and biomedical problems there is a need to identify parameters of the systems from experimental data. A typical example is the biochemical-kinetics systems describing oncogenic hyperplasia where the dynamical model is nonlinear and the number of the parameters to be identified can reach a few hundreds. Solving these large-scale identification problems by the local- or global-search methods can not be practical because of the complexity and prohibitive computing time. These difficulties can be overcome by application of the non-search techniques which are much less computation- demanding. The present work proposes key components of the corresponding mathematical formulation of the nonsearch paradigm. This new framework for the nonlinear large-scale parameter identification specifies and further develops the ideas of the well-known approach of A. Krasovskii. The issues are illustrated with a concise analytical example. The new results and a few directions for future research are summarized in a dedicated section.

keywords: nonlinear dynamic system, non-search parameter, identification, Krasovskii method, biochemical kinetics

1. Introduction

In many engineering problems there is a need to identify parameters of systems from experimental data. This is equally true for both simple and complex systems (the latter are common in biomedical technology). A typical example is the biochemical-kinetics systems describing oncogenic hyperplasia (e.g., [1] - [3]), the first and therefore inevitable stage in development of any solid tumor. Oncogenic hyperplasia is a complex process related to many different types of molecules interacting with hyperplastic cells. In particular, these

Please use the following format when citing this chapter:

Mamontov, E., and Koptioug, A., 2006, in IFIP International Federation for Information Processing, Volume 202, Systems, Control, Modeling and Optimization, eds. Ceragioli, F., Dontchev, A.,Furuta, H., Marti, K., Pandolfi, L., (Boston: Springer), pp. 269-278.

interactions include the autocrine mechanism where the transforming-growth-factor-α molecules bind to the epidermal-growth-factor receptors at the cell surfaces (e.g.,[4]).

Oncogenic hyperplasia is also implicated in many other proliferative diseases (e.g., vascular, gastrointestinal, endocrine). The existing detailed models for the above phenomenon are ordinary differential equation (ODE) systems with a hundred or more equations (e.g., see the 104 equations in [4], Tables 1-3). These systems usually include hundreds of parameters to be identified. In other words, if p is the vector of the parameters in the ODE system, then in the case of oncogenic hyperplasia

dimension m of vector p is on the order of a few hundreds. (1)

This feature is also common in many in biomedical and engineering systems. Under condition (1), the parameter identification (sometimes applied to biochemical kinetics [5], [6]) based on the local or global search can not be practical because of the complexity and prohibitive computing time. The problem can be resolved by application of the techniques much less demanding in terms of the required computations. One of them is the nonsearch parameter-identification strategy.

The purpose of the present work is to present the nonsearch paradigm (i.e. the basic aspects of the mathematical formulation) for dynamical systems that have property (1) and which are related to oncogenic hyperplasia. The latter inevitably leads to nonlinear systems (e.g., [4]). Subsequently, the above paradigm should correspondingly extend the nonsearch settings well known (e.g., [7]) for linear and simple systems.

2. Model

The present work considers the following nonlinear ODE system

$$dx/dt = f(t, x, p) \tag{2}$$

where $t \in \mathbf{R} = (-\infty, \infty)$ is the time, $x \in \mathbf{R}^n$ $(n \geq 1)$ is the state variable of the system under consideration, $p \in \mathbf{R}^m$ $(m \geq 1)$ is the vector of the system parameters discussed in Section 1, and f is the n -vector function sufficiently smooth on $\mathbf{R}^{n+m+1}$. As a rule, ODE (2) at every *fixed* p possesses the following two properties:

- it is asymptotically stable in the Lyapunov sense in the large;
- moreover, it has the unique generally nonstationary solution, say, $\varphi_s(t, p)$, usually called the *steady state* solution, which is *uniformly bounded* for *all* $t \in \mathbf{R}$.

The above stability implies that, if $\varphi(t, p, t_0, x_0)$ is the solution of ODE (2) with initial condition

$$x \mid_{t=t_0} = x_0 \tag{3}$$

where $t_0 \in \mathbf{R}$ and $x_0 \in \mathbf{R}^n$, then

$$\varphi_s(t, p) = \lim_{t_0 \to -\infty} \varphi(t, p, t_0, x_0) \, , \ \ p \in \mathbf{R}^m, \ \ x_0 \in \mathbf{R}^n. \tag{4}$$

A discussion and references on ODE (2) with the above features can be found in [8]. The generalization of $\varphi_s(\,\cdot, p)$ for the diffusion stochastic processes is a *nonstationary invariant* DSP (paper [9] summarizes the results on these invariant processes).

Remark 1. In many cases, solution $\varphi_s(\cdot, p)$ can be obtained by means of the limit relation (4). Alternatively, one can determine $\varphi_s(t, p)$ as the unique solution of certain *finite* (i.e. nondifferential) equation (see [8]). The simplest approximation for it is

$$f(t, x, p) = 0 \, , t \in \mathbf{R} \, , \tag{5}$$

which is the quasi-stationary version of ODE (2).

The next section proposes the necessary condition for the parameter identification to be meaningful.

3. Identification and uniqiuiness of parameters.

Identification of the parameter vector p in ODE (2) is commonly implemented in the following way. One obtains the measurement data $\overline{x}_s$ for the steady-state solution $\varphi_s(t,\cdot)$ of (2), and then determines the *actual* value p_s of p (i.e. the value corresponding to the measured time dependence) by solving equation

$$\overline{x}_s = \varphi_s(t, p_s) \, , \ t \in \mathbf{R} \, , \tag{6}$$

for p_s.

Remark 2. The above extraction of p_s from the *steady-state* solution makes p_s independent of initial value x_0 in (3).

System (6) of n equations is assumed to be *uniquely* solvable for m entries of vector p_s, no matter what the values n and m are. Theory of nonlinear-system identification not always can deal with the question whether this is possible or not. To fill the gap, one can prove the following proposition.

Proposition 1. If function φ_s is sufficiently smooth and equation (6) is solvable for p_s, then the equation has the only solution, if for any $p_I, p_{II} \in \mathbf{R}^m$

such that $p_I \neq p_{II}$, vectors $\int\limits_0^1 \frac{\partial \varphi_s(t,p_I+k(p_{II}-p_I))}{\partial p_i} \cdot dk, i=1,...,m$, as functions of t, are linearly independent for ***all*** $t \in \mathbf{R}$ (scalars p_i are the i-th entries of vector p).

The proof is based on the notion of a linear independence of vectors and representation

$$\varphi_s(t,p_{II}) - \varphi_s(t,p_I) = (p_{II}-p_I) \int\limits_0^1 \{\partial \varphi_s[t,p_I + k(p_{II}-p_I)]/\partial p\}\, dk,$$

which results from the well-known formula (e.g. [10]). Criteria for the *uniqueness in parameters* similar to Proposition 1 are of a fundamental importance because it is impossible to consider the ODE-related identification in a meaningful way without a criterion of this kind.

4. Nonsearch approach to the parameter identification

In the *nonlinear* nonsearch approach (e.g. [11]), one obtains the measurement data $\overline{x}$ on the solution of initial-value problem (2), (3) where p is equal to its actual value p_s, i.e.

$$\overline{x} = \varphi(t, p_s, t_0, x_0) \ , \tag{7}$$

$$d\overline{x}/dt = f(t, \overline{x}, p_s) \ . \tag{8}$$

To identify the parameter vector, i.e. to determine p_s, one implements the nonsearch procedure below.

Remark 3. The nonsearch procedure:

- allows p in ODE (2) to depend on time t;
- describes the t -dependent p with a dedicated ODE, say,

$$dp/dt = g(t,x,p) \tag{9}$$

  with initial condition

$$p\,|_{t=t_0} = p_0 \ , \tag{10}$$

  where g is the m -vector function sufficiently smooth in $\mathbf{R}^{n+m+1}$ and $p_0 \in \mathbf{R}^m$;

- and ensures certain stability of ODE system (2), (9) which provides the behaviors

$$\lim_{t\to\infty}(x-\overline{x}) = 0 \ , \ \lim_{t\to\infty} p = p_s, \tag{11}$$

  of the solution (x,p) of a problem (2), (9), (3), 10) for any initial vector (x_0, p_0).

The above recipe yields p_s by the limit relation (11) thereby avoiding using any search technique.

This in principle resolves the problem associated with the *high- number- of-parameters feature* (1) typical in oncogenic hyperplasia (see Section 1).
In the above recipe, the stability is the key property to be assured. This can be done with the help of the theorem below.

Theorem 1. Let the following assumptions hold.

(A) Hypothesis of Proposition 1 is valid.

(B) ODE (2) is asymptotically stable in the Lyapunov sense in the large at any t-dependent sufficiently smooth p, and det $[\partial f(t,x,p)/\partial x] \neq 0$.

(C) State x relaxes faster than parameter p and there exists the integral manifold, say, M for the \slow" variable p.

(D) ODE (9) is asymptotically stable in the Lyapunov sense in manifold M, and vector $\lim |_{t-t_0 \to \infty} p$ exists and is finite.

Then the nonsearch-identification recipe in Remark 3 is applicable.

The proof is not complicated. We only mention that the second equality in (11) follows from assumptions (A) and (D), and obvious relation $\lim_{t\to\infty} \overline{x} = \overline{x}_s$, which stems from assumption (B) and equality (8).

Remark 4. In assumption (C) of Theorem 1, the manifold corresponds to the function $\varphi_s(t,p)$ in Section 1. It can be obtained as the solution of the finite equation resulting from the two-groups-of-components version of the method in reference [8], p.459 (cf., Remark 1). The simplest approximation for this equation is (5).

In the literature, there are no practical or even constructive criteria to verify or provide asymptotic stability in the large of general nonlinear ODEs. If function g in (9) is nonlinear, it is especially difficult to design this function such that assumption (D) of Theorem 1 is assured. This is possible only in some particular cases.

In the general case, certain features *critical* to assumption (D) can be provided by the ideas of A. Krasovskii [11]. This topic is discussed in the next section.

5. Specification of the Krasovskii method and new issues related to it

The main relation in the method of Krasovskii is {[11], eqn. (3), (5), (6)}

$$dp/dt = g(t,x,p) = -dH/dp \qquad (12)$$

where H {[11], pp. 1852, 1856} is the strictly convex function of $x - \overline{x}$ and dH/dp (cf., [11], eqn. (11)) is the derivative along trajectories in the \slow"-variable integral manifold M. Note, however, that the manifold idea in the Krasovskii paradigm is recognized but not formally developed (for instance, the corresponding derivation in reference [11], eqn. (12), (13), is not valid when p depends on t). Following the above settings, for H in (12) we consider the expression

$$H(t,x,p) = \frac{1}{2}\left[(x-\overline{x})^T B(t,\overline{x},p_s)(x-\overline{x}) + (p-p_s)^T A(t,\overline{x},p_s)(p-p_s)\right] \tag{13}$$

where both matrices $A(t,\overline{x},p_s)$ and $B(t,\overline{x},p_s)$ are symmetric and positive definite. In contrast to (13), Krasovskii's expression for H does not include either the second term in the brackets or the value p_s, which is unknown (cf., assumption (D) of Theorem 1). Vector p_s in (13) is replaced below with a self-consistent predictor which is in a certain sense more accurate than p.

The derivative in (12) along trajectories in the \slow"-variable integral manifold M is evaluated as

$$dH(t,x,p)/dp = (\partial x/\partial p)^T B(t,\overline{x},p_s)(x-\overline{x}) + A(t,\overline{x},p_s)(p-p_s)$$

where the term $\partial x/\partial p$ on the right-hand side is estimated according to Remark 4 and equation (5). This results (see (12)) in

$$dp/dt = -\{[C(t,x,p)]^T B(t,\overline{x},p_s)(x-\overline{x}) + A(t,\overline{x},p_s)(p-p_s)\}, \tag{14}$$

where

$$C(t,x,p) = -[\partial f(t,x,p)/\partial x]^{-1}\partial f(t,x,p)/\partial p\ . \tag{15}$$

Equation (14) (see also (15)) generalizes the well-known Krasovskii equation {[11], eqn. (16)}.

Reformulating (14) to answer the question if assumption (D) of Theorem 1 holds, one applies:

- the first equality in (11), which stems from assumption (B) of Theorem 1 and
- the approximate expression $x - \overline{x} = C(t,\overline{x},p_s)(p-p_s)$, which stems from Remark 4, equation (5), and the first equality in (11).

The resulting form of (14) is

$$dp/dt = -\{A(t,\overline{x},p_s) + [C(t,\overline{x},p_s)]^T B(t,\overline{x},p_s)C(t,\overline{x},p_s)\}(p-p_s) \tag{16}$$

This is the specific version of the parameter ODE (9).

Remark 5. The last term on the right-hand side of (14) is missing in the Krasovskii equation {[11], eqn. (16)}. If it is not included in equation (16), this equation cannot be asymptotically stable when $m > n$ (since, in this case, matrix $[C(t, \overline{x}, p_s)]^T B(t, \overline{x}, p_s) C(t, \overline{x}, p_s)$ is singular, even if matrix $C(t, \overline{x}, p_s)$ is of the full rank, i.e. its rank is n).

Equation (16) is a linear ODE with time-dependent coefficients. Subsequently, its stability can be analyzed with the corresponding methods of the stability theory. Matrices $A(t, \overline{x}, p_s)$ and $B(t, \overline{x}, p_s)$ are to be determined to assure, *firstly*, the stability and *secondly*, the fact that vector p in ODE (16) relaxes slower than vector x in ODE (2) (which is required in assumption (C) of Theorem 1). We retain these topics for future research.

6. Time average of parameters as a self-consistent predictor of the actual values

Actual value p_s is obtained (see (11)) as the time limit of p from the solution of initial-value problem (2), (14), (3), (10). However, in the process of solving, p_s is unknown and hence can not be employed in ODE (14). For this reason, p_s in (14) is replaced with

$$q = \frac{1}{t - t_0} \int_{t_0}^{t} p \cdot ds \tag{17}$$

that results in

$$dp/dt = -\{A(t, \overline{x}, q) + [C(t, \overline{x}, q)]^T B(t, \overline{x}, q) C(t, \overline{x}, q)\}(p - q) \ . \tag{18}$$

Vector q is a *self- consistent* predictor for p_s. Indeed, in contrast to the well-known additional assumptions inherent in the parameter-*shift* technique (e.g. [7], Remark 8.4.1(ii) on p. 564), representation (17) involves only the time dependence of p without extra assumptions. Predictor q is also *more preferable* than any instantaneous value of p because of the following.

Within the treatment based on the Theorem 1, vector $\overline{x}$ in (18) is the measured value of the \fast"-variable x in ODE system (2), (18) whereas vector p in (18) is the \slow"variable in the system (c.f., assumption (C) of the Theorem 1). In view of the coupling of (18) with the \fast"-variable x, the \slow"variable p is generally prone to rapid but comparatively short-living changes. Subsequently, the advantage of the \time-average"form of (17) of predictor q for p_s is that it \filters out"fast variations in the time dependence of instantaneous value p.

Also note, that q defined by (17) is the solution of ODE

$$dq/dt = (p - q)/(t - t_0) \tag{19}$$

with initial condition

$$q\mid_{t=t_0}=p_0 \ . \tag{20}$$

This description is equivalent to (17) and, thus, can be used instead of it.

Initial-value problem (18), (19), (10), (20) is the model developed in the present work to determine actual value $p_s = \lim_{t\to\infty} p$ and to deliver it to the initial- value problem (2), (3). This model is intended to serve as a *computationally efficient alternative* to the search-based parameter identification. It is probably the *only practice-relevant option* in the high-number-of-parameters case (1). Also note that ODE (18) is the more specific form of ODE (9) of Theorem 1 derived within a generalization of the ideas of [11].

7. Example

The specific model below exemplifies the analyses in this section and Section 5. Let the identical molecules of concentration x and identical cells of concentration u be suspended in a dispersion medium (e.g., the blood plasma). Assume that the cells release certain molecules, the time-dependence $u(t)$ of u is known, and the molecules are disintegrated in the medium. Then these processes can be described with ODE (2), where $n = 1$ ($x_1 \equiv x$), $m = 2$, and $f(t,x,p) = -p_1x + p_2u(t)$, where p_1 and p_2 are the parameters. The problem is a determination of these parameters (generally dependent on the properties of the dispersion medium) from the measured data $\overline{x} = \overline{x}(t)$ (cf., (8)) on x.

According to the proposed treatment, the general model is system (2), (18), (19). In the example under consideration, the specific form of (2) is mentioned above. The form of (19) is fixed, independent of specific phenomena. The only issue is the specific form of (18). One can show (with the help of (15)) that this form is

$$\begin{pmatrix} {p_1}' \\ {p_2}' \end{pmatrix} = \left[\begin{pmatrix} A_1 & 0 \\ 0 & A_2 \end{pmatrix} + \frac{B}{p_1^2}\begin{pmatrix} (\overline{x}(t))^2 & -u(t)\overline{x}(t) \\ -u(t)\overline{x}(t) & (u(t))^2 \end{pmatrix}\right]\begin{pmatrix} p_1 - q_1 \\ p_2 - q_2 \end{pmatrix}$$

where scalars $A_1 > 0$ and $A_2 > 0$ are the diagonal entries of symmetric and positive definite matrix $A(t, \overline{x}, p_s)$, which is assumed constant, and scalar $B > 0$ is the 1x1 symmetric and positive definite matrix $B(t, \overline{x}, p_s)$. The mentioned scalars can be chosen as discussed in the text below Remark 5, in particular, involving the well-known results of stability theory (e.g. [12]]. However, to begin with, one can estimate them in line with the issues in ([7], Chapter 4).

Also note that the manifold in assertion (C) of Theorem 1 exists and is described as $x = \int_{-\infty}^{t}\left\{exp\left[-\int_s^t (p_1(\nu))^{-1}d\nu\right]p_2(s)u(s)\right\}ds$. The condition when this manifold is the one for "slow" vector (p_1, p_2) can be obtained by means of results of [8].

The above example illustrates the proposed innovative method, based upon the ideas of Theorem 1. The new approach presents nontrivial extension of the well+known techniques available in the literature (e.g., [7], Chapter 4).

8. Concluding remarks and directions for further research

The main new results developed in the present work include:

- *Proposition 1* (on the uniqueness in the parameters);
- *Theorem 1* (on the qualitative features of the model for the nonsearch identification of parameters of nonlinear systems);
- *three new issues* on a specification of the Krasovskii method:
  - the multi-aspect role of the \slow"-variable manifold,
  - the dissipative term in the ODE for the parameter vector,
  - the self-consistent predictor for the actual values of parameters;
- a generic *model for the nonsearch identification of nonlinear-system parameters (initial-value problem (2), (18)*, (19), (3), (10), (20)).

A few directions for future research are: (1) application of the present results to numerical parameter identification in biochemical kinetics of oncogenic hyperplasia in can-cer and other proliferative diseases, (2) generalization to Itô's stochastic nonlinear differential equations taking into account the influence of random noises (this can make use of the analytical-numerical methods developed for high-dimensional nonlinear stochastic systems [13]), and (3) extension of the stability-based nonsearch optimization/identifica-tion to new areas such as self-navigated truly autonomous (fully GPS-independent) robots capable of delivering light load, or intercepting mobile targets (e.g., [14]). Future rigorous analysis of the mathematical formulation proposed in this work will contribute to specification of the details which can facilitate practical application of the present approach.

References

[1] A. V. Koptioug, E. Mamontov. Toward prevention of hyperplasia in oncogeny and other proliferative diseases: The role of the cell genotoxicity in the model-based strategies. Abstracts: *7th Ann. Conf.: Functional Genomics -From Birth to Death.* Göteborg, Sweden, August 19-20, 2004.

[2] A. V. Koptioug, E. Mamontov, Z. Taib, M. Willander. The phase-transition morphogenic model for oncogeny as a genotoxic homeostatic dysfunction: Interdependence of modeling, advanced measurements, and numerical simulation. Abstracts: *ICSB2004, 5th Int. Conf. Systems Biology.* Heidelberg, Germany, October 9-13, 2004.

[3] K. Psiuk-Maksymowicz, E. Mamontov. The time-slice method for rapid solving the Cauchy problem for nonlinear reaction-diffusion equations in the competition of home-orhesis with genotoxically activated oncogenic hyperplasia. Abstracts: *The European Conference on Mathematical and Theoretical Biology.* Dresden, Germany, July 18-22, 2005.

[4] J. H. Miller, F. Zheng. Large-scale simulations of cellular signaling processes. *Parallel Computing* 30: 1137-1149, 2004.

[5] J. Yen, J. C. Liao, B. Lee, D. Randolph. A hybrid approach to modeling metabolic systems using a genetic algorithm and simplex method. *IEEE Trans. Systems, Man, and Cybernetics* 28(2): 173-191, 1998.

[6] C.-Y. F. Huang, J. E. Ferrel, Jr. Ultrasensitivity in the mitogen-activated protein kinase cascade. *Proc. Natl. Acad. Sci. USA* 93(September): 10078-10083, 1996.

[7] P. A. Ioannou, J. Sun. *Robust Adaptive Control.* Prentice- Hall, Upper Saddle River, NJ, USA, 1996.

[8] Y. V. Mamontov, M. Willander. Asymptotic method of finite equation for bounded solutions of nonlinear smooth ODEs. *Mathematica Japonica* 46(3): 451-461, 1997.

[9] E. Mamontov. Nonstationary invariant distributions and the hydrodynamic- style generalization of the Kolmogorov-forward/Fokker-Planck equation. *Appl. Math. Lett.* 18(9): 976-982, 2005.

[10] J. M. Ortega, W. C. Rheinboldt. *Iterative Solution of Nonlinear Equations in Several Variables.* Academic Press, New York, 1970.

[11] A. A. Krasovskii. Optimal algorithms in identification problem with an adaptive model. *Automation and Remote Control* (May 10): 1851-1857, 1977.

[12] N. Rouche, P. Habets, and M. Laloy. *Stability Theory by Liapunov's Direct Method.* Springer–Verlag, 1977.

[13] Y. V. Mamontov, M. Willander. *High-Dimensional Nonlinear Diffusion Stochastic Processes. Modeling for Engineering Applications.* World Scientific, Singapore, 2001.

[14] E. Mamontov, A. Koptioug, M. Mångård, K. Marti. *Asymptotic trajectory matching in self-navigation of autonomous manless interceptors: Nonsearch method and a formulation of the functional optimization of the stability of random systems.* Abstracts: *5th MATHMOD Vienna Conf., Special Session "Stochastic Optimization Methods", Chair: K. Marti.* Austria, Vienna, Vienna Univ. of Technol., 7-9 February, 2006. http://www.mathmod.at/.

DETECTION OF A RIGID INCLUSION IN AN ELASTIC BODY: UNIQUENESS AND STABILITY

A. Morassi[1] and E. Rosset[2]
[1] *Università degli Studi di Udine, Dipartimento di Georisorse e Territorio, Udine, Italy, antonino.morassi@uniud.it* [2] *Università degli Studi di Trieste, Dipartimento di Matematica e Informatica, Trieste, Italy, rossedi@univ.trieste.it*

Abstract We state uniqueness and stability results for the inverse problem of determining a rigid inclusion inside an isotropic elastic body Ω, from a single measurement of traction and displacement taken on the boundary of Ω.

Keywords: inverse problems, linearized elasticity, rigid inclusion.

1. Introduction

In this paper we consider the inverse problem of identifying a rigid inclusion inside an elastic body from boundary measurements of traction and displacement. This kind of problems arises, for instance, in non-destructive testing for damage assessment of mechanical specimens, which are possible defective due to the presence of interior rigid inclusions. More precisely, let the elastic body be represented by a bounded domain Ω in $\mathbf{R}^2$, or $\mathbf{R}^3$, inside which a possible unknown rigid inclusion D is present. Our aim is to identify D by applying a traction field φ at the boundary $\partial\Omega$ and by measuring the induced displacement field on a portion $\Sigma \subset \partial\Omega$.

Working within the framework of the linearized elasticity, where $\mathcal{C}$ denotes the known elasticity tensor of the material, the displacement field u satisfies the following boundary value problem

$$\begin{cases} \mathrm{div}(\mathcal{C}\nabla u) = 0, & \text{in } \Omega \setminus \overline{D}, \\ (\mathcal{C}\nabla u)\nu = \varphi, & \text{on } \partial\Omega, \\ u_{|\partial D} \in \mathcal{R}, \end{cases} \tag{1}$$

coupled with the *global equilibrium condition*

$$\int_{\partial D} (\mathcal{C}\nabla u)\nu \cdot r = 0, \quad \text{for every } r \in \mathcal{R}, \tag{2}$$

Please use the following format when citing this chapter:

Morassi, A., and Rosset, E., 2006, in IFIP International Federation for Information Processing, Volume 202, Systems, Control, Modeling and Optimization, eds. Ceragioli, F., Dontchev, A., Furuta, H., Marti, K., Pandolfi, L., (Boston: Springer), pp. 279-284.

where $\mathcal{R}$ denotes the linear space of the infinitesimal rigid displacements $r(x) = c + Wx$, where c is any constant n-vector and W is any constant skew $n \times n$ matrix. We shall assume $\mathcal{C}$ strongly convex and of Lamé type. Problem (1)-(2) admits a solution $u \in H^1(\Omega \setminus \overline{D})$, which is unique up to an infinitesimal rigid displacement. In order to specify a unique solution, we shall assume in the sequel the following normalization condition

$$u = 0 \quad \text{on } \partial D. \tag{3}$$

Therefore, the inverse problem consists in determining the unknown rigid inclusion D, appearing in problem (1)–(3), from a single pair of Cauchy data $\{u, (\mathcal{C}\nabla u)\nu\}$ on $\partial\Omega$.

The indeterminacy of the displacement field u and the consequent arbitrariness of the normalization (3) which we have chosen, lead to the following formulation of the uniqueness and stability issues.

Uniqueness issue.

Given two solutions u_i to (1)–(3) when $D = D_i$, $i = 1, 2$, satisfying

$$(\mathcal{C}\nabla u_i)\nu = \varphi, \quad \text{on } \partial\Omega, \tag{4}$$

$$u_1 - u_2{}_{|_\Sigma} \in \mathcal{R}, \tag{5}$$

does $D_1 = D_2$ hold?

Stability issue.

Given two solutions u_i to (1)–(3) when $D = D_i$, $i = 1, 2$, satisfying

$$(\mathcal{C}\nabla u_i)\nu = \varphi, \quad \text{on } \partial\Omega, \tag{6}$$

$$\min_{r \in \mathcal{R}} \|(u_1 - u_2) - r\|_{L^2(\Sigma)} < \epsilon, \quad \textit{for some } \epsilon > 0, \tag{7}$$

to evaluate the rate at which the Hausdorff distance between D_1 and D_2 tends to zero as ϵ tends to zero.

In this note we present the uniqueness and stability results obtained in [4], to which we refer for the proofs.

In particular, we have obtained uniqueness under the assumption that ∂D is of C^1 class, see Theorem 1. The proof of uniqueness is mainly based on the weak unique continuation principle for solutions to the Lamé system (first established by Weck [5]), the uniqueness for the corresponding Cauchy problem, (see, for instance, [2] and [3]), and geometrical arguments related to the structure of the linear space $\mathcal{R}$ which involve different techniques according to the space dimension.

This inverse problem is severely ill-posed from the point of view of stability and, therefore, only a weak rate of convergence, under some a priori information on the unknown boundary ∂D, can be expected. In [4] a constructive stability estimate of log-log type, assuming $C^{1,\alpha}$ regularity of ∂D, $0 < \alpha \leq 1$, has been proved, see Theorem 2 for a precise statement.

The key ingredients to prove stability are quantitative versions of the unique continuation principle, precisely: a three spheres inequality for solutions to the Lamé system, which was obtained in [1]; stability estimates for solutions to the Cauchy problem, obtained in [3]; a stability estimate of continuation from the interior for a mixed boundary value problem obtained in [4]. The complications of geometrical character arising in the proof of uniqueness, due to the general form of the condition (7), become significantly harder in the stability context, see the geometrical Lemma 7.1 in [4].

2. A priori information

In the sequel we use the following notation.

Let Ω be a bounded domain in $\mathbf{R}^n$. Given k, α, with $k \in \mathbf{N}$, $0 < \alpha \leq 1$, we say that a portion S of $\partial\Omega$ is of *class* $C^{k,\alpha}$ *with constants* ρ_0, $M_0 > 0$, if, for any $P \in S$, there exists a rigid transformation of coordinates under which we have $P = 0$ and

$$\Omega \cap B_{\rho_0}(0) = \{x = (x', x_n) \in B_{\rho_0}(0) \quad | \quad x_n > \psi(x')\},$$

where $x' = (x'_1, ..., x'_{n-1}) \in \mathbf{R}^{n-1}$ and ψ is a $C^{k,\alpha}$ function on $B'_{\rho_0}(0) = B_{\rho_0}(0) \cap \{x_n = 0\} \subset \mathbf{R}^{n-1}$ satisfying

$$\psi(0) = 0,$$

$$\nabla\psi(0) = 0, \quad \text{when } k \geq 1,$$

$$\|\psi\|_{C^{k,\alpha}(B'_{\rho_0}(0))} \leq M_0.$$

When $k = 0$, $\alpha = 1$, we also say that S is of *Lipschitz class with constants* ρ_0, M_0.

We denote by $\mathbf{M}^{m\times n}$ the space of $m \times n$ real valued matrices and by $\mathcal{L}(X, Y)$ the space of bounded linear operators between Banach spaces X and Y. When $m = n$, we shall also denote $\mathbf{M}^n = \mathbf{M}^{n\times n}$.

For every $n \times n$ matrix A and for every $\mathcal{C} \in \mathcal{L}(\mathbf{M}^n, \mathbf{M}^n)$, we use the following notation:

$$(\mathcal{C}A)_{ij} = \sum_{k,l=1}^{n} C_{ijkl}A_{kl}. \tag{8}$$

Let us introduce the linear space of the infinitesimal rigid displacements

$$\mathcal{R} = \left\{r(x) = c + Wx,\ c \in \mathbf{R}^n,\ W \in \mathbf{M}^n,\ W + W^T = 0\right\}, \tag{9}$$

where x is the vector position of a generic point in $\mathbf{R}^n$.

Let us list here the a priori information needed for stability.

i) A priori information on the domain.

We assume that Ω is a bounded domain in $\mathbf{R}^n$ such that

$$|\Omega| \leq M_1, \tag{10}$$

for some $M_1 > 0$, where $|\Omega|$ denotes the Lebesgue measure of Ω.

For the sake of simplicity, we assume in the sequel that

$$\partial\Omega \text{ is connected,} \tag{11}$$

but we emphasize that all the results stated in the next Section continue to hold in the general case when (11) is removed.

We assume that Ω contains an open, connected, rigid inclusion D such that

$$\Omega \setminus \overline{D} \text{ is connected,} \tag{12}$$

$$\partial D \text{ is connected,} \tag{13}$$

and

$$\text{dist}(D, \partial\Omega) \geq \rho_0. \tag{14}$$

Moreover, we assume that we can select an open portion Σ within $\partial\Omega$ (representing the portion of the boundary where measurements are taken) such that for some $P_0 \in \Sigma$

$$\partial\Omega \cap B_{\rho_0}(P_0) \subset \Sigma. \tag{15}$$

Regarding the regularity of the boundaries, given α, M_0, $0 < \alpha \leq 1$, $M_0 > 0$, we assume that

$$\partial\Omega \text{ is of } class\ C^{1,\alpha}\ with\ constants\ \rho_0, M_0, \tag{16}$$

$$\partial D \text{ is of } class\ C^{1,\alpha}\ with\ constants\ \rho_0, M_0, \tag{17}$$

and, moreover, that

$$\Sigma \text{ is of } class\ C^{2,\alpha}\ with\ constants\ \rho_0, M_0. \tag{18}$$

ii) Assumptions about the boundary data.

On the Neumann data φ appearing in problem (1) we assume that

$$\varphi \in H^{-\frac{1}{2}}(\partial\Omega, \mathbf{R}^n), \quad \varphi \not\equiv 0, \tag{19}$$

the compatibility condition

$$\int_{\partial\Omega} \varphi \cdot r = 0, \quad \text{for every } r \in \mathcal{R}, \tag{20}$$

and that, for a given constant $F > 0$,

$$\frac{\|\varphi\|_{H^{-\frac{1}{2}}(\partial\Omega,\mathbf{R}^n)}}{\|\varphi\|_{H^{-1}(\partial\Omega,\mathbf{R}^n)}} \leq F. \tag{21}$$

iii) Assumptions about the elasticity tensor.

We assume that the elastic material is *isotropic*, that is the elasticity tensor field $\mathcal{C} = \mathcal{C}(x) \in \mathcal{L}(\mathbf{M}^n, \mathbf{M}^n)$ has components $\mathcal{C}_{ijkl}$ given by

$$\mathcal{C}_{ijkl}(x) = \lambda(x)\delta_{ij}\delta_{kl} + \mu(x)(\delta_{ki}\delta_{lj} + \delta_{li}\delta_{kj}), \quad \text{for every } x \in \overline{\Omega}, \tag{22}$$

where $\lambda = \lambda(x)$ and $\mu = \mu(x)$ are the *Lamé moduli*. Moreover we assume that the *Lamé moduli* satisfy the $C^{1,1}$ regularity condition

$$\|\mu\|_{C^{1,1}(\overline{\Omega})} + \|\lambda\|_{C^{1,1}(\overline{\Omega})} \leq M. \tag{23}$$

and the *strong convexity* condition

$$\mu(x) \geq \alpha_0, \quad 2\mu(x) + n\lambda(x) \geq \beta_0, \quad \text{for every } x \in \overline{\Omega}, \tag{24}$$

where M, α_0, β_0 are given positive constants.

We shall refer to the set of constants α, ρ_0, M_0, M_1, F, α_0, β_0, M as to the *a priori data.*

By standard variational arguments it is easy to see that problem (1)–(3) admits a unique solution $u \in H^1(\Omega \setminus \overline{D}, \mathbf{R}^n)$ such that

$$\|u\|_{H^1(\Omega\setminus\overline{D},\mathbf{R}^n)} \leq C\|\varphi\|_{H^{-\frac{1}{2}}(\partial\Omega,\mathbf{R}^n)}, \tag{25}$$

where $C > 0$ only depends on α_0, β_0, ρ_0, M_0 and M_1.

3. Main results

Theorem 1 (Uniqueness)

Let Ω be a bounded domain satisfying (11) and having Lipschitz boundary. Let $D_i, i = 1, 2$, be two domains compactly contained in Ω, having C^1 boundary and satisfying (12) and (13). Moreover, let Σ be an open portion of $\partial\Omega$ of class $C^{2,\alpha}$. Let $u_i \in H^1(\Omega \setminus \overline{D_i}, \mathbf{R}^n)$ be the solution to (1)–(3), when $D = D_i$, $i = 1, 2$, let (19), (20) be satisfied and let the elasticity tensor $\mathcal{C}$ of Lamé type, with Lamé moduli λ and μ of $C^{1,1}$ class satisfying $\mu > 0$, $2\mu + n\lambda > 0$ in $\overline{\Omega}$. If we have

$$(u_1 - u_2)_{|\Sigma} \in \mathcal{R}, \tag{26}$$

then

$$D_1 = D_2. \tag{27}$$

Theorem 2 (Stability)

Let Ω be a domain satisfying (10), (11) and (16). Let D_i, $i = 1, 2$, be two connected open subsets of Ω satisfying (12), (13), (14) and (17). Moreover, let Σ be an open portion of $\partial\Omega$ satisfying (15) and (18). Let $u_i \in H^1(\Omega \setminus \overline{D_i}, \mathbf{R}^n)$ be the solution to (1)–(3), when $D = D_i$, $i = 1, 2$, and let (19)–(24) be satisfied. If, given $\epsilon > 0$, we have

$$\|u_1 - u_2 - \overline{r}\|_{L^2(\Sigma,\mathbf{R}^n)} = \min_{r\in\mathcal{R}} \|u_1 - u_2 - r\|_{L^2(\Sigma,\mathbf{R}^n)} \le \epsilon, \tag{28}$$

then we have

$$d_{\mathcal{H}}(\partial D_1, \partial D_2) \le \omega\left(\frac{\epsilon}{\|\varphi\|_{H^{-\frac{1}{2}}(\partial\Omega,\mathbf{R}^n)}}\right) \tag{29}$$

and

$$d_{\mathcal{H}}(\overline{D_1}, \overline{D_2}) \le \omega\left(\frac{\epsilon}{\|\varphi\|_{H^{-\frac{1}{2}}(\partial\Omega,\mathbf{R}^n)}}\right), \tag{30}$$

where ω is an increasing continuous function on $[0, \infty)$ which satisfies

$$\omega(t) \le C(\log|\log t|)^{-\eta}, \quad \text{for every } t,\ 0 < t < e^{-1}, \tag{31}$$

and $C, \eta, C > 0, 0 < \eta \le 1$, are constants only depending on the a priori data.

Here $d_{\mathcal{H}}$ denotes the Hausdorff distance between bounded closed sets of $\mathbf{R}^n$.

References

[1] G. Alessandrini, A. Morassi. Strong unique continuation for the Lamé system of elasticity. *Comm. Partial Differential Equations* 26:1787-1810, 2001.

[2] M. Eller, V. Isakov, G. Nakamura, D. Tataru. Uniqueness and stability in the Cauchy problem for Maxwell and elasticity systems. In *Nonlinear Partial Differential Equations*, Vol. 16, College de France Seminar. Chapman and Hill/CRC, 2000.

[3] A. Morassi, E. Rosset. Stable determination of cavities in elastic bodies. *Inverse Problems* 20:453-480, 2004.

[4] A. Morassi, E. Rosset. Uniqueness and stability in determining a rigid inclusion in an elastic body. Submitted, 2005.

[5] N. Weck. Außenraumaufgaben in der Theorie stationärer Schwingungen inhomogener elasticher Körper. *Math. Z.* 111:387-398, 1969.

SHAPE OPTIMIZATION OF CONTACT PROBLEMS WITH SLIP RATE DEPENDENT FRICTION

A. Myśliński [1]

[1] *System Research Institute, Warsaw, Poland, myslinsk@ibspan.waw.pl*

Abstract This paper deals with the formulation of a necessary optimality condition for a shape optimization problem of a viscoelastic body in unilateral dynamic contact with a rigid foundation. The contact with Coulomb friction is assumed to occur at a portion of the boundary of the body. The contact condition is described in velocities. The friction coefficient is assumed to be bounded and Lipschitz continuous with respect to a slip velocity. The evolution of the displacement of the viscoelastic body in unilateral contact is governed by a hemivariational inequality of the second order. The shape optimization problem for a viscoelastic body in contact consists in finding, in a contact region, such shape of the boundary of the domain occupied by the body that the normal contact stress is minimized. It is assumed, that the volume of the body is constant. Using material derivative method, we calculate the directional derivative of the cost functional and we formulate a necessary optimality condition for this problem.

keywords: dynamic unilateral problem, shape optimization, necessary optimality condition

1. Introduction

This paper deals with the formulation of a necessary optimality condition for a shape optimization problem of a viscoelastic body in unilateral dynamic contact with a rigid foundation. The contact with a given Coulomb friction [2, 3] is assumed to occur at a portion of the boundary of the body. The contact condition is described in velocities. Usually [2, 3] this contact problem is considered either with a constant or suitable small functional friction coefficient depending on spatial variables. Numerous experiments indicate [5, 7], that in the study of many frictional processes (stick - sleep motions, earthquake modelling, etc.) the friction coefficient has to be considered variable during the slip. The evolution of the body in the unilateral contact is described by a variational inequality of the second order [2, 4]. Under the assumption that the friction coefficient is bounded and Lipschitz continuous with respect to the sliding velocity, the

Please use the following format when citing this chapter:

Myśliński, A., 2006, in IFIP International Federation for Information Processing, Volume 202, Systems, Control, Modeling and Optimization, eds. Ceragioli, F., Dontchev, A., Furuta, H., Marti, K., Pandolfi, L., (Boston: Springer), pp. 285-295.

existence of solutions to the viscoelastic dynamic contact problems is shown in [3, 5]. For regularity result with respect to time variable see [7].

The shape optimization problem for a viscoelastic body in unilateral contact consists in finding, in a contact region, such shape of the boundary of the domain occupied by the body that the normal contact stress is minimized. It is assumed that the volume of the body is constant. Shape optimization of static contact problems was considered, among others, in [4, 11]. The necessary optimality conditions for shape optimization of dynamic contact problems were formulated in [6, 9].

In this paper we study this shape optimization problem for a viscoelastic body in unilateral dynamical contact with friction coefficient bounded and Lipschitz dependent on a slip rate. Assuming small friction coefficient and suitable regularity of data it can be shown [3] that the solution to dynamic contact problem is enough regular to differentiate it with respect to a parameter. Using material derivative method [11], we calculate the directional derivative of the cost functional and we formulate a necessary optimality condition for this problem. The present paper extends authors results contained in [9].

The following notation will be employed: $\Omega \subset R^2$ will denote the bounded domain with Lipschitz continuous boundary Γ. The time variable will be denoted by t and the time interval $I = (0, T), T > 0$. By $H^k(\Omega)$, $k \in (0, \infty)$, we will denote the Sobolev space of functions having derivatives in all directions of the order k belonging to $L^2(\Omega)$ [1]. For an interval I and a Banach space B, $L^p(I; B)$, $p \in [1, \infty)$, denotes the usual Bochner space [3]. $u_t = du/dt$ and $u_{tt} = d^2u/dt^2$ will denote first and second order derivatives, respectively, with respect to t of function u. u_{tN} and u_{tT} will denote normal and tangential components, respectively, of function u_t. $Q = I \times \Omega$, $\gamma_i = I \times \Gamma_i$, $i = 1, 2, 3$, where Γ_i are pieces of the boundary Γ.

2. Contact problem formulation

Consider deformations of a viscoelastic body occupying domain $\Omega \subset R^2$. The boundary Γ of domain Ω is Lipschitz continuous. The body is subject to body forces $f = (f_1, f_2)$. Moreover surface tractions $p = (p_1, p_2)$ are applied to a portion Γ_1 of the boundary Γ. We assume that the body is clamped along the portion Γ_0 of the boundary Γ and that the contact conditions are prescribed on the portion Γ_2 of the boundary Γ. Moreover $\Gamma_i \cap \Gamma_j = \emptyset$, $i \neq j$, $i, j = 0, 1, 2$, $\Gamma = \bar{\Gamma}_0 \cup \bar{\Gamma}_1 \cup \bar{\Gamma}_2$.

Let us denote by $u = (u_1, u_2)$, $u = u(t, x)$, $x \in \Omega$, $t \in [0, T]$, $T > 0$ the displacement of the body and by $\sigma = \{\sigma_{ij}(u(t, x))\}$, $i, j = 1, 2$, the stress field in the body. We shall consider viscoelastic bodies obeying Kevin - Voigt law [2–4]:

$$\sigma_{ij}(u) = c^0_{ijkl}(x)e_{kl}(u) + c^1_{ijkl}(x)e_{kl}(u_t) \quad x \in \Omega, \tag{1}$$

where $e_{kl}(u) = \frac{1}{2}(u_{k,l} + u_{l,k})$, $i, j, k, l = 1, 2$, $u_{k,l} = \partial u_k / \partial x_l$. The summation convention over repeated indices [2] is used throughout the paper. $c^0_{ijkl}(x)$ and $c^1_{ijkl}(x)$, $i, j, k, l = 1, 2$, are components of elasticity and viscoelasticity tensors. It is assumed that elements c^0_{ijkl} and c^1_{ijkl} satisfy usual symmetry, boundedness and ellipticity conditions [2–4]. The displacement field u is a solution of the linear viscoelastic system [2]:

$$u_{tti} - \sigma_{ij}(u(x))_{,j} = f_i(x), \quad (t, x) \in (0, \mathcal{T}) \times \Omega, \quad i, j = 1, 2, \tag{2}$$

where $\sigma_{ij}(x)_{,j} = \partial \sigma_{ij}(x) / \partial x_j$, $i, j = 1, 2$. There are given the following boundary conditions for $i, j = 1, 2$:

$$u_i(x) = 0 \text{ on } (0, \mathcal{T}) \times \Gamma_0, \text{ and } \sigma_{ij}(x) n_j = p_i \text{ on } (0, \mathcal{T}) \times \Gamma_1, \tag{3}$$

as well as the contact and friction conditions on the boundary $(0, \mathcal{T}) \times \Gamma_2$,

$$u_{tN} \leq 0, \quad \sigma_N \leq 0, \quad u_{tN} \sigma_N = 0, \tag{4}$$

$$u_{tT} = 0 \quad \Rightarrow \mid \sigma_T \mid \leq \mathcal{F} \mid \sigma_N \mid, \tag{5}$$

$$u_{tT} \neq 0 \quad \Rightarrow \quad \sigma_T = -\mathcal{F} \mid \sigma_N \mid \frac{u_{tT}}{\mid u_{tT} \mid}. \tag{6}$$

Here we denote [2]: $u_N = u_i n_i$, $\sigma_N = \sigma_{ij} n_i n_j$, $(u_T)_i = u_i - u_N n_i$, $(\sigma_T)_i = \sigma_{ij} n_j - \sigma_N n_i$ $i, j = 1, 2$, $n = (n_1, n_2)$ is the unit outward versor to the boundary Γ. $\mathcal{F} = \mathcal{F}(x, u_t) = \mathcal{F}(u_t)$ denotes a friction coefficient. The initial conditions are:

$$u_i(0, x) = u_0 \quad u_{ti}(0, x) = u_1, \quad i = 1, 2, \quad x \in \Omega, \tag{7}$$

where u_0 and u_1 are given functions. We shall consider problem (2) – (7) in the variational form. Assume,

$$\begin{aligned} f \in H^{1/4}(I; (H^1(\Omega; R^2))^*) \cap L^2(Q; R^2), \\ p \in L^2(I; (H^{1/2}(\Gamma_1; R^2))^*), \\ u_0 \in H^{3/2}(\Omega; R^2) \quad u_1 \in H^{3/2}(\Omega; R^2), \quad u_{1|\Gamma_2} = 0, \end{aligned} \tag{8}$$

are given. The space $L^2(Q; R^2)$ and the Sobolev spaces $H^{1/2}$ $(\Gamma_1; R^2)$ as well as $H^{1/4}(I; (H^1(\Omega; R^2))^*)$ are defined in [1]. Note, that from (8) as well as from Sobolev Imbedding Theorem [1] it follows that u_0 and u_1 in (7) are continuous on the boundary of cylinder Q. The friction coefficient $\mathcal{F} : \Gamma_2 \times R_+ \to R_+$, $\mathcal{F} \in L^\infty(\Gamma_2; R^2)$ has the following properties:

$$\mid \mathcal{F}(x, u_1) - \mathcal{F}(x, u_2) \mid \leq L \mid u_1 - u_2 \mid \quad \forall u_1, u_2 \in [0, +\infty) \text{ a.e. } x \in \Gamma_2, \tag{9}$$

$$0 \leq \mathcal{F}(x, u) \leq \mathcal{F}_0 \text{ a.e. } x \in \Gamma_2, \quad \forall u \in R_+, \tag{10}$$

and the function $x \to \mathcal{F}(x, u)$ is measurable for $u \in R_+$. We shall write $\mathcal{F} = \mathcal{F}(u_t) = \mathcal{F}(x, u_t)$. Let us introduce:

$$F = \{z \in L^2(I; H^1(\Omega; R^2)) \,:\, z_i = 0 \text{ on } (0,T) \times \Gamma_0\,, i = 1,2\}, \quad (11)$$

and the kinematically admissible set

$$K = \{z \in F \,:\, z_t = 0 \text{ on } (0,T) \times \Gamma_0\,,\ z_{tN} \le 0 \quad \text{on } (0,T) \times \Gamma_2\,\}. \quad (12)$$

The problem (1) - (7) is equivalent to the following variational problem [3]: find a function $u \in L^\infty(I; H^1(\Omega; R^2)) \cap H^{1/2}(I; L^2(\Omega; R^2)) \cap K$ such that $u_t \in L^\infty(I; L^2(\Omega; R^2)) \cap H^{1/2}(I; L^2(\Omega; R^2)) \cap K$ and $u_{tt} \in L^\infty(I; H^{-1}(\Omega; R^2)) \cap (H^{1/2}(I; L^2(\Omega; R^2)))^*$ satisfying the following inequality,

$$\begin{aligned} &\int_Q u_{tti}(v_i - u_{ti})dxd\tau + \int_Q \sigma_{ij}(u)e_{ij}(v_i - u_{ti})dxd\tau + \\ &\int_{\gamma_2} \mathcal{F} \mid \sigma_N(u) \mid (\mid v_T \mid - \mid u_{tT} \mid)dxd\tau \ge \int_Q f_i(v_i - u_{ti})dxd\tau + \\ &\int_{\gamma_1} p_i(v_i - u_{ti})dxd\tau \quad \forall v \in H^{1/2}(I; H^1(\Omega; R^2)) \cap K. \end{aligned} \quad (13)$$

The existence of solutions to system (1) - (7) was shown in [3, Theorem 5.1.8, p. 286]:

THEOREM 1 *Assume : (i) condition (8) holds, (ii) Γ_2 is of class C^β, $\beta > 2$, (iii) tensors $c^\eta_{ijkl}(x)$, $\eta = 0, 1$, are Hölder continuous with respect to x with $\beta' > 1/2$, (iv) the friction coefficient $\mathcal{F}$ is bounded. Than there exists a weak solution to the problem (1) - (7).*

Proof. The proof is based on penalization of the inequality (11), friction regularization and employment of localization and shifting technique due to Lions and Magenes. For details of the proof see [3].. ◇

REMARK 2 Assuming that $\mathcal{F}$ satisfies the conditions (9), (10) and using the same arguments as in proof of Theorem 2.1 in [5] one can prove that the solution to (11) is unique.

For the sake of brevity we shall consider the contact problem with the prescribed friction, i.e., we shall assume $g_N \in L^2(I; (H^{1/2}(\Gamma_2))^\star)$ is a given normal traction and

$$\mid \sigma_N \mid = g_N, \quad \mathcal{F}(u_t) \mid g_N \mid = \sigma_T \le 1. \quad (14)$$

The condition (6) is replaced by the following one,

$$u_{tT}\sigma_T + \mathcal{F} g_N \mid u_{tT} \mid = 0, \quad \mid \sigma_T \mid \le 1 \quad \text{on } I \times \Gamma_2. \quad (15)$$

Let us introduce the set,

$$\Lambda = \{\lambda \in L^\infty(I; L^2(\Gamma_2)) \ : \mid \lambda \mid \leq 1 \quad \text{on } I \times \Gamma_2\}. \tag{16}$$

Taking into account (15) the system (11) takes the form: find $u \in K$ and $\lambda \in \Lambda$ such that,

$$\begin{gathered}\int_Q u_{tti}(v_i - u_{ti}) dx d\tau + \int_Q \sigma_{ij}(u) e_{ij}(v_i - u_{ti}) dx d\tau - \\ \int_{\gamma_2} \mathcal{F} g_N \lambda_T (v_T - u_{tT}) dx d\tau \geq \int_Q f_i(v_i - u_{ti}) dx d\tau + \\ \int_{\gamma_1} p_i(v_i - u_{ti}) dx d\tau \quad \forall v \in H^{1/2}(I; H^1(\Omega; R^2)) \cap K,\end{gathered} \tag{17}$$

$$\int_{\gamma_2} \sigma_T u_{tT} ds d\tau \leq \int_{\gamma_2} \mathcal{F} g_N \lambda_T u_{tT} ds d\tau \quad \forall \lambda_T \in \Lambda. \tag{18}$$

3. Formulation of the shape optimization problem

Consider a family $\{\Omega_s\}$ of the domains Ω_s depending on a parameter s. The domain Ω_s we shall consider as an image of a reference domain Ω under a smooth mapping T_s. To describe the transformation T_s we shall use the speed method [11]. Let us denote by $V(.,.) \ : \ [0, \vartheta) \times R^2 \to R^2$ enough regular vector field depending on the parameter $s \in [0, \vartheta), \vartheta > 0$:

$$V(s,.) \in C^2(R^2, R^2) \ \forall s \in [0, \vartheta), \ \ V(., x) \in C([0, \vartheta), R^2) \ \forall x \in R^2. \tag{19}$$

Let $T_s(V) \ : \ R^2 \ni X \ \to \ x(t, X) \ \in \ R^2$ denotes the family of mappings depending on a velocity field V. The vector function $x(., X) = x(.)$ satisfies the systems of ordinary differential equations:

$$\frac{d}{d\tau} x(\tau, X) = V(\tau, x(\tau, X)), \tau \in [0, \vartheta), \quad x(0, X) = X \in R. \tag{20}$$

The family of domains $\{\Omega_s\}$ depending on parameter $s \in [0, \vartheta), \vartheta > 0$, is defined as follows: $\Omega_0 = \Omega$, and $\Omega_s = T_s(\Omega)(V) = \{x \in R^2 \ : \ \exists X \in R^2$ such that, $x = x(s, X)$, where the function $x(., X)$ satisfies equation (18) for $0 \leq \tau \leq s\}$.

Consider problem (17) - (18) in the domain Ω_s. Let F_s, K_s, Λ_s be defined, respectively, by (9, (10), (16) with Ω_s rather than Ω. We shall write $u_s = u(\Omega_s)$, $\sigma_s = \sigma(\Omega_s)$. The problem (17) - (18) in the domain Ω_s takes the form: find $u_s \in L^\infty(I; H^1(\Omega_s; R^2)) \cap H^{1/2}(I; L^2(\Omega_s; R^2)) \cap K$ such that $u_{ts} \in L^\infty(I; L^2(\Omega_s; R^2)) \cap H^{1/2}(I; L^2(\Omega_s; R^2)) \cap K$ and $u_{tts} \in$

$L^\infty(I; H^{-1}(\Omega_s; R^2)) \cap (H^{1/2}(I; L^2(\Omega_s; R^2)))^*$ and $\lambda_s \in \Lambda_s$ such that,

$$\int_{Q_s} u_{ttsi} v_i dx d\tau + \int_{Q_s} \sigma_{ij}(u_s) e_{ij}(v_i - u_{tsi}) dx d\tau -$$
$$\int_{\gamma_{s2}} \mathcal{F}_s g_N \lambda_s (v_T - u_{tsT}) dx d\tau \geq \int_{Q_s} f_i (v_i - u_{tsi}) dx d\tau + \qquad (21)$$
$$\int_{\gamma_{s1}} p_i (v_i - u_{tsi}) dx d\tau \quad \forall v \in H^{1/2}(I; H^1(\Omega_s; R^2)) \cap K,$$

$$\int_{\gamma_{s2}} \sigma_{sT} u_{tsT} ds d\tau \leq \int_{\gamma_{s2}} \mathcal{F}_s g_N \lambda_s u_{tsT} ds d\tau \quad \forall \lambda_s \in \Lambda_s. \qquad (22)$$

Let us formulate the optimization problem. By $\hat{\Omega} \subset R^2$ we denote a hold - all domain such that $\Omega_s \subset \hat{\Omega}$ for all $s \in [0, \vartheta), \vartheta > 0$ and by $P_{\hat{\Omega}}(\Omega)$ a finite perimeter [11] of a domain Ω in $\hat{\Omega}$. Let M be an auxiliary set M determined as $M = \{\phi \in L^\infty(I; H_0^2(\hat{\Omega}; R^2) \ : \phi \leq 0 \text{ on } I \times \hat{\Omega}, \ \| \phi \|_{L^\infty(I; H_0^2(\hat{\Omega}; R^2)} \leq 1\}$. Introduce, for a given $\phi \in M$, the following cost functional:

$$J_\phi(u_s) = J_\phi^1(u_s) + P_{\hat{\Omega}}(\Omega), \quad \text{where } J_\phi^1(u_s) = \int_{\gamma_{s2}} \sigma_{sN} \phi_{tNs} dz d\tau, \qquad (23)$$

and ϕ_{tNs} and σ_{sN} are normal components of ϕ_{ts} and σ_s, respectively, depending on a parameter s. The cost functional (23) depends on the solution $(u_s, \lambda_s) \in K_s \times \Lambda_s$ of the system (21) - (22).
We shall consider such family of domains $\{\Omega_s\}$, that every Ω_s, $s \in [0, \vartheta)$, $\vartheta > 0$, has a constant volume $c > 0$, i.e., every Ω_s belongs to the constraint set U given by:

$$U = \{\Omega_s \ : \int_{\Omega_s} dx = c\}. \qquad (24)$$

The set U is assumed to be nonempty. The shape optimization problem has the form:

For a given $\phi \in M$, find a domain Ω_s occupied by the body,
minimizing the cost functional (23) subject to $\Omega_s \in U$. (25)

The goal of the shape optimization problem (22) is to find such boundary Γ_2 of the domain Ω occupied by the body that the normal contact stress is minimized. It is known [4] that the normal contact stress attains peak in the contact area. This stress can be significantly reduced and uniformly distributed when the bodies in cantact have an optimal shape. The finite perimeter term in (23) is added to ensure the existence of solutions to the shape optimization problem (22). This term implies the compactness of the set (24) in L^2 topology of its characteristic functions. For detailed discussion concerning the conditions ensuring the existence of optimal solutions to shape optimization problems see [4, 11].

4. Shape derivatives of contact problem solution

In order to calculate Euler derivative (42) of the cost functional (23) we have to determine material $(\dot{u}, \dot{\lambda}) \in F \times \Lambda$ and shape derivatives $(u', \lambda') \in F \times \Lambda$ of a solution $(u_s, \lambda_s) \in K_s \times \Lambda_s$ of the system (21)–(22). Recall the notion of the material derivative [11]:

DEFINITION 3 *The material derivative $\dot{u} \in F$ of the function $u_s \in F_s$ at a point $X \in \Omega$ is determined by :*

$$\lim_{s \to 0} \| [(u_s \circ T_s) - u_0]/s - \dot{u} \|_F = 0, \tag{26}$$

where $u_0 \in F$, $u_s \circ T_s \in F$ is an image of function $u_s \in F_s$ in the space F under the mapping T_s.

Recall [11], that if the shape derivative $u' \in F$ of the function $u_s \in F_s$ exists, then it holds:

$$u' \stackrel{def}{=} \dot{u} - \nabla u V(0), \tag{27}$$

where $\dot{u} \in F$ is material derivative of the function $u_s \in F_s$. Taking into account Definition 3 we can calculate the material derivative of a solution to the system (21), (22):

LEMMA 4 *The material derivatives $(\dot{u}, \dot{\lambda}) \in K_1 \times \Lambda$ of a solution $(u_s, \lambda_s) \in K_s \times \Lambda_s$ to the system (21) – (22) are determined as a unique solution to the following system:*

$$\begin{aligned} &\int_Q \{(\dot{u}_{tt}\eta + u_{tt}\dot{\eta} + u_{tt}\eta div V(0) + (DV(0)u)_{tt}\eta + u_{tt}(DV(0)\eta) \\ &\qquad -\dot{f}\eta - f\dot{\eta} + (\sigma_{ij}(u)e_{kl}(\eta) - f\eta)div V(0)\} dx d\tau - \\ &\int_{\gamma_1} (\dot{p}\eta + p\dot{\eta} + p\eta D) dx d\tau - \int_{\gamma_2} \{(\frac{\partial \mathcal{F}}{\partial u_t}\dot{u}_t g_N \lambda \eta_{tT} + \mathcal{F} g_N \dot{\lambda} \eta_{tT}) + \\ &\qquad (\mathcal{F} g_N \lambda \dot{\eta}_{tT} + \mathcal{F} g_N \lambda \eta_{tT} D)\} dx d\tau \geq 0 \quad \forall \eta \in K_1, \end{aligned} \tag{28}$$

$$\begin{aligned} &\int_{\gamma_2} ((\frac{\partial \mathcal{F}}{\partial u_t}\dot{u}_t g_N \lambda + \mathcal{F} g_N \dot{\lambda} - \mu) u_{tT} + (\mathcal{F} g_N \lambda - \dot{\mu}) u_{tT} + \\ &\qquad (\mathcal{F} g_N \lambda - \mu)\dot{u}_{tT} + \mathcal{F}\lambda u_{tT} D\} g_N dx d\tau \quad \forall \mu \in L_1, \end{aligned} \tag{29}$$

where V(0) = V(0,X), DV(0) denotes the Jacobian of the matrix V(0) and D is the expression

$$D = div\ V(0) - (DV(0)n, n). \tag{30}$$

Moreover the sets K_1 and L_1 are determined by

$$\begin{aligned} K_1 = \{\xi \in F\ :\ \xi = u - DVu\ \ on\ \gamma_0,\ \ \xi n \geq nDV(0)u\ \ on\ A_1, \\ \xi n = nDV(0)u\ \ on\ A_2\ \}, \end{aligned} \tag{31}$$

$$L_1 = \{\xi \in \Lambda \,:\, \xi \geq 0 \; on \; B_2, \; \xi \leq 0 \; on \; B_1, \; \xi = 0 \; on \; B_0 \; \}, \qquad (32)$$

where the sets A_i, B_i, $i = 0, 1, 2$ *are given by:* $A_0 = \{x \in \gamma_2 \,:\, u_{tN} = 0\}$, $A_1 = \{x \in A_0 \,:\, \sigma_N = 0\}$, $A_2 = \{x \in A_0 \,:\, \sigma_N < 0\}$, $B_0 = \{x \in \gamma_2 \,:\, \lambda_T = 1, \;\; u_{tT} \neq 0\}$, $B_1 = \{x \in \gamma_2 \,:\, \lambda_T = -1, \;\; u_{tT} = 0\}$, $B_2 = \{x \in \gamma_2 \,:\, \lambda_T = 1, \,:\, u_{tT} = 0\}$.

Proof: is based on approach proposed in [11]. First we transport the system (21) – (22) to the fixed domain Ω. Let $u^s = u_s \circ T_s \in F$, $u = u_0 \in F$, $\lambda^s = \lambda_s \circ T_s \in \Lambda$, $\lambda = \lambda_0 \in \Lambda$. Since in general $u^s \notin K(\Omega)$ we introduce a new variable $z^s = DT_s^{-1}u^s \in K$. Moreover $\dot{z} = \dot{u} - DV(0)u$ [11]. Using this new variable z^s as well as the formulae for transformation of the function and its gradient into reference domain Ω [11] we write the system (21) – (22) in the reference domain Ω. Using the estimates on time derivative of function u [11] the Lipschitz continuity of u and λ satisfying (21) - (22) with respect to s can be proved. Applying to this system the result concerning the differentiability of solutions to variational inequality [11] we obtain that the material derivative $(\dot{u}, \dot{\lambda}) \in K_1 \times \Lambda$ satisfies the system (28) - (29). Moreover from the ellipticity condition of the elasticity coefficients by a standard argument [11] it follows that $(\dot{u}, \dot{\lambda}) \in K_1 \times \Lambda$ is a unique solution to the system (28) - (29). $\diamond$

Assume

$$\nabla u V(0) \in F, \quad \nabla \lambda_T V(0) \in \Lambda, \qquad (33)$$

where the spaces F and Λ are determined by (9) and (16) respectively. Integrating by parts system (28), (29) and taking into account (27), (33) we obtain the similar system to (28) - (29) determining the shape derivative $(u', \lambda_T') \in F \times L$ of the solution $(u_s, \lambda_{sT}) \in K_s \times L_s$ of the system (21) - (22):

$$\int_Q [u_{tt}'\eta + u_{tt}\eta' + (DV(0) +^* DV(0))u_{tt}\eta] dx d\tau +$$

$$\int_\gamma u_{tt}\eta V(0) n dx d\tau + \int_Q \sigma_{ij}(u') e_{kl}\eta dx d\tau -$$

$$\int_{\gamma_2} \{\frac{\partial \mathcal{F}}{\partial u_t} u_t' g_n \lambda \eta_{tT} + \mathcal{F} g_N \lambda' \eta_{tT} + \mathcal{F} g_N \lambda \eta_{tT}'\} dx d\tau +$$

$$I_1(u_t, \eta) + I_2(\lambda, u, \eta) \geq 0 \quad \forall \eta \in N_1, \qquad (34)$$

$$\int_{\gamma_2} [\frac{\partial \mathcal{F}}{\partial u_t} u_t' g_n \lambda u_{tT} u_{tT}' (\mu - \mathcal{F} g_N \lambda) - u_{tT} \mathcal{F} g_N \lambda'] dx d\tau + \qquad (35)$$

$$I_3(u, \mu - \lambda) \geq 0 \quad \forall \mu \in L_1, \qquad (36)$$

$$N_1 = \{\eta \in F \,:\, \eta = \lambda - DuV(0), \;\; \lambda \in K_1\}, \qquad (37)$$

$$I_1(\varphi, \phi) = \int_\gamma \{\sigma_{ij}(\varphi) e_{kl}\phi - f\phi -$$

$$((\nabla pn)\phi + (p\nabla\phi)n + p\phi\kappa)V(0)n\}dxd\tau, \tag{38}$$

$$I_2(\mu,\varphi,\phi) = \int_{\gamma_2} \{\frac{\partial\mathcal{F}}{\partial\varphi_t}\nabla\varphi_t g_n \mu\varphi_{tT} + (\nabla\mu\mathcal{F}g_N)n\nabla\phi +$$
$$\mathcal{F}g_N\mu(\nabla(\nabla\varphi n))\varphi + \mathcal{F}g_N\mu\nabla\varphi_{tT}H + \mathcal{F}g_N\mu\nabla\varphi n\}V(0)ndxd\tau, \tag{39}$$

$$I_3(\varphi,\mu-\lambda) = \int_{\gamma_2} [\frac{\partial\mathcal{F}}{\partial\varphi_t}\nabla\varphi_t g_n\mu + (\varphi n)(\mu - \mathcal{F}g_N\lambda) + \varphi(\nabla\mu n) -$$
$$\varphi(\nabla\mathcal{F}g_N\lambda n) + \varphi(\mu - \mathcal{F}g_N\lambda)\kappa]V(0)ndxd\tau. \tag{40}$$

where κ denotes a mean curvature of the boundary Γ [11].

5. Necessary optimality condition

Our goal is to calculate the directional derivative of the cost functional (23) with respect to the parameter s. We will use this derivative to formulate a necessary optimality condition for the optimization problem (22). First, recall from [11] the notion of Euler derivative of the cost functional depending on a domain Ω:

DEFINITION 5 *Euler derivative $dJ(\Omega;V)$ of the cost functional J at a point Ω in the direction of the vector field V is given by:*

$$dJ(\Omega;V) = \limsup_{s\to 0}[J(\Omega_s) - J(\Omega)]/s. \tag{41}$$

By direct application of this Definition we get the form of the directional derivative $dJ_\phi(u;V)$ of the cost functional (23):

LEMMA 6 *The directional derivative $dJ_\phi(u;V)$ of the cost functional (23), for $\phi \in M$ given, at a point $u \in K$ in the direction of vector field V is determined by:*

$$dJ_\phi(u;V) = \int_Q [u'_{tt}\eta + u_{tt}\eta' + (DV(0) +^* DV(0))u_{tt}\eta]dxd\tau +$$
$$\int_\gamma u_{tt}\eta V(0)ndxd\tau + \int_Q (\sigma'_{ij}e_{kl}(\phi)dxd\tau +$$
$$\int_\Gamma (\sigma_{ij}e_{kl}(\phi) - f\phi)V(0)nds - \int_{\Gamma_1} [(\nabla p\phi V(0) +$$
$$p\nabla\phi V(0) + p\phi D)]ds - \int_{\Gamma_2} \lambda'\phi_T ds + I_1(u,\phi) - I_2(\lambda,u,\phi), \tag{42}$$

where σ' is a shape derivative of the function σ_s with respect to s. ∇p is a gradient of function p with respect to x. Moreover $V(0) = V(0,X)$, ϕ_T and

σ_T are tangent components of ϕ and σ, respectively, as well as D is given by (30). $DV(0)$ denotes the Jacobian of the matrix $V(0)$.

In order to eliminate the shape derivative (u', λ') from (42) we introduce an adjoint state $(r, q) \in K_2 \times L_2$ defined as follows:

$$\int_Q r_{tt}\zeta dx d\tau + \int_Q \sigma_{ij}(\zeta) e_{kl}(\phi + r) dx d\tau +$$
$$\int_{\gamma_2} \frac{\partial \mathcal{F}}{\partial u_t} g_N r_{tT}(q - \lambda)\zeta dx d\tau = 0, \quad \forall \zeta \in K_2, \tag{43}$$

with

$$r(T, x) = 0, \quad r_t(T, x) = 0,$$

$$\int_{\gamma_2} \mathcal{F} g_N (r_{tT} + \phi_{tT} - u_{tT})\delta dx d\tau = 0, \quad \forall \delta \in L_2, \tag{44}$$

where $K_2 = \{\zeta \in K_1 : \zeta n = 0 \text{ on } A_0\}$ and $L_2 = \{\delta \in \Lambda : \delta = 0 \text{ on } A_0 \cap B_0\}$. Since $\phi \in M$ is a given element, then by the same arguments as used to show the existence of the solution $(u, \lambda) \in K \times L$ to the system (21)-(22) we can show the existence of the solution $(r, q) \in K_2 \times L_2$ to the system (43) - (44). From (42), (34), (35), (43), (44) we obtain:

$$dJ_\phi(u; V) = I_1(u, \phi + r) + I_2(\lambda, u, \phi + r) + I_3(u, q - \lambda). \tag{45}$$

The necessary optimality condition has a standard form [2, 9, 11]:

THEOREM 7 *Let $\Omega^\star \subset U$ be an optimal solution to the problem (22). Then there exists Lagrange multiplier $\mu \in R$ such that for all vector fields V determined by (17), (18) the following condition holds:*

$$dJ_\phi(u^\star; V) + \mu \int_{\Gamma^\star} V(0) n ds + dP_{\hat{\Omega}}(\Omega^\star; V) \geq 0, \tag{46}$$

where $dJ_\phi(u^\star; V)$ is given by (45), $\Gamma^\star$ denotes the boundary of the optimal domain $\Omega^\star$, $u^\star$ is a solution to the state system (17) - (18) in $\Omega^\star$, and $dP_{\hat{\Omega}}(\Omega^\star; V)$ denotes Euler derivative of $P_{\hat{\Omega}}(\Omega)$ at $\Omega^\star$ in a direction V.

6. Conclusions

The directional derivative of the cost functional with respect to the perturbation of the domain occupied by a viscoelastic body in unilateral contact with a rigid foundation was calculated and the necessary optimality condition for the shape optimization problem for the dynamical contact problem was formulated. This shape optimization problem has been numerically solved. Preliminary numerical results can be found in [10].

References

[1] R.A. Adams. *Sobolev Spaces* , Academic Press, New York, 1975.

[2] G. Duvaut and J.L. Lions. *Les inequations en mecanique et en physique*, Dunod, Paris, 1972.

[3] C. Eck, J. Jarusek and M. Krbec. *Unilateral Contact Problems. Variational Methods and Existence Theorems* , CRC Press, Boca Raton, Florida, USA, 2005.

[4] J. Haslinger, P. Neittaanmaki. *Finite Element Approximation for Optimal Shape Design. Theory and Application.*, John Wiley& Sons, 1988.

[5] I. Ionescu. Viscosity Solutions for Dynamic Problems with Slip - Rate Dependent Friction, *Quarterly of Mathematics*, LX:461 - 476, 2002.

[6] J. Jarušek, M. Krbec, M. Rao, J. Sokołowski. Conical Differentiability for Evolution Variational Inequalities, University of Nancy I, 2002.

[7] K.L. Kuttler, M. Schillor. Regularity of Solutions to a Dynamic Frictionless Contact Problem with Normal Compliance, *Nonlinear Analysis*, 59:1063 - 1075, 2004.

[8] K.L. Kuttler, M. Schillor, Dynamic Contact with Signorini's Condition and Slip Rate Dependent Friction, *Electronic Journal of Differential Equations*, 1 - 21, 2004.

[9] A. Myśliński. Shape Optimization for Dynamic Contact Problems, *Discussiones Mathematicae, Differential Inclusions, Control and Optimization*, 20:79 - 91, 2000.

[10] A. Myśliński. Augmented Lagrangian Techniques for Shape Optimal Design of Dynamic Contact Problems, CD - ROM Proceedings of the Fourth World Congress on Structural and Multidisciplinary Optimization, G. Chen, Y. Gu, S. Liu, Y. Wang eds., Liaoning Electronic Press, Shenyang, China, 2001.

[11] J. Sokołowski and J.P. Zolesio. *Introduction to Shape Optimization. Shape Sensitivity Analysis.* Springer, Berlin, 1992.

SECOND ORDER OPTIMALITY CONDITIONS FOR CONTROLS WITH CONTINUOUS AND BANG-BANG COMPONENTS

N.P. Osmolovskii [1] and H. Maurer[2]
[1] *Systems Research Institute, Polish Academy of Sciences, ul. Newelska 6, 01-447 Warszawa, Poland and University of Podlasie in Siedlce, 3 Maja, Siedlce, Poland, nikolai@osmolovskii.msk.ru**,
[2] *Wilhelms–Universität Münster, Institut für Numerische und Angewandte Mathematik, Einsteinstr. 62, D–48149 Münster, Germany, maurer@math.uni-muenster.de*

Abstract Second order necessary and sufficient optimality conditions for bang-bang control problems in a very general form have been obtained by the first author. These conditions require the positive (semi)-definiteness of a certain quadratic form on the finite-dimensional critical cone. In the present paper we formulate a generalization of these results to optimal control problems where the control variable has two components: a continuous unconstrained control appearing nonlinearly and a bang-bang control appearing linearly and belonging to a convex polyhedron. Many examples of control of this kind may be found in the literature.

keywords: bang–bang control, Pontryagin minimum principle, second order necessary and sufficient conditions, critical cone, quadratic form, strengthened Legendre condition

1. Introduction

The classical quadratic optimality conditions for an optimization or optimal control problem with constraints require that the second variation of the associated Lagrangian be nonnegative (positive) in all critical directions. Some general results along these lines were given in [5], [6], [7], [8], [9].

In the pure bang–bang case, where all control components appear linearly in the problem, second order necessary and sufficient optimality conditions have been obtained in Milyutin and Osmolovskii [4] in a very general form. In [2], [3] we have developed for this case numerical methods of testing the

*Paper partially supported by NSh-304.2003.1 and RFBR 04-01-00482.

Please use the following format when citing this chapter:

Osmolovskii, N.P., and Maurer, H., 2006, in IFIP International Federation for Information Processing, Volume 202, Systems, Control, Modeling and Optimization, eds. Ceragioli, F., Dontchev, A., Furuta, H., Marti, K., Pandolfi, L., (Boston: Springer), pp. 297-307.

positive definiteness of the quadratic form. A different approach to second order sufficient conditions for bang-bang controls was given in [1].

In this paper we formulate both necessary and sufficient quadratic optimality conditions for optimal control problems with a control variable having two components: a continuous unconstrained control appearing nonlinearly and a bang-bang control appearing linearly and belonging to a convex polyhedron. Such a type of control problem arises in many applications. Due to space restrictions we present only a summary of results on necessary and sufficient conditions.

In Section 2, we give a statement of the general control problem with continuous and bang-bang control components (main problem). We formulate the minimum principle (first order necessary optimality condition) and the notions of a Pontryagin and a bounded-strong local minimum. In Section 3, we present second order optimality conditions, both necessary and sufficient, for these two types of minimum in the main problem.

2. Control problem on a non–fixed time interval

2.1 The main problem

Let $x(t) \in \mathbf{R}^{d(x)}$ denote the state variable and let $u(t) \in \mathbf{R}^{d(u)}$, $v(t) \in \mathbf{R}^{d(v)}$ be the two types of control variables in the time interval $t \in [t_0, t_1]$ with a non–fixed initial time and final times t_0 and t_1. The following optimal control problem (1)–(4) will be referred to as the *main problem*:

$$\text{Minimize} \qquad \mathcal{J}(t_0, t_1, x(\cdot), u(\cdot), v(\cdot)) = J(t_0, x(t_0), t_1, x(t_1)) \tag{1}$$

subject to the constraints

$$\dot{x}(t) = f(t, x(t), u(t), v(t)), \quad u(t) \in U, \quad (t, x(t), v(t)) \in \mathcal{Q}, \tag{2}$$

$$\begin{aligned} &F(t_0, x(t_0), t_1, x(t_1)) \le 0, \quad K(t_0, x(t_0), t_1, x(t_1)) = 0, \\ &(t_0, x(t_0), t_1, x(t_1)) \in \mathcal{P}. \end{aligned} \tag{3}$$

The control variable u appears linearly in the system dynamics,

$$f(t, x, u, v) = a(t, x, v) + B(t, x, v)u\,, \tag{4}$$

whereas the control variable v appears nonlinearly in the dynamics in a sense which will be made more precise later. Here, F, K, a are column-vector functions, B is a $d(x) \times d(u)$ matrix function, $\mathcal{P} \subset \mathbf{R}^{2+2d(x)}$, $\mathcal{Q} \subset \mathbf{R}^{1+d(x)+d(v)}$ are open sets and $U \subset \mathbf{R}^{d(u)}$ is a convex polyhedron. The functions J, F, K are assumed to be twice continuously differentiable on $\mathcal{P}$ and the functions a, B are twice continuously differentiable on $\mathcal{Q}$. The dimensions of F, K are denoted by $d(F)$, $d(K)$. By $\Delta = [t_0, t_1]$ we shall denote the interval of control. We shall use the abbreviations

$$x_0 = x(t_0), \; x_1 = x(t_1), \; p = (t_0, x_0, t_1, x_1).$$

A process

$$\Pi = \{(x(t), u(t), v(t)) \mid t \in [t_0, t_1]\}$$

is said to be *admissible*, if $x(\cdot)$ is absolutely continuous, $u(\cdot)$, $v(\cdot)$ are measurable bounded on Δ and the triple of functions $(x(t), u(t), v(t))$ together with the end-points $p = (t_0, x(t_0), t_1, x(t_1))$ satisfies the constraints (2),(3).

DEFINITION 1 *The process Π affords a Pontryagin local minimum, if there is no sequence of admissible processes $\Pi^n = \{(x^n(t), u^n(t), v^n(t)) \mid t \in [t_0^n, t_1^n]\}$, $n = 1, 2, \ldots$ such that the following properties hold with $\Delta^n = [t_0^n, t_1^n]$:*

(a) $\mathcal{J}(\Pi^n) < \mathcal{J}(\Pi) \quad \forall\, n$ *and* $t_0^n \to t_0,\ t_1^n \to t_1$ *for* $n \to \infty$;

(b) $\max\limits_{\Delta^n \cap \Delta} |x^n(t) - x(t)| \to 0$ *for* $n \to \infty$;

(c) $\int\limits_{\Delta^n \cap \Delta} |u^n(t) - u(t)|\, dt \to 0, \quad \int\limits_{\Delta^n \cap \Delta} |v^n(t) - v(t)|\, dt \to 0$ *for* $n \to \infty$;

(d) there exists a compact set $C \subset \mathcal{Q}$ (which depends on the choice of the sequence) such that for all sufficiently large n, we have $(t, x^n(t), v^n(t)) \in C$ a.e. on Δ^n.

For convenience, let us formulate an equivalent definition of the Pontryagin minimum.

DEFINITION 2 *The process Π affords a Pontryagin local minimum, if for each compact set $C \subset \mathcal{Q}$ there exists $\varepsilon > 0$ such that $\mathcal{J}(\Pi') \geq \mathcal{J}(\Pi)$ for all admissible processes $\Pi' = \{(x'(t), u'(t), v'(t)) \mid t \in [t_0', t_1']\}$ such that*

(a) $|t_0' - t_0| < \varepsilon, \quad |t_1' - t_1| < \varepsilon$;

(b) $\max\limits_{\Delta' \cap \Delta} |x'(t) - x(t)| < \varepsilon$, *where* $\Delta' = [t_0', t_1']$;

(c) $\int\limits_{\Delta' \cap \Delta} |u'(t) - u(t)|\, dt < \varepsilon; \quad \int\limits_{\Delta' \cap \Delta} |v'(t) - v(t)|\, dt < \varepsilon$;

(d) $(t, x'(t), v'(t)) \in C$ *a.e. on* Δ'.

2.2 First order necessary optimality conditions

Let

$$\Pi = \{(x(t), u(t), v(t)) \mid t \in [t_0, t_1]\}$$

be a fixed admissible process such that the control $u(t)$ is a piecewise constant function and the control $v(t)$ is a continuous function on the interval $\Delta =$

$[t_0, t_1]$. In order to make the notations simpler we do not use such symbols and indices as zero, hat or asterisk to distinguish this trajectory from others. Denote by

$$\theta = \{\tau_1, \ldots, \tau_s\}, \quad t_0 < \tau_1 < \ldots < \tau_s < t_1$$

the finite set of all discontinuity points (jump points) of the control $u(t)$. Then $\dot{x}(t)$ is a piecewise continuous function whose points of discontinuity belong to θ, and hence $x(t)$ is a piecewise smooth function on Δ. Henceforth, we shall use the notation

$$[u]^k = u^{k+} - u^{k-}$$

to denote the jump of function $u(t)$ at the point $\tau_k \in \theta$, where

$$u^{k-} = u(\tau_k - 0), \quad u^{k+} = u(\tau_k + 0)$$

are the left hand and the right side hand values of the control $u(t)$ at τ_k, respectively. Similarly, we denote by $[\dot{x}]^k$ the jump of the function $\dot{x}(t)$ at the point τ_k.

Let us formulate a first-order necessary condition for optimality of the process Π in the form of the Pontryagin minimum principle. To this end we introduce the Pontryagin or Hamiltonian function

$$H(t, x, \psi, u, v) = \psi f(t, x, u, v) = \psi a(t, x, v) + \psi B(t, x, v)u, \tag{5}$$

where ψ is a row-vector of dimension $d(\psi) = d(x)$ while x, u, f, F and K are column-vectors. The factor of the control u in the Pontryagin function will be called the *switching function for the* u*-component*

$$\sigma(t, x, \psi, v) = \psi B(t, x, v) \tag{6}$$

which is a row vector of dimension $d(u)$. Denote the end-point Lagrange function by

$$l(\alpha_0, \alpha, \beta, p) = \alpha_0 J(p) + \alpha F(p) + \beta K(p), \quad p = (t_0, x_0, t_1, x_1),$$

where α and β are row-vectors with $d(\alpha) = d(F)$, $d(\beta) = d(K)$, and α_0 is a number. We introduce a tuple of Lagrange multipliers

$$\lambda = (\alpha_0, \alpha, \beta, \psi(\cdot), \psi_0(\cdot))$$

such that

$$\psi(\cdot) : \Delta \to \mathbf{R}^{d(x)}, \quad \psi_0(\cdot) : \Delta \to \mathbf{R}^1$$

are continuous on Δ and continuously differentiable on each interval of the set $\Delta \setminus \theta$. In the sequel, we shall denote first or second order partial derivatives by subscripts referring to the variables.

Denote by M_0 the set of the normalized tuples λ satisfying the minimum principle conditions for the process Π:

$$\alpha_0 \geq 0, \quad \alpha \geq 0, \quad \alpha F(p) = 0, \quad \alpha_0 + \sum \alpha_i + \sum |\beta_j| = 1, \quad (7)$$

$$\dot{\psi} = -H_x, \quad \dot{\psi}_0 = -H_t \quad \forall t \in \Delta \setminus \theta, \quad (8)$$

$$\psi(t_0) = -l_{x_0}, \quad \psi(t_1) = l_{x_1}, \quad \psi_0(t_0) = -l_{t_0}, \quad \psi_0(t_1) = l_{t_1}, \quad (9)$$

$$H(t, x(t), \psi(t), u, v) \leq H(t, x(t), \psi(t), u(t), v(t))$$

$$\text{for all } t \in \Delta \setminus \theta, \ u \in U, \ v \in \mathbf{R}^{d(v)} \text{ such that } (t, x(t), v) \in \mathcal{Q}, \quad (10)$$

$$H(t, x(t), \psi(t), u(t), v(t)) + \psi_0(t) = 0 \quad \forall\, t \in \Delta \setminus \theta. \quad (11)$$

The derivatives l_{x_0} and l_{x_1} are taken at the point $(\alpha_0, \alpha, \beta, p)$, where $p = (t_0, x(t_0), t_1, x(t_1))$, while the derivatives H_x, H_t are evaluated at the point $(t, x(t), \psi(t), u(t), v(t))$ for $t \in \Delta \setminus \theta$. The condition $M_0 \neq \emptyset$ constitutes the first order necessary condition for a Pontryagin minimum of the process Π which is the so called Pontryagin minimum principle, cf., e.g., Milyutin, Osmolovskii [4].

THEOREM 3 *If the process Π affords a Pontryagin minimum, then the set M_0 is nonempty. The set M_0 is a finite-dimensional compact set and the projector $\lambda \mapsto (\alpha_0, \alpha, \beta)$ is injective on M_0.*

In the sequel, it will be convenient to use the simple abbreviation (t) for indicating all arguments $(t, x(t), \psi(t), u(t), v(t))$, e.g.,
$H(t) = H(t, x(t), \psi(t), u(t), v(t))$ and $\sigma(t) = \sigma(t, x(t), \psi(t), v(t))$.

Let $\lambda = (\alpha_0, \alpha, \beta, \psi(\cdot), \psi_0(\cdot)) \in M_0$. From condition (11) it follows that $H(t)$ is a continuous function. In particular, we have $[H]^k = H^{k+} - H^{k-} = 0$ for each $\tau_k \in \theta$, where $H^{k-} := H(\tau_k, x(\tau_k), \psi(\tau_k), u(\tau_k - 0), v(\tau_k))$, $H^{k+} := H(\tau_k, x(\tau_k), \psi(\tau_k), u(\tau_k + 0), v(\tau^k))$. We shall denote by H^k the common value of H^{k-} and H^{k+}:

$$H^k := H^{k-} = H^{k+}.$$

The equalities $[H]^k = 0$, $[\psi]^k = 0 \ \forall\, t^k \in \theta$ constitute the Weierstrass–Erdmann conditions for broken extremals. We formulate one more condition of this type which is important for the statement of second-order conditions for extremal with jumps in the control. Namely, for $\lambda \in M_0$ and $\tau_k \in \theta$ consider the function

$$\begin{aligned}(\Delta_k H)(t) &= H(t, x(t), \psi(t), u^{k+}, v(\tau_k)) - H(t, x(t), \psi(t), u^{k-}, v(\tau_k)) \\ &= \sigma(t, x(t), \psi(t), v(\tau_k))\, [u]^k. \end{aligned} \quad (12)$$

LEMMA 4 *For each $\lambda \in M_0$ the following equalities hold*

$$\frac{d}{dt}(\Delta_k H)|_{t=\tau_k - 0} = \frac{d}{dt}(\Delta_k H)|_{t=\tau_k + 0}, \quad k = 1, \ldots, s.$$

Consequently, for each $\lambda \in M_0$ the function $(\Delta_k H)(t)$ has a derivative at the point $\tau_k \in \theta$. Define the quantity

$$D^k(H) = -\frac{d}{dt}(\Delta_k H)(\tau_k).$$

Then the minimum condition (10) implies the following property:

LEMMA 5 *For each $\lambda \in M_0$ the following conditions hold:*

$$D^k(H) \geq 0, \quad k = 1, \ldots, s. \tag{13}$$

The value $D^k(H)$ can also be written in the form

$$\begin{aligned} D^k(H) &= -H_x^{k+} H_\psi^{k-} + H_x^{k-} H_\psi^{k+} - [H_t]^k \\ &= \dot{\psi}^{k+} \dot{x}^{k-} - \dot{\psi}^{k-} \dot{x}^{k+} + [\psi_0]^k, \end{aligned}$$

where H_x^{k-} and H_x^{k+} are the left- and the right-hand values of the function $H_x(t)$ at τ_k, respectively, $[H_t]^k$ is the jump of the function $H_t(t)$ at τ_k, etc.

2.3 Integral cost function, unessential variables, strong minimum

It is well known that any control problem with a cost functional in integral form

$$\mathcal{J} = \int_{t_0}^{t_1} f_0(t, x(t), u(t), v(t))\, dt \tag{14}$$

can be brought to the canonical form (1) by introducing a new state variable y defined by the state equation

$$\dot{y} = f_0(t, x, u, v), \quad y(t_0) = 0\,. \tag{15}$$

This yields the cost function $\mathcal{J} = y(t_1)$. The control variable u is assumed to appear linearly in the function f_0,

$$f_0(t, x, u, v) = a_0(t, x, v) + B_0(t, x, v)u\,. \tag{16}$$

The component y is called an *unessential* component in the augmented problem. The general definition of an unessential component is as follows.

DEFINITION 6 *The state variable x_i, i.e., the i–th component of the state vector x is called* unessential *if the function f does not depend on x_i and if the functions F, J, K are affine in $x_{i0} = x_i(t_0)$ and $x_{i1} = x_i(t_1)$.*

In the following, let $\underline{x}$ denote the vector of all essential components of state vector x.

DEFINITION 7 *We say that the process* Π *affords a bounded-strong minimum if there is no sequence of admissible processes* $\Pi^n = \{(x^n(t), u^n(t), v^n(t)) \mid t \in [t_0^n, t_1^n]\}$, $n = 1, 2, \ldots$, *such that*

(a) $\mathcal{J}(\Pi^n) < \mathcal{J}(\Pi)$,

(b) $t_0^n \to t_0, \quad t_1^n \to t_1, \quad x^n(t_0) \to x(t_0)$ *(*$n \to \infty$*),*

(c) $\max_{\Delta^n \cap \Delta} |\underline{x}^n(t) - \underline{x}(t)| \to 0$ *(*$n \to \infty$*), where* $\Delta^n = [t_0^n, t_1^n]$,

(d) there exists a compact set $\mathcal{C} \subset \mathcal{Q}$ *(which depends on the choice of the sequence) such that for all sufficiently large n, we have* $(t, x^n(t), v^n(t)) \in \mathcal{C}$ *a.e. on* Δ^n.

An equivalent definition has the form.

DEFINITION 8 *The process* Π *affords a bounded-strong minimum, if for each compact set* $\mathcal{C} \subset \mathcal{Q}$ *there exists* $\varepsilon > 0$ *such that* $\mathcal{J}(\Pi') \geq \mathcal{J}(\Pi)$ *for all admissible processes* $\Pi' = \{(x'(t), u'(t), v'(t)) \mid t \in [t_0', t_1']\}$ *such that*

(a) $|t_0' - t_0| < \varepsilon, \quad |t_1' - t_1| < \varepsilon, \quad |x'(t_0) - x(t_0)| < \varepsilon$;

(b) $\max_{\Delta' \cap \Delta} |\underline{x}'(t) - \underline{x}(t)| < \varepsilon$, *where* $\Delta' = [t_0', t_1']$;

(c) $(t, x'(t), v'(t)) \in \mathcal{C}$ *a.e. on* Δ'.

The *strict* bounded-strong minimum is defined in a similar way, with the non-strict inequality $\mathcal{J}(\Pi') \geq \mathcal{J}(\Pi)$ replaced by the strict one and the process Π' required to be different from Π.

3. Quadratic necessary and sufficient optimality conditions

In this section, we shall formulate a quadratic necessary optimality condition of a Pontryagin minimum (Definition 1) for given control process Π. A strengthening of this quadratic condition yields a quadratic sufficient condition of a bounded-strong minimum (Definition 7). These quadratic conditions are based on the properties of a quadratic form on the so-called critical cone whose elements are first order variations along a given process Π. The main results of this section (Theorems 9 and 12) are due to Osmolovskii.

3.1 Critical cone

For a given process Π we introduce the space $\mathcal{Z}^2(\theta)$ and the *critical cone* $\mathcal{K} \subset \mathcal{Z}^2(\theta)$. Denote by $P_\theta W^{1,2}_{d(x)}(\Delta, \mathbf{R}^{d(x)})$ the space of piecewise continuous functions $\bar{x}(\cdot) : \Delta \to \mathbf{R}^{d(x)}$, which are absolutely continuous on each interval of the set $\Delta \setminus \theta$ and have a square integrable first derivative. By $L^2(\Delta, \mathbf{R}^{d(v)})$ we denote the space of square integrable functions $\bar{v}(\cdot) : \Delta \to \mathbf{R}^{d(v)}$. For each $\bar{x} \in P_\theta W^{1,2}_{d(x)}(\Delta, \mathbf{R}^{d(x)})$ and for $\tau_k \in \theta$ we set

$$\bar{x}^{k-} = \bar{x}(\tau_k - 0), \quad \bar{x}^{k+} = \bar{x}(\tau_k + 0), \quad [\bar{x}]^k = \bar{x}^{k+} - \bar{x}^{k-}.$$

Let $\bar{z} = (\bar{t}_0, \bar{t}_1, \xi, \bar{x}, \bar{v})$, where $\bar{t}_0, \bar{t}_1 \in \mathbf{R}^1$, $\xi \in \mathbf{R}^s$, $\bar{x} \in P_\theta W^{1,2}_{d(x)}(\Delta, \mathbf{R}^{d(x)})$, $\bar{v} \in L^2(\Delta, \mathbf{R}^{d(v)})$. Thus,

$$\bar{z} \in \mathcal{Z}^2(\theta) := \mathbf{R}^2 \times \mathbf{R}^s \times P_\theta W^{1,2}_{d(x)}(\Delta, \mathbf{R}^{d(x)}) \times L^2(\Delta, \mathbf{R}^{d(v)}).$$

For each $\bar{z}$ we set

$$\tilde{x}_0 = \bar{x}(t_0) + \bar{t}_0 \dot{x}(t_0), \quad \tilde{x}_1 = \bar{x}(t_1) + \bar{t}_1 \dot{x}(t_1), \quad \tilde{p} = (\bar{t}_0, \tilde{x}_0, \bar{t}_1, \tilde{x}_1). \tag{17}$$

The vector $\tilde{p}$ is considered as a column vector. Note that $\bar{t}_0 = 0$, respectively, $\bar{t}_1 = 0$ holds for a *fixed* initial time t_0, respectively, final time t_1. Denote by $I_F(p) = \{i \in \{1, \ldots, d(F)\} \mid F_i(p) = 0\}$ the set of indices of all active endpoint inequalities $F_i(p) \le 0$ at the point $p = (t_0, x(t_0), t_1, x(t_1))$. Denote by $\mathcal{K}$ the set of all $\bar{z} \in \mathcal{Z}^2(\theta)$ satisfying the following conditions:

$$J'(p)\tilde{p} \le 0, \quad F_i'(p)\tilde{p} \le 0 \ \forall\, i \in I_F(p), \quad K'(p)\tilde{p} = 0, \tag{18}$$

$$\dot{\bar{x}}(t) = f_x'(t, x(t), u(t), v(t))\bar{x}(t) + f_v'(t, x(t), u(t), v(t))\bar{v}(t), \tag{19}$$

$$[\bar{x}]^k = [\dot{x}]^k \xi_k, \quad k = 1, \ldots, s, \tag{20}$$

where $p = (x(t_0), t_0, x(t_1), t_1)$, $[\dot{x}] = \dot{x}(\tau_k + 0) - \dot{x}(\tau_k - 0)$.

It is obvious that $\mathcal{K}$ is a convex cone with finitely many faces in the space $\mathcal{Z}^2(\theta)$. The convex cone $\mathcal{K}$ is called the *critical cone.*

3.2 Quadratic necessary optimality conditions

Let us introduce a quadratic form on the critical cone $\mathcal{K}$ defined by the conditions (18)-(20). For each $\lambda \in M_0$ and $\bar{z} \in \mathcal{K}$ we set

$$\Omega(\lambda, \bar{z}) = \langle A\tilde{p}, \tilde{p}\rangle + \sum_{k=1}^{s} (D^k(H)\xi_k^2 - [\dot{\psi}]^k \bar{x}^k_{\mathrm{av}} \xi_k)$$

$$+ \int_{t_0}^{t_1} (\langle H_{xx}(t)\bar{x}(t), \bar{x}(t)\rangle + 2\langle H_{xv}(t)\bar{v}(t), \bar{x}(t)\rangle + \langle H_{vv}(t)\bar{v}(t), \bar{v}(t)\rangle)\, dt, \tag{21}$$

where

$$\begin{aligned}\langle A\tilde{p},\tilde{p}\rangle &= \langle l_{pp}\tilde{p},\tilde{p}\rangle + 2\dot{\psi}(t_0)\tilde{x}_0\bar{t}_0 + (\dot{\psi}_0(t_0) - \dot{\psi}(t_0)\dot{x}(t_0))\bar{t}_0^2 \\ &\quad -2\dot{\psi}(t_1)\tilde{x}_1\bar{t}_1 - (\dot{\psi}_0(t_1) - \dot{\psi}(t_1)\dot{x}(t_1))\bar{t}_1^2, \end{aligned} \tag{22}$$

$$l_{pp} = l_{pp}(\alpha_0,\alpha,\beta,p), \quad p = (t_0, x(t_0), t_1, x(t_1)), \quad \bar{x}^k_{\text{av}} = \frac{1}{2}(\bar{x}^{k-} + \bar{x}^{k+}),$$

$$H_{xx}(t) = H_{xx}(t, x(t), \psi(t), u(t), v(t)), \quad \text{etc.}$$

Note that the functional $\Omega(\lambda, \bar{z})$ is linear in λ and quadratic in $\bar{z}$. Also note that for a problem on a fixed time interval $[t_0, t_1]$ we have $\bar{t}_0 = \bar{t}_1 = 0$ and, hence, the quadratic form (22) reduces to $\langle A\tilde{p},\tilde{p}\rangle = \langle l_{pp}\tilde{p},\tilde{p}\rangle$. The following theorem gives the main second order necessary condition of optimality.

THEOREM 9 *If the process Π affords a Pontryagin minimum, then the following Condition $\mathcal{A}$ holds: the set M_0 is nonempty and*

$$\max_{\lambda\in M_0} \Omega(\lambda,\bar{z}) \geq 0 \quad \textit{for all} \quad \bar{z} \in \mathcal{K}.$$

This theorem follows from Theorem 2.3 in [7].

We call Condition $\mathcal{A}$ the necessary quadratic condition, although it is truly quadratic only if M_0 is a singleton.

3.3 Quadratic sufficient optimality conditions

A natural strengthening of the necessary Condition $\mathcal{A}$ turns out to be a sufficient optimality condition not only for a Pontryagin minimum, but also for a bounded-strong minimum; cf. Definition 7. Denote by M_0^+ the set of all $\lambda \in M_0$ satisfying the conditions:

(a) $H(t, x(t), \psi(t), u, v) < H(t, x(t), \psi(t), u(t), v(t))$, for all $t \in \Delta \setminus \theta$, $u \in U$, $v \in \mathbf{R}^{d(v)}$, such that $(t, x(t), v) \in \mathcal{Q}$ and $(u, v) \neq (u(t), v(t))$;

(b) $H(\tau_k, x(\tau_k), \psi(\tau_k), u, v) < H^k$ for all $\tau_k \in \theta$, $u \in U$, $v \in \mathbf{R}^{d(v)}$ such that $(\tau_k, x(\tau_k), v) \in \mathcal{Q}$, $(u, v) \neq (u(\tau_k - 0), v(\tau_k))$, $(u, v) \neq (u(\tau_k + 0), v(\tau^k))$, where $H^k := H^{k-} = H^{k+}$.

Let $\operatorname{Arg}\max_{u'\in U} \sigma u'$ be the set of points $v \in U$ where the maximum of the linear function $\sigma u'$ is attained

DEFINITION 10 *For a given admissible process Π with a piecewise constant control $u(t)$ and continuous control $v(t)$ we shall say that $u(t)$ is a strict bang-bang control, if the set M_0 is nonempty and there exists $\lambda \in M_0$ such that*

$$Arg \max_{u'\in U} \sigma(t)u' = [u(t-0), u(t+0)],$$

where $[u(t-0), u(t+0)]$ *denotes the line segment spanned by the vectors* $u(t-0), u(t+0)$.

If $\dim(u) = 1$, then the strict bang-bang property is equivalent to $\sigma(t) \neq 0$ $\forall t \in \Delta \backslash \theta$. If the set M_0^+ is nonempty, then, obviously, $u(t)$ is a strict bang-bang control.

DEFINITION 11 *An element* $\lambda \in M_0$ *is said to be strictly Legendrian if the following conditions are satisfied*

(a) *for each* $t \in \Delta \setminus \theta$ *the quadratic form*

$$\langle H_{vv}(t, x(t), \psi(t), u(t), v(t))\bar{v}, \bar{v}\rangle$$

is positive definite in $\mathbf{R}^{d(v)}$;

(b) *for each* $\tau_k \in \theta$ *the quadratic form*

$$\langle H_{vv}(\tau_k, x(\tau_k), \psi(\tau_k), u(\tau_k - 0), v(\tau_k))\bar{v}, \bar{v}\rangle$$

is positive definite in $\mathbf{R}^{d(v)}$;

(c) *for each* $\tau_k \in \theta$ *the quadratic form*

$$\langle H_{vv}(\tau_k, x(\tau_k), \psi(\tau_k), u(\tau_k + 0), v(\tau_k))\bar{v}, \bar{v}\rangle$$

is positive definite in $\mathbf{R}^{d(v)}$.

(d) $D^k(H) > 0$ *for all* $\tau_k \in \theta$.

Denote by $\mathrm{Leg}_+(M_0^+)$ the set of all strictly Legendrian elements $\lambda \in M_0^+$ and put

$$\bar{\gamma}(\bar{z}) = \bar{t}_0^2 + \bar{t}_1^2 + \langle \xi, \xi\rangle + \langle \bar{x}(t_0), \bar{x}(t_0)\rangle + \int_{t_0}^{t_1} \langle \bar{v}(t), \bar{v}(t)\rangle \, dt.$$

THEOREM 12 *Let the following Condition* $\mathcal{B}$ *be fulfilled for the process* Π :

(a) *the set* $\mathrm{Leg}_+(M_0^+)$ *is nonempty;*

(b) *there exists a nonempty compact set*

$$M \subset \mathrm{Leg}_+(M_0^+)$$

and a number $C > 0$ *such that*

$$\max_{\lambda \in M} \Omega(\lambda, \bar{z}) \geq C\bar{\gamma}(\bar{z}) \quad \textit{for all } \bar{z} \in \mathcal{K}.$$

Then Π *is a strict bounded-strong minimum.*

If the set $\mathrm{Leg}_+(M_0^+)$ is nonempty and $\mathcal{K}=\{0\}$ then (b) is fulfilled automatically. This is a first order sufficient optimality condition of a strict bounded-strong minimum. Let us emphasize that there is only a minimal gap between the necessary condition $\mathcal{A}$ and the sufficient condition $\mathcal{B}$!

References

[1] A.A. Agrachev, G. Stefani, P.L. ZezzaStrong optimality for a bang–bang trajectory. *SIAM J. Control and Optimization* 41:991–1014, 2002.

[2] H. Maurer, N.P. Osmolovskii. Second order sufficient conditions for time optimal bang–bang control problems. *SIAM J. Control and Optimization* 42:2239–2263, 2004.

[3] H. Maurer, N.P. Osmolovskii. Second order optimality conditions for bang–bang control problems. *Control & Cybernetics* 32:555-584, 2003.

[4] A.A. Milyutin, N.P. Osmolovskii. *Calculus of Variations and Optimal Control.* Translations of Mathematical Monographs, Vol. 180, American Mathematical Society, Providence, 1998.

[5] N.P. Osmolovskii. High-order necessary and sufficient conditions for Pontryagin and bounded-strong minima in the optimal control problems. *Dokl. Akad. Nauk SSSR, Ser. Cybernetics and Control Theory* 303: 1052–1056, 1988; English transl., *Sov. Phys. Dokl.* 33, N. 12:883–885, 1988.

[6] N.P. Osmolovskii. Quadratic conditions for nonsingular extremals in optimal control (A theoretical treatment). *Russian J. of Mathematical Physics* 2:487–516, 1995.

[7] N.P. Osmolovskii. Second order conditions for broken extremal. In: *Calculus of variations and optimal control.* (Technion 1998), A. Ioffe, S. Reich and I. Shafir, eds., Chapman and Hall/CRC, Boca Raton, Florida, 198-216, 2000.

[8] N.P. Osmolovskii. Second-order sufficient conditions for an extremum in optimal control. *Control and Cybernetics* 31 803-831, 2002.

[9] N.P. Osmolovskii. Quadratic optimality conditions for broken extremals in the general problem of calculus of variations. *Journal of Math. Science* 123: 3987-4122, 2004.

References

ON A VARIATIONAL PROBLEM OF ULAM

S. Villa[1]
[1]*DIMA, Università di Genova, Genova, Italy, villa@dima.unige.it*

Abstract Using the concept of well-posedness under perturbations we give an answer to the following question posed by Stanislaw Ulam : "when it is true that solutions of two problems in the calculus of variations which corresponds to "close" physical data must be close to each other?"("A collection of mathematical problems", 1960).

keywords:well-posedness, integral functionals, bounded Hausdorff convergence

1. Introduction

The origin of the problem studied in this paper can be found in the following question posed by Stanislaw Ulam in 1960 in the book "A collection of mathematical problems" (see [14]).

"If we consider a typical elementary problem, that of finding an extremum y_0 of

$$I(y) = \int_a^b F(x, y(x), y'(x))\, dx \tag{1}$$

[...] a question arises, namely, the conditions which guarantee that for every $\epsilon > 0$ there exists a $d > 0$ such that for all sufficiently "regular" $G(x, y, z)$ with $|G - F| < d$, there exists a minimum y_1 for

$$J(y) = \int_a^b G(x, y(x), y'(x))\, dx, \tag{2}$$

where $|y_1 - y_0| < \epsilon$.

We assume merely the proximity of F and G and nothing is assumed about the proximity of their partial derivatives, occurring in the Lagrangian equations.

Speaking descriptively, the question is: when it is true that solutions of two problems in the calculus of variations which corresponds to "close" physical data must be close to each other?"

Please use the following format when citing this chapter:

Villa, S., 2006, in IFIP International Federation for Information Processing, Volume 202, Systems, Control, Modeling and Optimization, eds. Ceragioli, F., Dontchev, A., Furuta, H., Marti, K., Pandolfi, L., (Boston: Springer), pp. 309-318.

The aim of the paper is to give an answer to this question in the more general case of multiple integrals defined by

$$J(u) = \int_\Omega L(x, u(x), Du(x))\, dx \tag{3}$$

on a suitable Sobolev space and to locate it in an appropriate context.

The first remark is that in this form the question has always a negative answer; in fact Bobylev showed in [4] that whatever the functional I defined by (1) under study would be, it is possible to destroy all its minimum points by disturbing to be as small as wished.

EXAMPLE 1 *[[4]] Let $L : [0,1] \times \mathbb{R} \to [0, +\infty)$ be a three times differentiable function and let us consider the associated integral functional*

$$J(u) = \int_0^1 L(x, u(x), u'(x))\, dx,$$

where $u \in C^1(0,1)$ and $u(0) = u(1) = 0$. Let us suppose that J is Tikhonov well-posed with unique minimizer $u_0 = 0$.

Consider the one-parameter family of functionals

$$J_\epsilon(u) = \int_0^1 \left[L(x, u(x), u'(x)) + \epsilon \cos\left(\frac{u'(x)}{\epsilon}\right)\right] dx.$$

For a sufficiently small ϵ the functional J_ϵ has no minimum points in the unit ball

$$B = \{u(x) \in C^1(0,1) \mid \|u\| \le 1\}.$$

Anyway Bobylev showed that it is possible to restrict the class of admissible perturbations in order to ensure that every perturbed problem has a solution. In fact he showed that considering only strictly convex and sufficiently regular integrands, the problem of Ulam is solved positively.

The same approach is developed in [13], where it is proved the existence of minimizers for problems in a neighborhood of a given one, if a uniform strictly subdifferentiability property is satisfied.

Since we do not want to impose convexity, we choose another way to treat this problem, taking into account the stability of approximate minimizers instead of that of the minimizers. In fact, since the given functional I has a minimizer, it follows from the proximity of F and G that the infimum of J (defined by (2)) is finite. Therefore for every $\epsilon > 0$, surely there exist functions y_ϵ such that $J(y_\epsilon) < \inf J + \epsilon$. So the question posed by Ulam can be modified into

"when it is true that approximate solutions of two problems in the calculus of variations which corresponds to "close" physical data must be close to each other?"

Clearly in the case that the minimizers of close problems exist, an affirmative answer to this question implies an affirmative answer to the original one.

This question is intimately related to the well-posedness properties of the minimization problem of the functional I as Ulam himself wrote in his book: "Affirmative theorems of this sort would ensure the stability of the solutions of physical problems even with respect to "hidden" parameters.[...] It seems desirable in many mathematical formulations of physical problems to add still another requirement to the well-known desiderata of Hadamard of existence, uniqueness and continuity of the dependence of solutions on the initial parameters.

Specifically one should have a stability in the strongest sense illustrated by us above: the solutions should vary continuously even when the operator itself is subject to "small" variations".

This is exactly the point of view that we adopt throughout this work. Let us better explain what we mean. Given an optimization problem, arising in the calculus of variations or not, it is possible to see it as an element of a "variational system" according to the definition given by Rockafellar and Wets in [11].

More precisely we consider a fixed space $\mathcal{A}$ of parameters, endowed with a convergence structure, and we associate to each parameter an optimization problem. Moreover we of course assume that the problem we are starting from corresponds to a fixed $a_0 \in \mathcal{A}$. In this way we generate a family of problems having the same structure of the starting one.

Remarkable examples are the cases in which the space of parameters is the space of admissible initial data or the space of integrands satisfying some given properties.

Once that the given minimization problem is embedded in a suitable family, we can speak of well-posedness under perturbations of it. This concept, introduced by Zolezzi in [20], rigorously expresses the generalization of the Hadamard's idea of well-posedness mentioned by Ulam, in the more general form, which takes into account not only the behavior of the minimizers, but also that of the approximate minimizers (see Definition 3), requiring the continuous dependence of the approximate solutions on the parameters.

This notion is the right instrument to reformulate the problem of Ulam in a better way. The final version of the problem is therefore:

It is possible to find sufficient conditions on the integrand which guarantees well-posedness under perturbations of the functional I?

In this new form the example proposed by Bobylev is no more a counterexample (see Example 10), since we are considering approximate minimizers. Moreover it is somehow misleading in the sense that we explain below.

Looking at Example 1 it turns out in fact that the existence of a minimizer is a very unstable property if we consider perturbations of the integrand with respect to the uniform metric (which is a very strong one) but Ioffe and Zaslavski proved that for several variational systems (among them integral functionals parameterized by the integrands) well-posedness under perturbations is a generic property (with respect to a suitable topology).

This result is not completely satisfactory since it only tells us that there are a lot of well-posed problems without giving a method to find them.

In the literature there are not many constructive results on this subject, and the existing ones often require strong regularity properties on the integrand and convexity (see [22, 20, 4, 12]). Therefore our aim is to find some weaker sufficient conditions still guaranteeing well-posedness of a given problem of the calculus of variations.

The last thing that we want to point out is the choice of the distance between two problems. The uniform convergence of the integrands proposed by Ulam is too strong to treat this kind of problems.

The right choice seems to be one of the so called "variational convergences", as Γ, Mosco and bounded Hausdorff convergence. We choose the last one because it gives us the useful link between Tikhonov well-posedness and well-posedness under perturbations that we mentioned above (see Theorem 4), and moreover gives the possibility of finding quantitative bounds on the distance between minima and minimizers of two given problems.

The paper is organized as follows: the answer to the question analyzed here will be given in Section 3, whereas Section 2 contains some prelimanaries, as the introduction of the well-posedness concepts and the relations between them.

The last section is devoted to some dominated convergence theorems for the bounded Hausdorff convergence of integral functionals.

The proofs of the results presented here can be found in the papers [16, 17, 15, 18].

2. Well-posedness concepts and preliminaries

We start introducing the definitions that we need. We only review the well-posedness concepts whereas for the definition of bounded Hausdorff (bH) convergence we refer to the book [3] and to the papers [1, 2].

DEFINITION 2 *We say that a function $J : X \to \mathbb{R} \cup \{+\infty\}$ is Tikhonov well-posed if it satisfies the following conditions:*

a) there exists a unique global minimizer u_0 of J;

b) if u_h is any minimizing sequence, i.e. a sequence such that $J(u_h) \to J(u_0)$, then $u_h \to u_0$.

For a more general definition, and for several characterizations of Tikhonov well-posedness, we refer to [6].

Now we consider also a convergence space $\mathcal{A}$ and a fixed point $f \in \mathcal{A}$. We are given the proper extended real-valued functions

$$J : X \to (-\infty, +\infty], \quad I : \mathcal{A} \times X \to (-\infty, +\infty]$$

such that

$$J(u) = I(f, u), \; u \in X.$$

The corresponding value function is given by

$$V(g) = \inf\{I(g, u) \,|\, u \in X\}, \; g \in \mathcal{A}.$$

DEFINITION 3 *([21]) The problem of minimizing J on X is called well-posed under perturbations (with respect to the embedding defined by I) iff*

1. *$V(g) > -\infty$ for all $g \in \mathcal{A}$,*
2. *there exists a unique global minimizer u_0 for J,*
3. *for every sequences $f_h \to f$ in $\mathcal{A}$ and $u_h \in X$ such that*

$$I(f_h, u_h) - V(f_h) \to 0 \;\; as \;\; h \to +\infty \tag{4}$$

we have $u_h \to u_0$ in X.

Sequences satisfying condition (4) are called asymptotically minimizing sequences.

Let $L(X) = \{G : X \to \mathbb{R} \cup \{+\infty\}$, bounded from below, proper and l.s.c.$\}$ and $\psi : X \to \mathbb{R} \cup \{+\infty\}$ such that $\lim_{\|u\| \to +\infty} \psi(u) = +\infty$. Set

$$\mathcal{A} = \{G \in L(X) \,|\, G(u) \geq \psi(u) \text{ for each } u\}, \tag{5}$$

endowed with the bH-convergence, and define

$$I : \mathcal{A} \times X \to (-\infty, +\infty]$$

$$I(G, u) = G(u). \tag{6}$$

The following result is a rewriting in different terms of Theorem 3.5 of [16].

THEOREM 4 *Let X be a Banach space, $\mathcal{A}$ and I defined by (5) and (6). Assume that $J \in \mathcal{A}$ is Tikhonov wellposed. Then J is wellposed under perturbations with respect to the embedding defined by I.*

Theorem 4 cannot be extended without further restrictions on the space $\mathcal{A}$ of admissible perturbations putting on $\mathcal{A}$ a weaker convergence as Example 4.9 of [10] shows and this is why we choose bH-convergence to study this problem.

We conclude this section recalling the definition of strict convexity at a point in order to illustrate Theorem 6. In the sequel, given a function $L : \mathbb{R}^n \to [0, +\infty]$ we will denote by L^{**} its convex regularization.

In the following we denote by L_r the function defined on $\mathbb{R}^n$ by setting $L_r = L$ on the closed ball $\overline{B}(u_0, r)$ with center u_0 and radius r and $L_r = +\infty$ otherwise.

DEFINITION 5 *Let U be a closed convex subset of $\mathbb{R}^n$ and $u_0 \in U$. We say that a function $L : \mathbb{R}^n \to \mathbb{R} \cup \{+\infty\}$ is* convex at u_0 *with respect to the set U if*

$$\sum_{i=1}^{m} c_i L(v_i) \geq L(u_0) \tag{7}$$

for every $m > 0$, for every $v_i \neq u_0$ and $c_i > 0$ such that $v_i \in U$, $\sum_{i=1}^{m} c_i = 1$ and $\sum_{i=1}^{m} c_i v_i = u_0$.

If inequality (7) is always strict, we say that the function L is strictly convex at u_0 *with respect to U.*

Clearly, there are many examples of convex functions at a point which are not globally convex; for instance every function is convex at its minimum point, if it has one.

3. Well-posedness results

Following the idea described in in the Introduction, we give an answer to Ulam's question in the following way: given a normal integrand (see [7]) we find some conditions on it guaranteeing Tykhonov well-posedness of the associated integral functional.

More precisely, consider $L : \Omega \times \mathbb{R}^{nm} \to [0, +\infty]$ a normal integral and define the associated integral functional

$$J(u) = \int_{\Omega} L(x, Du(x))\, dx \tag{8}$$

on the space $W_0^{1,1}(\Omega; \mathbb{R}^m)$. The key result of this section is the following theorem which enable us to prove Corollary 9.

THEOREM 6 *Let L and J be defined as in Corollary 6. Suppose that J is coercive having a unique minimum point $u_0 \in W_0^{1,1}(\Omega; \mathbb{R}^m)$.*

Moreover assume that $v \mapsto L(x, v)$ is strictly convex at the point $Du_0(x)$ for almost every $x \in \Omega$. Then J is Tikhonov wellposed in $W_0^{1,1}(\Omega; \mathbb{R}^m)$.

Equivalently, if $u_h \in W_0^{1,1}(\Omega; \mathbb{R}^m)$ is any minimizing sequence, then:

$$|u_h - u_0 \,| \to 0 \text{ in } W_0^{1,1}(\Omega; \mathbb{R}^m).$$

We skip the proof of Theorem 6, which can be found in [17], and we just state the basic tools in order to prove it. The first is Lemma 7, which gives a characterization of the points of strict convexity in terms of their geometric properties.

LEMMA 7 *Let $L : \mathbb{R}^n \to [0, +\infty]$ be lower semicontinuous and proper. Then:*

1. *If $(u_0, L^{**}(u_0))$ is an extreme point of $\operatorname{epi} L^{**}$ then L is strictly convex at u_0.*
2. *If L is strictly convex at u_0, then $(u_0, L(u_0))$ is an extreme point of $\operatorname{epi}(L_r)^{**}$ for all $r > 0$.*

The second lemma is a generalization to the nonconvex setting of Lemma 3 of [19] which can be applied thanks to a semicontinuity result (see Lemma 4.3 of [17]).

LEMMA 8 *Let $L : \Omega \times \mathbb{R}^{mn} \to [0, +\infty]$ be a normal integrand and $u_h, u_0 \in W^{1,1}(\Omega; \mathbb{R}^m)$. Suppose that the function $v \mapsto L(x, v)$ is convex at the point $Du_0(x)$ for almost every $x \in \Omega$.*

If $|u_h - u_0|_{L^1} \to 0$, $Du_h \rightharpoonup Du_0$ and $J(u_h) \to J(u_0)$ then:

$$(x \mapsto L(x, Du_h(x))) \rightharpoonup (x \mapsto L(x, Du_0(x)))$$

in $L^1(\Omega; \mathbb{R})$.

As a consequence of Theorems 4 and 6, we get the following Corollary, which establishes a criterium of well-posedness under perturbations for nonconvex integral functionals.

COROLLARY 9 *Let $L, L_h : \Omega \times \mathbb{R}^{nm} \to [0, +\infty]$ be normal integrands and let J be defined as (8). Suppose that J is coercive having a unique minimum point $u_0 \in W_0^{1,1}(\Omega; \mathbb{R}^m)$.*

Moreover assume that $v \mapsto L(x, v)$ is strictly convex at the point $Du_0(x)$ for almost every $x \in \Omega$. Consider the sequence J_h defined by $J_h(u) := \int_\Omega L_h(x, Du(x))\, dx$ on $W_0^{1,1}(\Omega, \mathbb{R}^m)$. Assume that

(i) $J_h \to J$ with respect to the bH-convergence on $W_0^{1,1}(\Omega; \mathbb{R}^m)$;

(ii) J_h is equicoercive.

Then:

$$|u_h - u_0| \to 0 \text{ in } W_0^{1,1}(\Omega; \mathbb{R}^m)$$

for every asymptotically minimizing sequence u_h. In other words the problem of minimizing J is well-posed under equicoercive perturbations with respect to the bH-convergence.

Referring again to the comments in the Introduction, in the next example we show in which sense Corollary 9 gives an answer to the Ulam's question.

EXAMPLE 10 *Let us consider the functionals*

$$J(u) = \int_0^1 (u(x)^2 + u'(x)^4)\, dx$$

and

$$J_h(u) = \int_0^1 (u(x)^2 + (u'(x)^2 - \frac{1}{h})^2)\, dx,$$

with $u \in W_0^{1,4}(0,1)$.

It is easy to show that the perturbed functionals J_h *do not have a minimizer for any* h*; in fact in this case we do not have uniform strict convexity of the perturbed functionals. According to Example 1, this means that even if* J *is strictly convex and Tykhonov well-posed it is possible to perturb it with perturbations as small as wished and destroy the existence of a minimizer.*

Anyway, since J *is Tykhonov well-posed and the sequence* J_h *is convergent with respect to the bounded Hausdorff topology (see Remark 11), every asymptotically minimizing sequence strongly converges.*

This shows how the concept of well-posedness under perturbations allow us to avoid the requirement of the existence of the minimizer of the perturbed problems.

4. Dominated convergence theorems

In order to obtain well-posedness results it is useful to obtain criteria which guarantee that a sequence of lower semicontinuous functions converges in the bH-sense to a certain function (see for instance Example 10).

The only known result on this subject is about quadratic functionals of elliptic type and can be found in [5].

In this section we study some classes of integral functionals, and we prove that, under suitable hypotheses, convergence in different senses of the sequence of the integrands is a sufficient condition to assure convergence of the sequence of the associated integrals (with respect to the bounded Hausdorff convergence). The first result is the following remark. The proof is elementary, but we note that sequences of normal integrands satisfying condition (9) are said to be convergent modulo given growth or with respect to the τ_p topology and this family of topologies was introduced in [9] to prove the so called antirelaxation theorem, which establishes the genericity of well-posed problems arising in the calculus of variations with respect to τ_p perturbations of the integrand.

REMARK 11 *Fix* $p \geq 1$ *and let* $L_h, L : \Omega \times \mathbb{R}^m \times \mathbb{R}^{nm} \to \mathbb{R}$ *be normal integrands.*

Assume that for every $\epsilon > 0$ *there exist* $N_\epsilon \in \mathbb{N}$, $\phi_\epsilon \in L^1(\Omega; \mathbb{R})$ *such that* $\|\phi_\epsilon\| \leq 1$ *and*

$$|L_h(x,u,z) - L(x,u,z)| \leq \epsilon(\phi_\epsilon(x) + |u|^p + |z|^p) \qquad (9)$$

for every $h > N_\epsilon$, *for every* $(u,z) \in \mathbb{R}^n \times \mathbb{R}^{nm}$ *and for almost every* $x \in \Omega$.

Let J, J_h *the associated integral functionals on* $W^{1,p}(\Omega; \mathbb{R}^m)$. *Then* $J_h \to J$ *with respect to bH-convergence in* $W^{1,p}(\Omega; \mathbb{R}^m)$.

It follows from this Remark that τ_p is the natural convergence to require on the integrands in order to obtain bH-convergence of the associated integrals. Nevertheless, for every $p \geq 1$, uniform convergence modulo given growth is stronger than uniform convergence on bounded subsets, so it would be useful to find a weaker convergence on the integrands still guaranteeing bH-convergence of the integral functionals.

This is the aim of Theorem 11, where we relax the hypotheses on convergence, requiring only pointwise convergence of the integrands, but we pay this weaker requirement imposing a restrictive growth condition (which cannot be weakened), and an equilipschitz property on bounded subsets for the integrands. Moreover we deal with functionals not depending explicitely on u. So in the next theorem, the functionals considered are of the form

$$J(u) := \int_\Omega L(x, Du(x))\, dx.$$

Let us state the assumption that we will make in the following theorem.

A) Let $L_h : \Omega \times \mathbb{R}^{nm} \to \mathbb{R}$ be a sequence of functions. Suppose that for every $R > 0$ there exists $L(\cdot, R) \in L^1(\Omega, \mathbb{R})$ such that

$$|L_h(x,z) - L_h(x,y)| \leq K(x,R)|z-y|$$

for almost every $x \in \mathbb{R}^n$, for every $z, y \in B_R(0)$ and for every $h \in \mathbb{N}$.

We note that this equilipschitz property is always satisfied by a sequence of equibounded finite convex functions.

THEOREM 12 *Let $L, L_h : \Omega \times \mathbb{R}^{nm} \to [0, +\infty)$ be normal integrands and assume that L, L_h satisfy assumption A). Let $p > 1$ and suppose that there exist $1 < q < p$, $r > \frac{p}{p-q}$ and $\Lambda \in L^r(\Omega, \mathbb{R})$ such that for every $z \in \mathbb{R}^{nm}$ and for almost every $x \in \Omega$.*

$$0 \leq L_h(x,z) \leq \Lambda(x)(|z|^q + 1). \tag{10}$$

If for all $z \in \mathbb{R}^{nm}$ $L_h(\cdot, z) \to L(\cdot, z)$ almost everywhere, then $J_h \to J$ with respect to the bounded Hausdorff convergence on $W^{1,p}(\Omega; \mathbb{R}^m)$.

We note that condition (10) cannot be weakened as Example 3.26 in [18] shows.

Other results on this subject and some consequences of Theorem 12 can be found in [15].

References

[1] H. Attouch and R. Wets. Quantitative stability of variational systems. I. The epigraphical distance. *Trans. Amer. Math. Soc.*, 328(2):695–729, 1991.

[2] H. Attouch and R. Wets. Quantitative stability of variational systems. II. A framework for nonlinear conditioning. *SIAM J. Optim.*, 3(2):359–381, 1993.

[3] G. Beer. *Topologies on closed and closed convex sets*, volume 268 of *Mathematics and its Applications*. Kluwer Academic Publishers Group, Dordrecht, 1993.

[4] N. A. Bobylëv. On a problem of s. ulam. *Nonlinear Anal.*, 24(3):309–322, 1995.

[5] Z. Chbani. Caractérisation de la convergence d'intégrales définies à partir d'opérateurs elliptiques par la convergence des coefficients. *Sém. Anal. Convexe*, 21:Exp. No. 12, 14, 1991.

[6] A. L. Dontchev and T. Zolezzi. *Well-posed optimization problems*, volume 1543 of *Lecture Notes in Mathematics*. Springer-Verlag, Berlin, 1993.

[7] I. Ekeland and R. Temam. *Convex analysis and variational problems*. North-Holland Publishing Co., Amsterdam, 1976. Translated from the French, Studies in Mathematics and its Applications, Vol. 1.

[8] J. Hadamard. Sur les problèmes aux dérivées partielles et leur signification physique. *Princeton University Bulletin*, pages 49–52, 1902.

[9] A. D. Ioffe and A. J. Zaslavski. Variational principles and well-posedness in optimization and calculus of variations. *SIAM J. Control Optim.*, 38(2):566–581 (electronic), 2000.

[10] R. Lucchetti. Some aspects of the connections between Hadamard and Tyhonov well-posedness of convex programs. *Boll. Un. Mat. Ital. C (6)*, 1(1):337–345, 1982.

[11] R. T. Rockafellar and R. J.-B. Wets. Variational systems, an introduction. In *Multifunctions and integrands (Catania, 1983)*, volume 1091 of *Lecture Notes in Math.*, pages 1–54. Springer, Berlin, 1984.

[12] M. A. Sychëv. Necessary and sufficient conditions in theorems of semicontinuity and convergence with a functional. *Mat. Sb.*, 186(6):77–108, 1995.

[13] M. A. Sychëv. On the continuous dependence of the solutions of the simplest variational problems on the integrand. *Siberian Math. J.*, 36(2):379–388, 1995.

[14] S. M. Ulam. *A collection of mathematical problems*. Interscience Tracts in Pure and Applied Mathematics, no. 8. Interscience Publishers, New York-London, 1960.

[15] S. Villa. Bounded hausdorff convergence of integral functionals and related well-posedness properties. *preprint*.

[16] S. Villa. AW-convergence and well-posedness of non convex functions. *J. Convex Anal.*, 10(2):351–364, 2003.

[17] S. Villa. Well-posedness of nonconvex integral functionals. *SIAM J. Control Optim.*, 43(4):1298–1312 (electronic), 2004/05.

[18] S. Villa. *Well-posed problems of the calculus of variations*. 2005.

[19] A. Visintin. Strong convergence results related to strict convexity. *Comm. Partial Differential Equations*, 9(5):439–466, 1984.

[20] T. Zolezzi. Well-posed optimization problems for integral functionals. *J. Optim. Theory Appl.*, 31(3):417–430, 1980.

[21] T. Zolezzi. Well-posedness criteria in optimization with application to the calculus of variations. *Nonlinear Anal.*, 25(5):437–453, 1995.

[22] T. Zolezzi. Wellposed problems of the calculus of variations for nonconvex integrals. *J. Convex Anal.*, 2(1-2):375–383, 1995.

NUMERICAL SOLUTION OF OPTIMAL CONTROL PROBLEMS WITH CONVEX CONTROL CONSTRAINTS

[1]D. Wachsmuth
[1]*Institut für Mathematik, TU Berlin, Str. des 17. Juni 136, D-1062 Berlin, Germany,*
wachsmut@math.tu-berlin.de

Abstract We study optimal control problems with vector-valued controls. In the article, we propose a solution strategy to solve optimal control problems with pointwise convex control constraints. It involves a SQP-like step with an imbedded active-set algorithm.

keywords: Optimal control, convex control constraints, set-valued mappings.

1. Introduction

In this article, we want to investigate solution strategies to solve optimal control problems with pointwise convex control constraints. In flow control, any control regardless of distributed or boundary control is a vector-valued function. That means, there are many possibilities to formulate control constraints. Here, we will study a general concept of such control constraints. As an example of an optimal control problem with vector-valued controls we chose the following problem of optimal distributed control of the instationary Navier-Stokes equations in two dimensions. To be more specific, we want to minimize the following quadratic objective functional:

$$J(y,u) = \frac{\alpha_T}{2}\int_\Omega |y - y_T|^2 dx + \frac{\alpha_Q}{2}\int_Q |y - y_Q|^2 dxdt + \frac{\gamma}{2}\int_Q |u|^2 dxdt \quad (1)$$

subject to the instationary Navier-Stokes equations

$$\begin{aligned} y_t - \nu\Delta y + (y\cdot\nabla)y + \nabla p &= u && \text{in } Q,\\ \operatorname{div} y &= 0 && \text{in } Q,\\ y(0) &= y_0 && \text{in } \Omega, \end{aligned} \quad (2)$$

and to the control constraints $u \in U_{ad}$ with U_{ad} defined by

$$U_{ad} = \{u \in L^2(Q)^2 : \ u(x,t) \in U(x,t) \text{ a.e. on } Q\}. \quad (3)$$

Please use the following format when citing this chapter:

Wachsmuth, D., 2006, in IFIP International Federation for Information Processing, Volume 202, Systems, Control, Modeling and Optimization, eds. Ceragioli, F., Dontchev, A., Furuta, H., Marti, K., Pandolfi, L., (Boston: Springer), pp. 319-327.

Here, Ω is a bounded domain in R^2, Q denotes the time-space cylinder $Q := \Omega \times (0, T)$. The set of admissible controls is generated by the set-valued mapping U, $U : Q \mapsto 2^{R^2}$. The conditions imposed on the various ingredients of the optimal control problem are specified in Sections 2.1 and 2.2, see assumptions (A) and (AU).

Several choices of the control constraint U are discussed in [13]. Let us only mention that the case of box-constraints is a special case of our general convex constraint. Optimal control problems with such control constraints are rarely investigated in literature. We refer to Bonnans [2], Dunn [3], Páles and Zeidan [7], and the authors article [14].

The plan of the article is as follows. In Section 2 we collect results concerning the solvability of the state equation and the optimal control problem. Necessary optimality conditions are stated in Section 3. Section 4 is devoted to the derivation of a solution strategy. There, a new active-set algorithm is described. Numerical experiments that confirm the efficiency of the proposed algorithm are presented in Section 5.

2. Statement of the optimal control problem

We define the spaces of solenoidal functions $H := \{v \in L^2(\Omega)^2 : \operatorname{div} v = 0\}$, $V := \{v \in H_0^1(\Omega)^2 : \operatorname{div} v = 0\}$. Further, we will work with the standard spaces of abstract functions from $[0, T]$ to a real Banach space X, $L^p(0, T; X)$, endowed with their natural norms.

2.1 The state equation

Before we start with the discussion of the state equation, we specify the requirements for the various ingredients describing the optimal control problem. In the sequel, we assume that the following conditions are satisfied:

$$
(\mathbf{A}) \begin{cases} 1 \;\; \Omega \subset R^2 \text{ is domain with Lipschitz boundary } \Gamma := \partial\Omega, \\ 2 \;\; y_0,\, y_T \in H,\, y_Q \in L^2(Q)^2, \\ 3 \;\; \alpha_T,\, \alpha_Q \geq 0,\, \gamma,\, \nu > 0. \end{cases}
$$

We introduce a linear operator $A : L^2(0, T; V) \mapsto L^2(0, T; V')$ by

$$
\int_0^T \langle (Ay)(t), v(t) \rangle_{V',V} dt := \int_0^T (y(t), v(t))_V dt,
$$

and a nonlinear operator B by

$$
\int_0^T \langle (B(y))(t), v(t) \rangle_{V',V} \, dt := \int_0^T \int_\Omega \sum_{i,j=1}^2 y_i(t) \frac{\partial y_j(t)}{\partial x_i} v_j(t) dx dt.
$$

Now, we define the notation of weak solutions for the instationary Navier-Stokes equations (2) in the Hilbert space setting. A function $y \in L^2(0,T;V)$ with $y_t \in L^2(0,T;V')$ is called weak solution of (2) if it satisfies

$$\begin{aligned} y_t + \nu A y + B(y) &= f \text{ in } L^2(0,T;V'), \\ y(0) &= y_0. \end{aligned} \tag{4}$$

Results concerning the solvability of (4) are standard, cf. [8] for proofs and further details.

2.2 Set-valued functions

Before we begin with the formulation of the optimal control problem with inclusion constraints, we will specify the notation and assumptions for the admissible set $U(\cdot)$. It is itself a mapping from the control domain Q to the set of subsets of R^2, a so-called *set-valued mapping*.

The controls are taken from the space $L^2(Q)^2$, so we have to impose at least some measurability conditions on the mapping U. In the sequel, we will work with measurable set-valued mappings, we refer to the textbook by Aubin and Frankowska [1]. Once and for all, we specify the requirements for the function U, which defines the control constraints.

(AU)

The set-valued function $U : Q \to R^2$ satisfies:

1. *U is a measurable set-valued function.*
2. *The images of U are non-empty, closed, and convex a.e. on Q. That is, the sets $U(x,t)$ are non-empty, closed and convex for almost all $(x,t) \in Q$.*
3. *There exists a function $f_U \in L^2(Q)^2$ with $f_U(x,t) \in U(x,t)$ a.e. on Q.*

The assumption (AU) is as general as possible. In the case that the set-valued function U is a constant function, i.e. $U(x,t) \equiv U_0$, we can give a simpler characterization: it suffices that U_0 is non-empty, closed, and convex. For further details, we refer to [14].

3. First-order necessary conditions

We will repeat the exact statement of the necessary optimality conditions for convenience of the reader.

THEOREM 1 (NECESSARY CONDITION) *Let $\bar{u}$ be locally optimal in $L^2(Q)^2$ with associated state $\bar{y} = y(\bar{u})$. Then there exists a unique Lagrange multiplier $\bar{\lambda} \in L^2(0,T;V)$ with $\bar{\lambda}_t \in L^{4/3}(0,T;V)$, which is the weak solution of the* *adjoint equation*

$$
\begin{aligned}
-\bar{\lambda}_t + \nu A\bar{\lambda} + B'(\bar{y})^*\bar{\lambda} &= \alpha_Q(\bar{y} - y_Q) \\
\bar{\lambda}(T) &= \alpha_T(\bar{y}(T) - y_T).
\end{aligned} \tag{5}
$$

Moreover, it holds

$$
(\gamma\bar{u} + \bar{\lambda}, u - \bar{u})_Q \geq 0 \quad \forall u \in U_{ad}. \tag{6}
$$

We can reformulate the variational inequality (6) as the projection representation of the optimal control

$$
\bar{u}(x,t) = \text{Proj}_{U(x,t)}\left(-\frac{1}{\gamma}\bar{\lambda}(x,t)\right) \qquad \text{a.e. on } Q. \tag{7}
$$

Secondly, another formulation of the variational inequality uses the normal cone $N_{U_{ad}}(\bar{u})$. Then inequality (6) can be written equivalently as

$$
\nu\bar{u} + \bar{\lambda} + N_{U_{ad}}(\bar{u}) \ni 0. \tag{8}
$$

It will allow us to write the optimality system as a generalized equation.

4. Solution strategy

Now, we are going to describe our strategy to solve optimal control problems with pointwise convex control constraints. The starting point of our considerations is Newtons method applied to generalized equations.

4.1 Generalized Newton's method

In the sequel, we want to apply Newton's method to the optimality system. This system can be written as a generalized equation. Let us define a function F by

$$
F(y,u,\lambda) = \begin{pmatrix} y_t + \nu Ay + B(y) \\ y(0) \\ -\lambda_t + \nu A\lambda + B'(y)^*\lambda \\ \lambda(T) \\ \gamma u + \lambda \end{pmatrix} - \begin{pmatrix} u \\ y_0 \\ \alpha_Q(y - y_Q) \\ \alpha_T(y(T) - y_T) \\ 0 \end{pmatrix}. \tag{9}
$$

Then a triple $(\bar{y}, \bar{u}, \bar{\lambda})$ fulfills the necessary optimality conditions consisting of the state equation (4), the adjoint equation (5), and the variational inequality (6) if and only if it fulfills the generalized equation

$$
F(\bar{y}, \bar{u}, \bar{\lambda}) + (0,0,0,0,N_{U_{ad}}(\bar{u}))^T \ni 0. \tag{10}
$$

Now, we will apply the generalized Newton method to equation (10). If the control constraints are box constraints then this method is equivalent to the SQP-method, see e.g. [9]. Given iterates (y_n, u_n, λ_n) we have to solve the linearized generalized equation: 0 should belong to the set

$$F(y_n, u_n, \lambda_n) + F'(y_n, u_n, \lambda_n)[(y - y_n, u - u_n, \lambda - \lambda_n)] + (0, 0, 0, 0, N_{U_{ad}}(\bar{u}))^T \tag{11}$$

in every step. It turns out that this equation is the optimality system of the linear-quadratic optimal control problem.

$$\min J_n(y, u) = \frac{\alpha_T}{2} |y(T) - y_d|_H^2 + \frac{\alpha_Q}{2} \|y - y_Q\|_2^2 - B(y - y_n)\lambda_n \tag{12}$$

subject to the linearized state equation

$$\begin{aligned} y_t + \nu A y + B'(y_n) y &= u + B(y_n) \\ y(0) &= y_0 \end{aligned} \tag{13}$$

and the control constraint

$$u \in U_{ad}. \tag{14}$$

Let us emphasize the following observation. In the generalized equation (10) only $N(x)$ represents the control constraint. And it is the only term that was *not* linearized in (11). Consequently, the subproblem (12)–(14) is subject to the same control constraint $u(x, t) \in U(x, t)$ as the original non-linear problem. That means, the control constraint is not linearized, even if it is written as an inequality like $u_1(x, t)^2 + u_2(x, t)^2 \leq \rho(x, t)^2$. This is a difference to the standard SQP-method for optimization problems with nonlinear inequalities. An inequality $g(x) \leq 0$ would be linearized to $g(x_n) + g'(x_n)(x - x_n) \leq 0$.

4.2 Active-set strategy

As for the box-constrained case we will use an active set algorithm. It is very similar to the well-known primal-dual active-set strategy. The algorithm we propose here tries to solve the projection representation of optimal controls given by (7).

The algorithm to solve the subproblem (12)–(14) works as follows. We denote the control iterates of step k by u^k. The state y^k is the solution of (13) with right-hand side u^k, and λ^k is the solution of the adjoint equation associated to (13).

Algorithm. Take a starting guess u^0 with associated state y^0 and adjoint λ^0. Set $k = 0$.

1 Given u^k, y^k, λ^k. Determine the active set $A^{k+1} = A^{k+1}(\lambda^k)$ by

$$A^{k+1} := \left\{ (x, t) : -\frac{1}{\gamma} \lambda^k(x, t) \notin U(x, t) \right\}.$$

2 Minimize the functional J_n given by (12) subject to the linearized state equation (13) and to the control constraints

$$\left.\begin{array}{lcl} u|_{A^{k+1}} & = & \mathrm{Proj}_{U(x,t)}\left(-\frac{1}{\gamma}\lambda^k(x,t)\right) \\ u|_{Q\setminus A^{k+1}} & & \text{free.} \end{array}\right\}$$

Denote the solution by $(\tilde{u},\, \tilde{y},\, \tilde{\lambda})$.

3 Project $u^{k+1} := \mathrm{Proj}_{U(x,t)}\left(-\frac{1}{\gamma}\tilde{\lambda}(x,t)\right)$, compute y^{k+1} and λ^{k+1}.

4 If $\left\|u^{k+1} - \mathrm{Proj}_{U_{ad}}\left(-\frac{1}{\gamma}\lambda^{k+1}\right)\right\|_2 < \epsilon \rightarrow$ ready,
else set $k := k+1$ and go back to 1.

In the first step of the algorithm, the active set is determined. The control constraint at a particular point (x,t) is considered active if $-1/\gamma\,\lambda^k(x,t)$ does not belong to $U(x,t)$. In the second step, a linear-quadratic optimization problem is solved. It involves no inequality constraints, since the control is fixed on the active set and the control is free on the inactive set. After that, the optimal adjoint state of that problem is projected on the admissible set to get the new control. The algorithm stops if the residual in the projection representation is small enough.

Let us compare the behaviour of our algorithm and the primal-dual method in the detection of the active sets for the simple box constraint $|u_i| \leq 1$, $i = 1,2$. Let us assume that $-1/\gamma\,\lambda^k(x_0,t_0) = (2,0)$, i.e. $-1/\gamma\,\lambda^k(x_0,t_0)$ is not admissible. Then in our method the point (x_0,t_0) will belong to the active set A^{k+1} in the next step. And at this point the value $(1,0)$ is prescribed for u^{k+1}: $u^{k+1}(x_0,t_0) = (1,0)$. In the primal-dual method, the point (x_0,t_0) will be added to the active set A_1^{k+1} associated to the inequality $|u_1| \leq 1$. It will not belong to the active set A_2^{k+1} for the second inequality $|u_2| \leq 1$ since this inequality was not violated. And for the inner problem we will get the constraints $u_1^{k+1}(x_0,t_0) = 1$ and $u_2^{k+1}(x_0,t_0)$ =free, that is the control is allowed to vary in tangential directions on the right side of the box!

5. Numerical results

Now, let us report about the performance of the proposed algorithm. To study the convergence speed We present results for an optimal control problem, where the solution is known.

5.1 Problem setting

The computational domain was chosen to be the unit square $\Omega = (0,1)^2$ with final time $T = 1$. We want to minimize the functional

$$\frac{1}{2}\int_0^1\int_\Omega |y(x,t) - y_d(x,t)|^2 dxdt + \frac{\gamma}{2}\int_0^1\int_\Omega |u(x,t)|^2 dxdt$$

subject to the instationary Navier-Stokes equations on $\Omega\times(0,1)$ with distributed control

$$\begin{aligned} y_t - \nu\Delta y + (y\cdot\nabla)y + \nabla p &= u + f && \text{in } Q,\\ \operatorname{div} y &= 0 && \text{in } Q,\\ y(0) &= y_0 && \text{in } \Omega. \end{aligned}$$

and subject to some control constraints to be specified later on. Let us construct a triple of state, control and adjoint, that satisfies the first-order optimality system. We chose as state and adjoint state

$$\bar{y}(x,t) = e^{-\nu t}\begin{pmatrix} \sin^2(\pi x_1)\sin(\pi x_2)\cos(\pi x_2) \\ -\sin^2(\pi x_2)\sin(\pi x_1)\cos(\pi x_1) \end{pmatrix},$$

$$\bar{\lambda}(x,t) = \left(e^{-\nu t} - e^{-\nu}\right)\begin{pmatrix} \sin^2(\pi x_1)\sin(\pi x_2)\cos(\pi x_2) \\ -\sin^2(\pi x_2)\sin(\pi x_1)\cos(\pi x_1) \end{pmatrix}.$$

Regardless of the choice of U_{ad}, the control is computed using the projection formula as

$$\bar{u} = \text{Proj}_{U_{ad}}\left(-\frac{1}{\gamma}\bar{\lambda}(x,t)\right).$$

The quantities f, y_0, y_d are now chosen in such a way that $\bar{y}$ and $\bar{\lambda}$ are the solutions of the state and adjoint equations, respectively.

5.2 Discretization

The continuous problem was discretized using Taylor-Hood finite elements with different mesh sizes. The grid consists of 4096 triangles with 8321 velocity and 2113 pressure nodes yielding a mesh size $h = 0.03125$. Further, we use the semi-implicit Euler scheme for time integration with a equidistant time discretization with step length $\tau = 0.000625$. The control is approximated by piecewise continuous functions in time and space.

We used the SQP-method without any globalization to solve the problem with box constraints. The constrained SQP-subproblems were solved by the primal-dual active-set method. This method is known to converge locally with super-linear convergence rate [4, 11] if a sufficient optimality condition holds. Under some strong assumptions it converges even globally [6].

To solve the optimal control problem with convex constraints, we employ the solution strategy proposed in the previous section: generalized Newton method with active-set strategy as inner loop. The quadratic subproblems of the active-set method were again solved by the CG algorithm.

5.3 Results

Now, let us report about the results for convex control constraints compared to the box-constraints considered above. The parameters of the example are set to $\gamma = 0.01$ and $\nu = 0.1$. We computed solutions for the box constraint

$$|u_i(x,t)| \leq 1.0$$

and for the convex constraint in polar coordinates

$$0 \leq u_r(x,t) \leq \psi(u_\phi(x,t)).$$

The function $\psi(\phi)$ is given as the spline interpolation with cubic splines and periodic boundary conditons of the function values in Table 1. It is chosen such that the admissible set is comparable to the box constrained case. The projection on that set was computed by Newtons method.

ϕ	0	$\frac{\pi}{4}$	$\frac{\pi}{2}$	$\frac{3}{4}\pi$	π	$\frac{5}{4}\pi$	$\frac{3}{2}\pi$	$\frac{7}{4}\pi$
$\psi(\phi)$	1.0	1.0	0.64	0.44	0.4	0.44	0.64	1.0

Table 1. Convex constraint

Let us compare briefly the convergence of the active set algorithms: the primal-dual active set method for the box constrained problem and the active set method proposed above to solve the convex constrained problem. In all our computations both methods showed a similar behaviour, which can be seen in Table 2. Although they solved optimal control problems with different control constraints, they needed almost the same number of outer (SQP/generalized Newton) and inner (active-set) iterations. Also the residual $u - \text{Proj}(-1/\gamma\,\lambda)$ depicted by 'Res' decreased in the same way. So, we can say that our algorithm is as efficient as the primal dual active set method applied to box constraints.

References

[1] J.-P. Aubin and H. Frankowska. *Set-valued analysis*. Birkhäuser, 1990.

[2] J. F. Bonnans. Second-order analysis for constrained optimal control problems of semilinear elliptic equations. *Appl. Math. Optim.*, 38:303–325, 1998.

[3] J. C. Dunn. Second-order optimality conditions in sets of L^∞ functions with range in a polyhedron. *SIAM J. Control Optim.*, 33(5):1603–1635, 1995.

[4] M. Hintermüller, K. Ito, and K. Kunisch. The primal-dual active set strategy as a semismooth Newton method. *SIAM J. Optim.*, 13:865–888, 2003.

Box constraints			Convex constraints		
SQP		PD-AS	gNewton		AS
It	$\|u^n - u^{n-1}\|_\infty$	Res	It	$\|u^n - u^{n-1}\|_\infty$	Res
1		0.1564	1		$0.9912 \cdot 10^{-1}$
		$0.4823 \cdot 10^{-1}$			$0.2327 \cdot 10^{-1}$
		$0.1128 \cdot 10^{-1}$			$0.3475 \cdot 10^{-2}$
2		$0.1128 \cdot 10^{-1}$	2		$0.3498 \cdot 10^{-2}$
	0.1221	$0.4971 \cdot 10^{-3}$		0.1311	$0.1976 \cdot 10^{-3}$
3		$0.6967 \cdot 10^{-3}$	3		$0.2530 \cdot 10^{-3}$
	$0.1148 \cdot 10^{-1}$	$0.1561 \cdot 10^{-5}$			$0.3741 \cdot 10^{-4}$
				$0.1186 \cdot 10^{-1}$	$0.7047 \cdot 10^{-5}$
4		$0.1891 \cdot 10^{-5}$	4		$0.7035 \cdot 10^{-5}$
	$0.4105 \cdot 10^{-6}$	$0.4643 \cdot 10^{-6}$			$0.1291 \cdot 10^{-5}$
				$0.2804 \cdot 10^{-3}$	$0.2308 \cdot 10^{-6}$

Table 2. Comparison of the algorithms

[5] M. Hinze. *Optimal and instantaneous control of the instationary Navier-Stokes equations.* Habilitation, TU Berlin, 2002.

[6] K. Kunisch and A. Rösch. Primal-dual active set strategy for a general class of constrained optimal control problems. *SIAM J. Optim.*, 13:321–334, 2002.

[7] Zs. Páles and V. Zeidan. Optimum problems with measurable set-valued constraints. *SIAM J. Optim.*, 11:426–443, 2000.

[8] R. Temam. *Navier-Stokes equations.* North Holland, Amsterdam, 1979.

[9] F. Tröltzsch. On the Lagrange-Newton-SQP method for the optimal control of semilinear parabolic equations. *SIAM J. Control Optim.*, 38:294–312, 1999.

[10] F. Tröltzsch and D. Wachsmuth. Second-order sufficient optimality conditions for the optimal control of Navier-Stokes equations. *ESAIM: COCV*, 12:93–119, 2006.

[11] M. Ulbrich. Constrained optimal control of Navier-Stokes flow by semismooth Newton methods. *Systems & Control Letters*, 48:297–311, 2003.

[12] D. Wachsmuth. Regularity and stability of optimal controls of instationary Navier-Stokes equations. *Control and Cybernetics*, 34:387–410, 2005.

[13] D. Wachsmuth. Optimal control problems with convex control constraints. Preprint 35-2005, Institut für Mathematik, TU Berlin, submitted, 2005.

[14] D. Wachsmuth. Sufficient second-order optimality conditions for convex control constraints. *J. Math. Anal. App.*, 2006. To appear.